BIOSTATS BASICS

BIOSTATS BASICS
A Student Handbook

James L. Gould   and   **Grant F. Gould**

Princeton University   *MIT/Comverse Network Systems*

with

BioStats Basics Online

an interactive tutorial and basic collection of statistical tests
including questions, glossary, and data sets

Grant F. Gould and **James L. Gould**

W. H. FREEMAN AND COMPANY
NEW YORK

Publisher: Susan Brennan
Executive Editor: Sara Tenney
Development Editor: Michelle Baildon
New Media and Supplements Editor: Joy Hilgendorf
Marketing Manager: John Britch
Project Editor: Mary Louise Byrd
Cover and Text Design: Diana Blume
Illustration Coordinator: Bill Page
Editorial Assistant: Katie Mergen
Illustrations: Fine Line Illustrations
Photo Editor: Patricia Marx
Production Coordinator: Paul W. Rohloff
Composition: Matrix Publishing Services, Inc.
Manufacturing: RR Donnelley & Sons

Library of Congress Cataloging-in-Publication Data

Gould, James L., 1945–
  BioStats basics : a student handbook / James L. Gould and
Grant F. Gould.
      p. cm.
"With BioStats Basics online, an interactive tutorial and basic
collection of statistical tests including questions, glossary, and
data sets."
  ISBN 0-7167-3416-8 (paperback)
  1. Biometry. I. Gould, Grant F. II. Title.
QH323.5 .G67 2001
570'.1'5195—dc21   2001004467

Printed in the United States of America

First Printing 2001

To our most inspiring teachers
Liz Cutler and Davida Phillips
and in memory of
Janet Elson and Anne Shepherd

Contents in Brief

PREFACE Origin and Overview   xvii

1 **Cause and Effect**   1

2 **Data**   13

3 **Binomial Distributions**   41

4 **Continuous Parametric Distributions: I**   79

5 **Continuous Parametric Distributions: II**   111

6 **Data Transformations**   139

7 **Multiple Parametric Distributions:
      ANOVA**   151

8 **Categorical Data**   189

9 **Nonparametric Continuous Data**   215

10 **Circular Data**   235

11 **Relationships between Variables I:
       Correlation and Regression**   255

12 **Relationships between Variables II:
       Multiple Regression**   289

13 **None of the Above**   309

14 **Once Over Lightly**   325

Answers to Exercises   335

Statistics Labs   361

Selected Statistical Tables   369

Glossary   405

Bibliography   417

Index   419

Contents

PREFACE Origin and Overview   xvii

    Problems and Balance   xvii

    Biostats Basics Online: A Tutorial-Based Approach   xviii

    Why Is This Book So Thin?   xix

    Examples and Exercises   xx

1   Cause and Effect   1

    1.1 Probability and Survival   2

    1.2 Probability and Research   5

    1.3 Recognizing Differences   9

    1.4 Probability and Math   11

    Points to Remember   11

2   Data   13

    2.1 Why Do We Need Data and Statistics?   13

    2.2 Types of Data   16

    2.3 Displaying Data   18

    2.4 Distribution Types   22

    2.5 Distribution Shapes   24

    2.6 Sampling   27

    2.7 Comparing Distributions   29

    Points to Remember   36

    Exercises   36

Tests covered in
this chapter:
- Computing equal-probability binomials
- Computing binomial probabilities when the alternative events are not equally likely
- Applying the Poisson analysis for rare binomial events

3   Binomial Distributions   41

3.1 What Kinds of Measurements Yield Binomial Distributions?   41

3.2 The Product Law and the Importance of Independence   44

3.3 Comparing Distributions   48

3.4 Probability: Was the Sample Drawn from the Null Distribution?   48

3.5 The Role of Sample Size   55

3.6 The Issue of Tails   57

3.7 Post Hoc Analysis   59

3.8 The Uses of Post Hoc Probabilities: Bayesian Logic   60

3.9 Binomials with Unequal Probabilities   63

3.10 Binomials with Estimated Probabilities   65

3.11 Looking at the End: The Poisson Distribution   65

Points to Remember   71

Exercises   72

More Than the Basics   76

Error and significance   76

From binomial to Poisson   77

Tests covered in
this chapter:
- Testing for normality
- Testing two-choice distributions
- The Gauss test
- The one-sample *t*-test

4   Continuous Parametric Distributions: I   79

4.1 What Is "Parametric"?   79

4.2 What Sorts of Measurements Yield Parametric Distributions?   80

4.3 Computing the Parameters of a Parametric Distribution   83

4.4 How Do We Know If a Distribution Is Parametric?   88

4.5 What Do Binomial and Parametric Distributions Have in Common?   91

4.6 Comparing a Parent and a Sample Distribution   93

Points to Remember   101

Exercises   102

More Than the Basics   106

From binomial to bell   106

Why two sets of symbols?   107

The parameters   107

The one-sample t-test   109

5   Continuous Parametric Distributions: II   111

5.1 Sample Size and Certainty   111

5.2 Comparing Sample Distributions: The Logic
of the *t*-Tests   112

5.3 Checking That Variances Are Equal   115

5.4 Other Reasons to Compare Variances   117

5.5 Comparing the Means of Paired versus
Unpaired Data   121

5.6 What Do You Do If the Standard Deviations
Are Not Similar?   123

5.7 The Standard Error and Confidence Intervals   126

Points to Remember   131

Exercises   132

More Than the Basics   135

Two-sample t-test   135

Two sample t-test with dissimilar SDs   136

Variance, additivity, and standard error   136

6   Data Transformations   139

6.1 Why Parametric Is Better   139

6.2 Nonparametric May Be Only Skin Deep   140

*Tests covered in
this chapter:*
• The *F*-test
• The paired *t*-test
• The unpaired *t*-test
• The unequal SD *t*-test

*Test covered in
this chapter:*
• Transformations

6.3 The Catch 141

6.4 Popular Transformations 141

6.5 Sample Means: The Ultimate Transformation? 143

6.6 Comparisons from Sample Means 145

Points to Remember 146

Exercises 146

More Than the Basics 148

Additivity 148

Guiding principles 148

Sample sets and SEs 149

7   Multiple Parametric Distributions: ANOVA   151

7.1 Why It's Illegal to Perform Multiple *t*-tests within a Data Set 151

7.2 Comparing Means without Comparing Means 154

7.3 But Which One Is Different? 163

7.4 Two-Way ANOVAs 165

7.5 Hoary Extensions of the Two-Way ANOVA 169

7.6 Other Beyond-the-Scope ANOVAs 172

7.7 What to Do with Multiple Pairwise Comparisons 174

Points to Remember 178

Exercises 180

More Than the Basics 182

The F-distribution 182

One-way ANOVA computation 183

The Tukey-Kramer Method 184

Two-way ANOVA computation 185

Additivity revisited 186

ANOVA III 186

8   Categorical Data   189

8.1 Where Do Categorical Data Come From?   189

8.2 Comparing a Sample to a Null Distribution   190

8.3 How Is the Goodness-of-Fit Test Different
    from the Binomial Test?   194

8.4 Applications of Chi-Square When Category
    or Binomial Probabilities Are Estimated   196

8.5 Chi-Square and the Quick-but-Dirty Approach   198

8.6 Comparing Two or More Sample Distributions   199

Points to Remember   205

Exercises   206

More Than the Basics   211

   Chi-square goodness-of-fit computation   211

   Chi-square independence computation   212

   Fisher's Exact Test   212

*Tests covered in
this chapter:*

- Chi-square goodness-
  of-fit test
- Chi-square indepen-
  dence test

9   Nonparametric Continuous Data   215

9.1 Why Nonparametric Tests Are Less Powerful   215

9.2 Testing Paired Two-Sample Data   217

9.3 Evaluating Grouped Multiple-Sample Data   220

9.4 Testing Unpaired Two-Sample Data   222

9.5 Evaluating Unpaired Multiple-Sample Data   224

Points to Remember   226

Exercises   227

More Than the Basics   232

   The signed-rank computation   232

   The Friedman computation   232

   The rank-sum computation   233

   The U-test   233

   The Kruskal-Wallis computation   233

*Tests covered in
this chapter:*

- The signed-rank test
- The Friedman test
- The rank-sum test
- The Kruskal-Wallis
  test

Tests covered in this chapter:

- The Rayleigh z-test
- The Rayleigh u-test
- The Watson-Williams test

10 Circular Data   **235**

10.1 Where Do Circular Distributions Come From? 235

10.2 Determining the Mean Bearing
        and Degree of Dispersion   236

10.3 Testing for Clustering   240

10.4 Testing a Specific Hypothesis   242

10.5 Comparing Two Samples   245

Points to Remember   248

Exercises   248

More Than the Basics   252

   Circular distributions   252

   The Rayleigh z-computation   252

   The Rayleigh u-test   252

   The Watson-Williams computation   253

Tests covered in this chapter:

- Nonparametric correlation analysis
- Parametric correlation analysis
- Linear regression analysis

11 Relationships between Variables I: Correlation and Regression   **255**

11.1 Correlation versus Cause and Effect:
        When to Draw the Line   255

11.2 Correlating Nonparametric Data   260

11.3 Parametric Correlation Analysis   265

11.4 Linear Regression Analysis   267

11.5 Which Hypothesis to Test?   273

11.6 What If the Data Are Nonlinear?   275

11.7 Cause and Effect?   277

Points to Remember   279

Exercises   280

More Than the Basics   286

   Calculating nonparametric correlation values   286

Additivity once more   286
Calculating parametric correlation values   286
Calculating linear regression values   287

12 Relationships between Variables II: Multiple Regression   289

12.1 How Multiple Variables Interact: Path Analysis   289
12.2 The Goal of Two-Variable Multiple Regression   294
12.3 Admit or Reject? The Problem   295
12.4 Sorting Out the Influence of Multiple Variables   298
12.5 Higher-Level Multiple Regression   302
Points to Remember   303
Exercises   304
More Than the Basics   306
Accounting for variance   306
Calculating multiple regression   307

13 None of the Above   309

13.1 The Quick-but-Dirty Approach   310
13.2 The Academic Approach   313
13.3 The Hard Way: Monte Carlo Simulations   314
13.4 Computers: From Zero to Null in 14 Hours   317
13.5 The Theory behind Statistics   318
Points to Remember   321
Exercises   322

14 Once Over Lightly   325

14.1 Distribution Types   325
14.2 Types of Data   326

Test covered in this chapter:
• Multiple regression

Test covered in this chapter:
• Monte Carlo analysis

14.3 Characterizing Distributions   326

14.4 Distribution Shapes   327

14.5 Statistical Tests   328

14.6 Which Test Is Appropriate?   329

Exercises   333

Answers to Exercises   **335**

Statistics Labs   **361**

Selected Statistical Tables   **369**

Glossary   **405**

Bibliography   **417**

Index   **419**

PREFACE Origin and Overview

This book has two origins. The first was an effort to rescue some important data on honey bee learning: one of us discovered, after the fact, a statistical flaw in an experimental design. The problem (like the one that initiated the study of probability by Pascal in 1654) involved the effect of stopping a series of events before their planned termination. This experience led both of us to look into the logic of the many statistical tests we had been applying mindlessly to data.

The second and more important instigation was the need to teach a survey of statistical methods, first to students in Princeton's animal behavior laboratory course, and then to biology majors, and finally to nonmajors—students in the other natural and social sciences. The analysis of complex data has become a fundamental component of every branch of biological studies, from animal behavior to ecology to genetics. We've now used this coordinated text and software at Princeton for fifteen years.

Problems and Balance

We wrote this book because choosing the right text for our needs was a problem: we could not find a book that focused on supplying students just enough understanding of probability to design efficient experiments and choose the appropriate statistical test. Texts that are heavily mathematical are excellent for many students, but ours were overwhelmed by derivations and numerical details that they would gladly have taken on faith. As our result, our emphasis in *BioStats Basics* rests on the *logic* behind choosing statistical tests rather than their mathematical derivations and pencil-and-paper computations.

Another problem for us was that most books focus heavily or exclusively on parametric analysis. Our students, however, often find themselves out of necessity collecting or analyzing nonparametric data; even when their data *might be* (or perhaps even *ought to be*) parametric, the sample sizes are frequently too small to allow them to be sure of this. Therefore, we include in this text a broad coverage of the two dozen most important tests for both parametric and nonparametric data.

A third problem was that most texts are so formal that, in their attempt to adhere to strict linguistic distinctions important to statisticians, the plethora of awkward and narrowly defined terms becomes confusing. The previous paragraph is a good example: in it we refer to nonparametric data, a class of distributions familiar to most researchers. But how many know (or would care) that technically there is no such thing as parametric and nonparametric data? There is parametric analysis (and normal data), and there is nonparametric analysis (and non-normal data). This is a counterproductive level of verbal precision to inflict on students just trying to learn how to design experiments and evaluate their results—and a level not met by most practicing scientists. In short, we wanted a more informal and practical presentation of probability and statistical testing.

We developed the computer aid *Biostats Basics Online* to help students around the difficulties encountered with conventional statistical programs. Though the packages work well, they are not coordinated with a clear introductory text; they presuppose that the student understands and actively remembers the appropriate context for the tests, as well as the limitations and presuppositions for each. The ability to input data and receive back a precise-looking number seems to mesmerize some students, while others find the undocumented blank columns, waiting to be filled with data, simply intimidating.

Biostats Basics Online: A Tutorial-Based Approach

Our original solution was a simple HyperCard-based tutorial that began with an intuitive introduction to probability (liberally supplied with user-controlled simulations) followed by "cards" that explained each test and provided type-in fields so that the student could enter data and have them analyzed. Having an easily used test juxtaposed with a basic explanation of its workings, logic, and conditions was a great success from the outset. We combined this tutorial with a two-hour lab in which students gathered data on the heights of the two sexes and then analyzed them using the five most common tests. They learned at once the differing degrees of power of these various statistical techniques. The version of *BioStats Basics Online* that accompanies this text is much expanded and allows students to save and manipulate data to a much greater extent. It is the evolutionary product of many cycles of feedback from students and teaching assistants; we are grateful to the hundreds of student comments that have shaped the present version, as well as to our colleagues Andy Dobson and Henry Horn (who have taught courses based on the text and software and who provided important suggestions). *BioStats Basics Online* can be downloaded from the text's Web site at www.whfreeman.com/gould.

Why Is This Book So Thin?

The present book is the result of our fifteen years of experience teaching statistics to students who need the right mix of theory and practice to design experiments and analyze results, and whose interest needs to be kept fully engaged at every step along the way. This is why we get into analyzing data and testing hypotheses very early, while we are still learning about probability. It also explains why we concentrate on the two dozen most useful tests, plus several methods of transforming data: for our audience, this is an almost complete inventory of what they are likely ever to need.

Finally, it accounts for the absence of step-by-step descriptions of how to perform each test with pencil and paper: we assume our students will use *BioStats Basics Online* or another software tool to perform the computations. Such examples as we do develop in detail (e.g., binomials, integrating the area under the Gaussian curve, chi-square, and two of the nonparametric tests) are chosen because they make a clear link between the data, the test methods, and the meaning of the statistics. They are reality checks of a sort, and, presented as they are in the context of an example, they are anything but tedious.

The chapter on Monte Carlo methods is more useful to students than the details of additional special-case ANOVAs. One result of this self-imposed limitation to the main tests is that we have been able to create a logical flow sheet of questions and conditions that guide the student to the correct test; this useful tool is found inside the back cover of the text, as well as in each chapter.

Our focus on the important over the encyclopedic, the practical over the theoretical, has made this more of a handbook than a conventional text. To learn statistics is not to read about statistics but to use these tools in practice. We designed the book to support such active usage. The size and spiral binding support the portability of the work. The easy-to-use test-choice tree found at the back of the text provides the reader with the logic connecting research questions, data collection, and statistical choices. This serves as a quick way for the user to choose which statistical test or tests are the appropriate tools to explore further in analyzing the particular biological question at hand. And each chapter is tabbed for quick reference.

We can say without much chance of contradiction that many professional statisticians will find the irreverent tone of the book offensive; our approach is useful in helping to demystify statistics for most students, but we can see that a more reverential commentary would better suit a number of our mathematically gifted colleagues. We also know from experience that departments that deal with a fairly narrow range of data types will find the breadth of our approach unnecessary. Some colleagues in social psychology, economics, and sociology, for instance, deal (or pretend to deal) exclusively with parametric data; for them a text that devotes itself to an in-depth treatment of t-tests and parametric correlations is far more appropriate than one that wanders into nonparametric tests, circular statistics, and Monte Carlo simulations. Experience has taught us, though, that the level and breadth of our presentation are ideal for most students in the life sciences.

Examples and Exercises

Our students come from a variety of departments in the social and natural sciences; and therefore we have drawn examples from a range of fields, but with an emphasis on experiments that are both interesting and accessible. Many are from senior honors thesis projects. *BioStats Basics Online* includes data sets from many experiments and observational studies, as well as those bottomless sources of statistical fodder, ecology, medicine, and baseball. Some of the end-of-the-chapter exercises use these data sets, while others depend on shorter sets found in the text itself. (All these are also are found on *BioStats Basics Online* for the students' convenience.)

Our examples are limited to those for which we can readily lay our hands on the original data. We would very much welcome interesting data from our readers. The best way to send it is by e-mail (with a description of the experiment) to Gould@Princeton.edu or to GGould@alum.MIT.edu.

Passive analysis of canned data is no substitute for analyzing data you have gathered yourself. Moreover, active data gathering provides firsthand experience with the difficulties, ambiguities, and critical choices that any researcher must face in formulating hypotheses, designing experiments, and organizing the results. We find, therefore, that where time allows, a well-conceived lab can greatly enrich a statistics course. "Statistics Labs," beginning on page 361, provides outlines of several lab projects that have worked well for us.

Cause and Effect

A table in an aviary used for testing hungry blue jays is decorated with 16 mealworms, laid out in a 4×4 array (Figure 1-1). Meal-worms—which are not worms at all but the brown, hard-skinned larvae of grain beetles—are normally a preferred food for blue jays. In this case, however, the blue jays are in for a nasty sur-prise. In the wild, many animals must learn which objects in the environment are edible and which are not. The aviary experi-ment we are describing is designed to study this process under

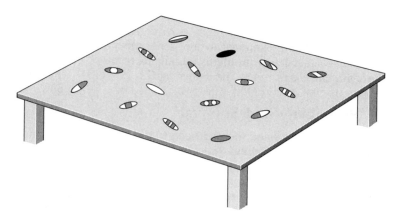

1-1 A testing table used for studying how birds learn to recognize toxic food. The 16 mealworms are painted with a variety of colors and patterns. Four of the mealworms have been dipped in a bitter solution. One or more elements of the colors or patterns is consistently correlated with unpleasant prey.

controlled conditions. Four of the mealworms have been dipped in quinine, an alkaloid with an extremely bitter taste. All have been decorated with nontoxic paints. Naïve birds will eat the painted but undipped mealworms, but they will spit out the quinine-treated ones and then work to get rid of the taste by drinking, vomiting, wiping their beaks, and so on.

The problem for these birds is to figure out, with the minimum number of unpleasant mistakes, what combination of markings predicts a bad taste while still consuming the maximum number of nontoxic prey. The problem for the researchers is to determine, with the minimum number of repetitions of the experiment, how quickly the birds find a solution. Both species are faced with the worry that a series of lucky or unlucky samplings by an individual bird may give an incorrect impression about either what is safe or how quickly the learning is occurring. For both, then, chance generates an unavoidable element of "noise" in the experiments, and can lead to mistaken deductions in the short run. On the other hand, sampling an infinite number of mealworms, or running an infinite number of tests, is an unacceptable price for the birds and the researchers alike.

This book looks at the ways researchers solve the intellectual puzzle of sorting out chance and causation. Both our desire and our ability to do so may owe a great deal to our evolutionary history as animals, when generation after generation of our ancestors faced problems analogous to those confronting the wary blue jays in the mealworm test.

1.1 Probability and Survival

We humans have a deep-seated need to discover the cause-and-effect relationships behind the events in our lives and in the world around us. The prosperity of astrologers suggests that this need is, if anything, stronger when there are apparently no causes to be found. This desire to understand the workings of the world is the fuel that powers the natural and social sciences. Its origin is doubtless in the need to learn from experience—but from a limited set of experiences. Even insects regularly infer cause and effect when they deduce which features are consistently associated with a food source and which are irrelevant.

This process of learning to recognize causal connections is called *classical conditioning.* Ivan Pavlov stumbled across it during his studies of digestion in dogs. When a normally meaningless cue regularly preceded the presentation of a food dish, dogs came to understand the connection and salivated in expectation of the food. Underlying this seemingly trivial behavioral change is the same processing that generates our detailed understanding of nature and one another. We now know that classical conditioning is an elaborate though unconscious exercise in probability analysis in which experience is compared with chance expectations. As animals encounter the same innately recognized stimulus again and again (the taste of food for dogs or bees or birds), they sort out which of the cues present predict the stimulus well, which are less useful, which do not correlate at all, and even which predict the absence of the stimulus. They are not perfect at this task: "superstitious" behaviors sometimes develop based on a few early coincidences an animal chances to experience. But the innate use of probability to sort out apparent cause and effect is much more efficient for many behaviors than a rigid automatic response to some limited range of innately recognized cues.

Bees apply this probabilistic approach to odors, colors, shapes (Figure 1-2), nearby landmarks, locations, times of day, and even to the process of working out how best to manipulate flowers so that they can harvest their food quickly—a kind of feedback-based learning called *operant conditioning.* Their ability to learn what does *not* predict food is seen when they avoid colors and shapes that have previously offered only water. Some species, including the herring gull, actually learn negative correlations *better* than positive ones; most animals, however, are like the herring gull's more optimistic cousin the laughing gull, and tend to learn and remember the positive correlations best.

Even before learning begins, gulls and honey bees use probabilities to guide their experimentation. For bees this means selecting which of the many stimuli they encounter to land on and explore as possible flowers. Violet-colored targets are selected as plausible candidates more often than yellow, and green targets are tried only rarely; likewise, visually complex candidates get

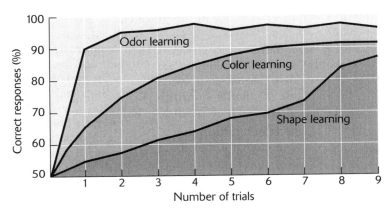

1-2 In a two-choice test the probability of a bee selecting the correct cue depends both on the number of times it has been fed on that cue previously (the number of trials) and the kind of cue (odors are learned much faster than shapes). Because this is a two-choice test, bees score about 50% correct by chance on the first visit, before any learning can occur. (We will explain in Chapter 2 that connecting the data points with lines in this kind of test is not technically correct; until then, however, we will indulge in this satisfying visual trick.)

more attention than simple shapes (Figure 1-3). Nevertheless, bees hedge their bets and occasionally land on simple green targets—that is, they introduce a degree of adaptive randomness into their experiments. Gull chicks, too, experiment in their attempts to solicit food, pecking at a variety of objects (including their feet and blades of grass). But most often by far they peck at vertical bars moving horizontally—the cue provided by the beaks of their parents when they are offering to feed the chicks.

The use of probability by animals is greatly underestimated, a bias no doubt fed by the widespread (but incorrect) impression that chance is synonymous with unpredictability. But the purpose of probability analysis is to sort out chance and apparent causation—to find the "signal" in the apparent chaos of "noisy" experience. An impressive example of the innate use of probability analysis that brings this strategy into the range of

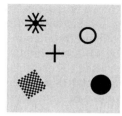

1-3 Although searching bees will land on any of these figures, they bias their "research" toward visually complex shapes. Thus the radial figure at the top receives more attention than the two circular figures, but bees land on the checkerboard in the lower left more often than on any of the other shapes.

our own experience is seen in human language acquisition. Infants are born with neural circuits that enable them to distinguish among the three dozen consonants found in human speech (standard English employs fewer than two dozen of these), and to parse vowel sounds in a way that factors out variability arising from differences between the pitch of one speaker's voice and another's. Separating a continuous stream of ordinary speech into distinct words (a feat still beyond computers) is a skill acquired at eight months of age when the brain automatically begins analyzing the probabilities of various combinations of sound transitions. Transitions between consonants within words are more common than transitions between consonants that end one word and begin another. Within just two minutes of exposure to nonsense syllables, human infants can extract the salient probabilities and make accurate guesses about word boundaries in any artificial language created by experimenters.

1.2 Probability and Research

Researchers, whether in the natural or the social sciences, use much more efficient and systematic versions of this trial-and-error process in their work. Though the details of the methods most appropriate to each field vary, they always require formulating clear, testable hypotheses about the apparent cause-

and-effect connections between events, and then testing these hypotheses by changing one factor at a time (Figure 1-4). If the results are consistent with the prediction derived from the hypothesis, they may go on to devise another test; if not, they reexamine the hypothesis itself. A truism of science is that researchers can never *prove* that a hypothesis is true; they can only exclude other hypotheses (and then only if the tests are well designed). The reason that a hypothesis cannot be proved is that no one can ever be sure it is the only explanation that can possibly account for the observations; this is why we've been

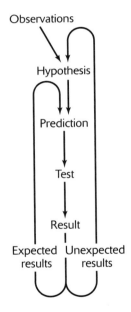

1-4 The researcher begins by formulating a hypothesis based on preliminary observations. The hypothesis is then used to make a prediction. A test is formulated with clear expectations of what will happen if the hypothesis is true versus the outcome if it is not; the not-true prediction is generally called a *null hypothesis*. The test is run and the result is compared to the predictions. If the result is consistent with the experimental prediction, new tests of the hypothesis are devised; if not, the hypothesis is reexamined and perhaps modified in preparation for a new round of testing.

careful to talk in terms of "apparent" causes. But by excluding (through repeated hypothesis testing) each of the plausible alternatives, a fair degree of confidence develops. Of course, that confidence can be misleading: the set of hypotheses that underlie Newtonian physics seemed to work well for two centuries, but Einstein showed the world that they weren't the whole story.

Let's look at this approach to research through the example of the toxic-food experiment that was in progress at the beginning of the chapter. The researcher may have begun with the casual and counterintuitive observation that brightly colored insects are consumed less often by many insectivorous birds than are their camouflaged cousins: monarch butterflies, milkweed bugs, and wasps, for instance, are ignored in favor of flies, moths, and caterpillars. What hypotheses spring to mind to account for this apparent avoidance of easy-to-see prey? It could be that the initial observation is simply wrong. Or, if avoidance is real, perhaps it is instinctive: the birds are programmed to avoid any brightly colored prey, or are provided with an innate, pictorial, enemies list. Or maybe the behavior depends on learning, either through personal experience or observation of the behavior of other birds. If experience is involved, perhaps the birds learn about toxic insects as a class (bright colors are bad news) or individually (yellow-and-black stripes on the abdomen are bad, as are orange-and-black wings). Of course, the avoidance might be partially or initially innate, with learning serving to modify the behavior.

You might begin with the hypothesis that the behavior is innate. How would you test that idea? You must have an alternative to compare your hypothesis to, a so-called ***null hypothesis***. (We'll discuss the null hypothesis in more detail later; for now we'll simply say that it is the hypothesis that the variable you're changing will not affect the results of your tests.) In the toxic-food experiment, the null hypothesis is that the birds have no innate information available. Now you need to decide how to compare these two hypotheses. You might settle on the strategy of decorating palatable prey like mealworms with the same markings as milkweed bugs, which blue jays usually pass up. You would doubtless want to offer unmarked mealworms so that even the hungriest birds would not be tempted by prey they "know" to be bad. But then these control mealworms will lack

The **null hypothesis** predicts that changing the variable under study will have no effect on the results. (See Chapter 2 for a more detailed discussion.)

any odor or texture that might be imparted by the paint; perhaps they should be given dull decorations. And if you are looking at innate knowledge, you'll need naïve birds. In short, you will need to think through all the relevant variables and be sure that all are held constant in the appropriate way except for the one (color) you *intend* to test.

There are still more decisions to be made: How should the prey be laid out so as to avoid biases? What kind of behavioral data should you collect? How much data will you need to gather? How can you determine whether any trend you observe is meaningful? As we will illustrate again and again, your knowledge of statistics can help you choose the optimum way of gathering data and, with a preliminary test, decide on the best sample size; as a result, you will be able to minimize the amount of work necessary for distinguishing between your hypotheses.

The efficient use of this systematic hypothesis-testing approach depends on understanding probability in order to design and analyze your tests for sorting out chance and causation. At the same time, it can run into other more subtle problems with probability. The most insidious temptation is to formulate a hypothesis *after* looking at the results; in fact, the odds of a certain result occurring are $1:1$ after it has occurred. In Chapter 3 we will consider in more detail when this formulation of a "post hoc hypothesis" can and cannot be used to identify cause and effect.

Even when the hypothesis is formulated in advance, there are potential worries and complications. As a researcher, you can often have more than one prediction of what may happen (in addition to the obligatory null hypothesis that the factor being changed is unimportant); how does that alter your analysis? Or it may be that you have previously collected data to compare with the new results, in which case the null hypothesis will instead be that the new data set will be essentially identical to the preexisting set. When you work with two sets of data—two "sample distributions"—you are faced with the worry that there may be chance variation in *both* of the samples you are seeking to compare, and thus two sources of "noise" (random variability) rather than one. Clearly, asking whether two distributions are different depends on how certain you are that each data set is a good representation of reality.

1.3 Recognizing Differences

Chance variation, whether in the sample you are dealing with or the one you may be comparing it to, is the major challenge in analyzing data. Because practical considerations limit the scope of an experiment to a finite number of tests, there is always the worry that the results obtained may not mean what they appear to. For instance, you may reasonably hypothesize that a flipped coin should turn up heads 50% of the time, and if in fact you were able to test it a very large number of times, the result would doubtless be very close to 50%. But if you test the coin only 10 times and find that it lands with the head side up 80% of the time, do you conclude the coin is biased, or that you were just "unlucky" in your first few tests? Is the anomalous outcome the result of chance or causation? How can you know without devoting inordinate amounts of time to testing? How much difference from the null prediction of a 50:50 outcome is worrisome?

Animals face this issue as well, and they have evolved so that their behavior balances the time spent in gathering "data" about food (or mates or danger) against the need to learn and respond quickly. In life-and-death situations they are even wired to jump to conclusions on the basis of a single experience. Rats, for instance, undergo "one-trial learning" when a novel food makes them ill, and thus they develop a better-safe-than-sorry phobia or superstition. A similar pattern is seen in humans undergoing chemotherapy.

Researchers also employ an intuitive sense of the likely and unlikely in choosing hypotheses and judging initial observations; indeed, we all do this in many aspects of our daily lives. But to evaluate experimental or observational results objectively, we must depend on probability theory and its associated statistical tests. Statistics allows us to quantify the probability that a particular result deviates only by chance from what we would expect if the variable we are studying has no effect. Now you might assume that this means that statistics can therefore tell us if a phenomenon—that is, the effect of a variable—is real. But as we said, statistics and the scientist's methodology can only prove hypotheses false: there is always some small probability that extreme results are the result of chance, as well as the much more common possibility that the variable you are studying is not the actual cause of the effect, but instead some secondary or correlated cause. Thus when we speak of chance and causation, we are referring to the probability that an effect has some cause as opposed to no cause (and is thus a chance result); if you are lucky in your thinking, the actual cause of the effect reflected in your data will be the possible cause you have in mind.

As we have said, a good understanding of the nature of chance and statistics also permits researchers to design clear-cut

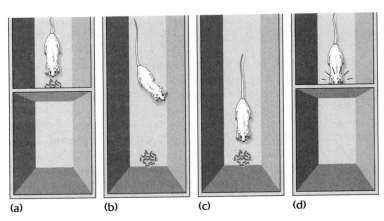

(a)   (b)   (c)   (d)

1-5 Habit formation—that is, operant conditioning—develops as animals correlate their behavior with its consequences. In this simple case, a rat has learned to choose a maze arm and run down it to a food dish **(a)**. In both training and testing the maze is well illuminated, and thus the food is clearly visible. **(b)** After numerous training trials the partition is removed and the food moved to the end of the arm. Now the rat reveals just how mindless its maze behavior has become: it stops where the food used to be and searches. **(c)** Eventually, after many more trials, the rat will run to the new location with no more than a slight break in stride at the original location. **(d)** If the partition and food are now returned to their original locations, the mostly reprogrammed rat will crash into the redeployed partition on the next trial. [Drawing based on J. Sparks, *The Discovery of Animal Behaviour* (Boston: Little, Brown, 1982), p. 106.]

experiments that minimize the number of tests and the amount of time necessary to complete a study. You can clearly improve on the innate in this regard: the literature of psychology is filled with examples of the illogical behavior animals can be forced into by their rough-and-ready systems for divining apparent cause and effect and generating efficient "habits" (Figure 1-5).

1.4 Probability and Math

Fortunately for all of us, the elements of probability necessary to an understanding of statistical procedures are fairly simple, much less abstruse than those steeped in statistical lore like to let on. And the basic rules for deciding what sort of tests are appropriate are also well within the grasp of ordinary mortals. Once you understand the different kinds of data that the world supplies to the curious, it usually becomes pretty obvious how they must be analyzed. Moreover, even the basic level of understanding we are aiming for in this book should allow the more adventurous to derive reasonable tests for novel kinds of data. After all, as you will see, some of the most commonly used tests were developed by trial and error to solve a pressing practical problem; the elegant but (in experimental practice) largely irrelevant mathematical derivations that fill the pages of probability texts were often discovered later.

Points to remember

✔ Statistical analysis exists to help distinguish chance variation from causal effects.

✔ Statistical analysis is used to ask how likely it is that any difference between one set of data and another set arose simply by chance.

✔ The null hypothesis in statistical analyses is that there is no difference between two sets of data beyond what would be expected from chance alone. Thus the null hypothesis predicts that the difference in experimental or other conditions that distinguishes the two sets will have no effect on the results.

Data

Life for researchers would be far simpler if, as was common even only 50 years ago, we could just collect our data and publish them, without worrying about statistical analyses. But experience has shown that we cannot trust unaided readers to estimate the reliability of a set of results unless the trend in the data is overwhelming. So, how are we to compensate for our species' probabalistic shortcomings with the least pain and suffering?

2.1 Why Do We Need Data and Statistics?

The essential job of statistics is to discriminate between chance and causation, and thus allow us to formulate and present a clear picture of the world. To do this two sets of data are generally compared. The default or **null hypothesis** is usually that one set of data was drawn at random from the other set—or rather, from all the instances of which the other set is a sample. Imagine a box with the heights of every adult female in America on individual slips of paper, and another box with the heights of every adult American male. If you draw one slip from one of the boxes, can you deduce which box the slip came from? If you draw 10 slips at random? Even if you know which box you are drawing from, how many randomly drawn slips would you need to draw in order to get a good estimate of the average height of males and females?

These questions are the province of statistical analysis. Statistics can never tell us for sure that you are drawing slips from the male box, for example. But it can tell you the exact

The **null hypothesis** predicts that the set of data being analyzed is drawn (that is, *sampled*) from the same large set of instances from which the other data sample is derived. Thus, allowing for chance variation in sampling, the two sets should be the same.

probability that you would draw this set of slips, or a set more extreme, by chance if you were to repeat the draw an infinite number of times. Statistics can also tell you how good an estimate of the average height on the slips in the box you are likely to have when you repeatedly draw this particular number of slips. Thus statistics tells us how unlikely a particular data set is, but it leaves it to us to decide if we draw the line between

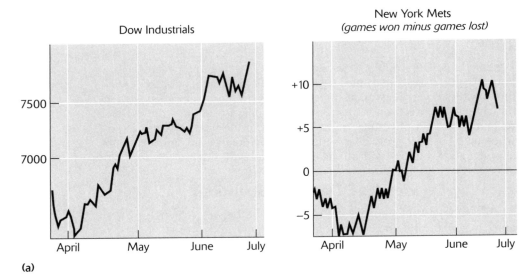

(a)

2-1 (a) This pair of graphs is from the *New York Times* of July 4, 1997; the graphs suggest a cause-and-effect relationship between the stock portfolios of Major League Baseball players and the performance of the team located closest to Wall Street. Since changes in the team record lag changes in the Dow Industrials by an average of five days, the stock market would logically have to be the cause. Not presented: records of the other 27 teams or data for the Mets in any other year. **(b)** Here is a clear relationship between two variables among biology students in a statistics class: number of syllables in an individual's given name and average grade point average (GPA) over the previous three years. Since each student had a name before taking any courses, the only plausible cause-and-effect inference would seem to be that name length enhances academic chance and causation at one in ten, or one in a hundred, or one in a thousand. And as we emphasized in the last chapter, even if the data clearly indicate a difference, statistics cannot tell us if the cause-and-effect relationship you had in mind when formulating hypotheses and experimental tests represents the true cause or is a secondary or correlated effect. (As a cautionary lesson, we offer Figure 2-1 for your consideration; the issues

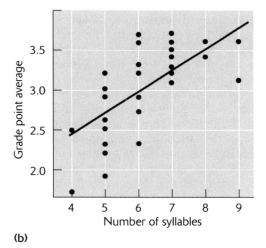

(b)

2-1 (*continued*) performance. Your personal experience may suggest that this is unlikely to be a real phenomenon, and impel you to look for some correlated- or secondary-cause effect.

When the data represent something more esoteric, however, the implausibility of the apparent cause may not be as obvious; even when the putative cause seems perfectly reasonable, showing a statistical relationship does not prove that the suspected cause is in fact the real one. (In this case, the correlation is a secondary consequence of female names being, on average, slightly longer than those of males, while the GPAs of female biology students are somewhat higher than those male GPAs. The ultimate causes of the name-length and GPA discrepancies are another matter entirely.)

surrounding cause-and-effect deductions are considered in more detail in Chapters 11 and 12.)

Statistical analysis requires understanding the ways in which data can be summarized and characterized. This chapter will introduce these essential preliminaries in preparation for the analyses that will be the subjects of the later chapters.

2.2 Types of Data

Data are nothing more than observations collected into a set. Each datum is one element of the data set. The nature of the data set determines which sorts of statistical approaches are appropriate. Thus it is essential for you to identify the kind of data to be tortured and otherwise manipulated before subjecting them to any math. Luckily, there are only three general types to distinguish among: categorical, continuous, and discrete.

Categorical data fall into mutually exclusive categories.

Categorical data are those that exist in mutually exclusive categories. For example, we frequently categorize humans as male or female (Figure 2-2a). Vertebrate animals may be categorized as fish, amphibians, reptiles, birds, or mammals; you have to be extremely well informed to quibble with these apparently mutually exclusive divisions. General Motors automobiles may be unambiguously categorized as labeled Chevrolet, Pontiac, and Buick, among others (we say "labeled" because they have so many internal parts in common). Statistical tests for categorical data are quite different from those for other kinds of data.

Continuous data are measured in terms of a continuously variable parameter.

Continuous data are those that measure a continuously subdivisible parameter. For example, human height is continuous because an individual may be 175.02 cm tall, or 157.03, or anything in between, not to mention shorter or taller (Figure 2-2b). Animal weights are continuous because any weight within the observed range of weights is possible. In fact, if your equipment is sensitive enough, you should be able to measure a decline in weight with each exhalation as water vapor is lost to the air.

Discrete data are measured in terms of a parameter that can have only certain discrete values.

Discrete data are those that exist as a continuum along which only certain values are possible. Thus the number of students in a class is discrete because the data can exist only as whole numbers (Figure 2-2c)—except in the minds of college

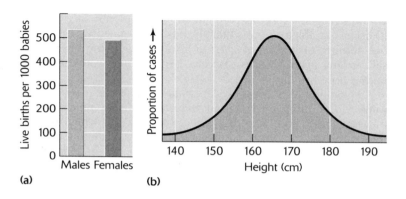

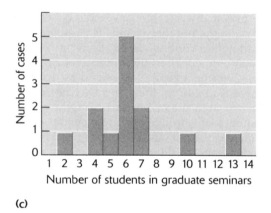

(c)

2-2 (a) Categorical data exist in one of two or more mutually exclusive categories; shown here is the slight (5%) male bias in live births. Because male infants survive less well than females, by age five the ratio approaches 50:50. **(b)** Continuous data can take any value in a range. Shown here is the distribution of adult female heights measured among members of a college staff. Any height that the measuring instrument is capable of resolving is possible. **(c)** Discrete data can take only certain values within its range. This distribution represents the number of students in various graduate seminars in Freeman University's biology department; all values are exact whole numbers.

2 Data

administrators, there are no fractional students or teachers (though our department officially has 12.17 faculty). Similarly, the number of brands of sport-utility vehicles being offered for sale in any given year is discrete, because it is constrained to be a whole number.

The difference between "continuous" and "discrete" is the same as the distinction made by the grammatically precise when they say "less" and "fewer": we have *less* rain, in which we are struck by *fewer* raindrops, because the former is a continuous measure (e.g., 1.61375 cc) while the latter is discrete (e.g., 43 drops).

Despite these differences, continuous data are often grouped into "bins" to create discrete data for convenience, especially when the data are to be plotted. For example, in analyzing height we might group, or "lump," all males of heights 174.50 cm to 175.49 cm into a discrete bin labeled "175 cm." If, as is frequently the case, the number of bins is large, the data (whether they were originally continuous or discrete) can be treated as essentially continuous; when the number of bins is small, however, the data will necessarily be treated as discrete.

The conversion can go the other way: were we to average class sizes across an academic department, the sum of the many discrete numbers of enrollees divided by the equally discrete number of classes is unlikely to result in a whole number. Data on average class sizes in different departments are now, as a result, continuous. But though discrete and continuous data can sometimes be manipulated from one sort into the other, categorical data remain resolutely categorical.

2.3 Displaying Data

A **distribution** is the way data are arranged with regard to a variable.

When you have a set of data, it usually forms a ***distribution*** of some sort. There is a deeply ingrained or encultured need in many of us to take a list of numbers and attempt to graph or otherwise plot it so we can visualize how the data are distributed. But this is not necessarily a simple matter. Were you to plot male heights, for instance, few of us would choose to place a point along a height axis for each male (Figure 2-3a). Instead, most of us would lump the data into bins and then represent

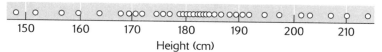

(a)

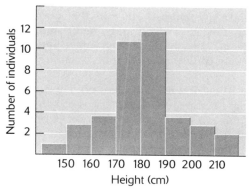

(b)

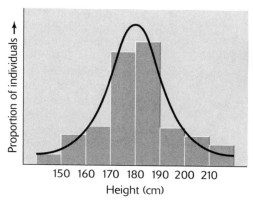

(c)

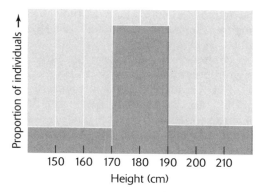

(d)

2-3 Heights of male graduates in a university biology department displayed in three ways: **(a)** true values are plotted along a height axis; **(b)** the values are lumped into ranges and plotted as a histogram; **(c)** a line is fitted by eye to the histogram (which is possible because the underlying data are continuous; **(d)** the values are lumped into histogram units with different *x*-axis widths.

the data distribution by using the number of individuals in each bin on the vertical (y) axis and cm of height on the horizontal (x) axis. Thus you would first obtain a different series of points, each representing the number of instances lumped into each bin (Figure 2-3b). Our mind's eye almost demands that points be connected by a line, either exact or idealized by being smoothed (Figure 2-3c). But since the data have been grouped into bins, the technically correct way of displaying them is as a *bar graph* (see Figure 2-3b).

The bar graph is a subset of a more general display strategy, the *histogram*. In a histogram, the area of each bar is proportional to the number of data points it represents. In the case of our bar graph for height, the width of each bar (2 cm of height) is the same in each case and the vertical extent corresponds to some discrete number of individuals in that bin. But were you to combine some bins at the end to smooth out the ragged extremes of the distribution, you would have a histogram. Now, however, the vertical extent of the bar will not necessarily correspond to a discrete number, since the *area* of the rectangle you create to group data must represent the number of individuals in the bin; the height of the bar is thus the number of individuals divided by the number of bin widths (see Figure 2-3c).

Because height is a continuous variable, we do little violence to the data when we represent the discrete distribution as a line (see Figure 2-3c). But consider the data in Figure 2-4a, which shows the number of chicks fledged per female as a function of the number of females nesting on one male redwing blackbird territory. The temptation to draw a line through these points is almost overwhelming (Figure 2-4b), but note that the values on the x axis are not bins but discrete values—there cannot be 3.3 nests on a territory. Thus the only way to plot these data is as a histogram (Figure 2-4c). But do not despair: in Chapter 11 we will discuss a circumstance in which a straight line *can* be drawn legitimately through this set of points.

Another way of displaying data is the *scatterplot,* in which two values are represented for each data point. Let's say you are looking for a relationship between math and verbal SAT scores. You might plot the mSAT score of each individual along the y axis and the vSAT along the x axis (Figure 2-5). Each point

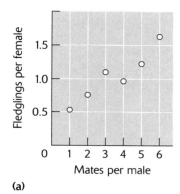

(a)

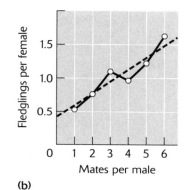

(b)

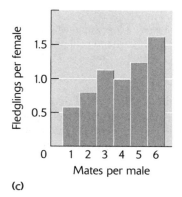

(c)

2-4 Number of redwing blackbird chicks fledged on territories with
different numbers of redwing nests (mates per male). **(a)** The data;
(b) the data connected by a solid line and approximated with a
dashed straight line; **(c)** the data plotted as a bar graph. Of these,
the solid line is the only plot that is clearly wrong, because it implies
that the data are continuous along the mates-per-male axis. The
dashed line is possible only if the fledgling data are distributed in
a particular way (discussed in Chapter 11). [Based on data in
C. H. Holm, *Ecology* 54, 356–65 (1973).]

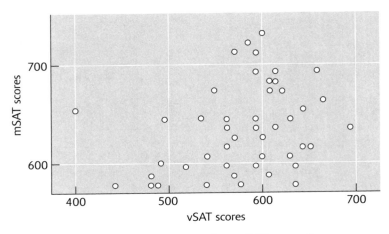

2-5 A scatterplot of vSAT scores against mSAT scores for one university's biology majors. Chapter 11 describes how to draw a best-fit line through data like these.

represents one individual's pair of scores; the pattern of points may suggest some relationship between these two variables. In Chapter 11 we will explain when it is legitimate to plot a best-fit line though this set of points, and how to draw that line and decide what it means.

2.4 Distribution Types

Now that you can recognize data types and plot their distribution, we can begin to reflect on what these distributions can actually do for us and our understanding of the phenomena we hope to analyze. Sometimes in this book we will be plotting distributions and analyzing them in isolation; the scatterplot of SAT scores could be dealt with in this way. More often, however, we will be comparing two (or more) distributions with each other.

There are three general types of distributions to be compared. One of these, almost inevitably, is the **sample distribution**—the data you are trying to interpret. You obtain the sample distribution by collecting some number of data points from

A **sample distribution** is a set of data sampled from the population of all possible instances of the data.

the larger collection of all possible data points; this large, all-inclusive (and often imaginary) collection is the ***parent distribution*** (or, to the statistically precise, the ***population***; but because "population" has a clear real-world meaning that can arise while working with data—studying a population of redwing blackbirds, for instance—we will avoid using it as a statistical term). Thus if you collect data on the heights of 25 American male college students, you have a sample distribution. If you then compare it to the distribution for all American males (or all male college students, or all college students), you are comparing this sample to a parent distribution. The parent distribution, since it includes all the data, is perfectly "known"—that is, in statistical parlance, there is no error in the sample beyond what might have been introduced in measuring the heights in the first place. In practice, a sample distribution itself can be so large that it can serve as a parent distribution because the likely chance error in a sample becomes very small as the sample size grows.

> A **parent distribution** (or **population**) is the collection of all possible instances of whatever is being measured.

2 Data

Another kind of distribution we might be comparing our sample to is a ***null distribution***, an idealized predicted distribution which, as a product of the mind, lacks even measurement error. (Statisticians prefer the term ***hypothetical distribution***, but again since we regularly deal with the real-world concept of hypotheses, it seems clearer to use the intuitively clear term "null distribution" instead.) Thus you might have created a mathematical model for growth that predicts a certain distribution of heights; you could then compare your sample distribution (or even the parent distribution) to this null distribution. The null distribution is sometimes confused with the crucial null hypothesis, especially when the null hypothesis is based on a null distribution.

> A **null distribution** (or **hypothetical distribution**) is an idealized predicted distribution.

> The **null hypothesis** is that the sample distribution being analyzed is drawn from the parent distribution, the null distribution, or the other sample distribution with which you are comparing it; thus there is no difference between the distributions beyond that which chance might create.

Now that we've distinguished among distribution types, we can refine our definition of the null hypothesis: the typical ***null hypothesis*** is that the sample distribution you obtain or observe is drawn from the parent distribution, the null distribution, or another sample distribution. To put it another way, the null hypothesis is that the variable, trend, or population you are looking at in your sample causes or creates *no difference* between the sample distribution on the one hand and the alternative distribution to which you wish to compare it on the other.

As we hinted in the previous paragraph, sample distributions are not always compared to either a parent or null distribution; at least as often you will be comparing two sample distributions with each other. Such comparisons generally involve two sets of data that differ in one particular. Figure 2-6, for instance, compares male and female shoe sizes: gender, obviously, is the factor that distinguishes the two distributions. In this case, the null hypothesis is that the two sample distributions are drawn from the same parent distribution. Whatever the comparison, then, the null hypothesis is that there is no difference between the parent distributions being contrasted.

2.5 Distribution Shapes

It is convenient to refer to these various distributions by their shapes, and in fact the shape often tells us something about how a distribution can be analyzed. Two of the distributions we have

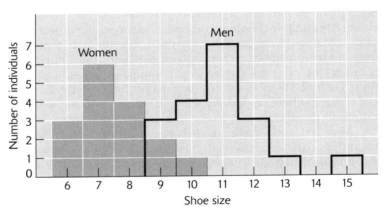

2-6 The two distributions compared here are male and female shoe sizes in a statistics class, with male sizes converted to equivalent female sizes so that all sizes can be plotted on the same axis.

looked at in this chapter are commonly called ***normal distributions***; we will also call these ***parametric distributions*** for no better reason than that most scientists, but not statisticians, use the terms interchangeably. As we will discuss later, it is technically possible to consider any distribution parametric, but to clumsy scientists like ourselves, "normal" and "parametric" are synonymous; and because normal has a clear real-world meaning while "parametric" does not, we will use the latter despite its high quibble value among those seriously into statistics. The data contributing to a parametric distribution are commonly referred to as ***parametric data***. Figures 2-3 and 2-6 have the characteristic ***bell-curve*** shape of classic normal (parametric) distributions. The mSAT and vSAT data are also parametric, though this is not obvious in the scatterplot (see Figure 2-5). Parametric data, as we will see, are suitable for a variety of powerful statistical analyses.

> A **normal** (or **parametric**) **distribution** is one whose data are distributed in a characteristic **bell-curve** pattern.

    Nonparametric data form ***nonparametric distributions*** that are not bell shaped. (Statisticians prefer the term ***nonnormal data***.) These distributions may, like those of parametric data, be symmetrical, but be too peaked in the center or too flat to qualify as parametric. More often, nonparametric data are skewed or have more than one peak. Figure 2-7a shows the skewed times recorded from chickens feigning death in the presence of an apparent predator; Figure 2-7b shows the bimodal (two-peaked) curve representing American heights—a mixture of two parametric curves, one for males and another for females. Discrete data with very few bins are necessarily nonparametric, even if the distribution is symmetrical. Categorical data (e.g., brand of car) are obviously nonparametric because even with many categories there is no logic for placing them in any particular order along the x axis.

> A **nonnormal** (or **nonparametric**) **distribution** is one whose data are distributed in any manner other than the bell-curve pattern characteristic of parametric distributions.

    Finally, there are ***circular distributions***. The data that generate these distributions arise from measurement along an axis that repeats itself periodically, such as direction (where $0°$ is the same as $360°$), time of day, day of the week/month/year, and so on. The data in Figure 2-8 represent the directional estimates of undergraduates attempting to point to campus after being displaced several miles along an indirect route while blindfolded. (They were clueless.)

> A **circular distribution** is one whose data are taken from a variable whose value repeats itself.

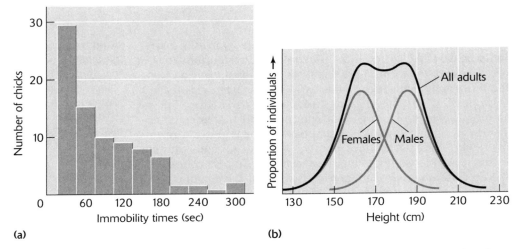

(a)

(b)

2-7 (a) The distribution of immobility times (death feigning) among 79 chicks tested under control conditions; this distribution is skewed to the left. **(b)** A distribution of adult height (parents of college students). The curve is bimodal (two-peaked) because it mixes two different distributions: that for males, centered on 180 cm, and that for females, centered on 165 cm.

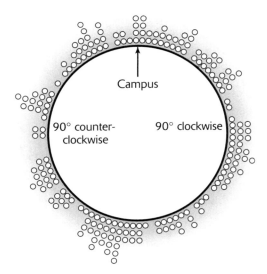

2-8 The estimates by 74 students of the direction to the campus from one or more of 13 sites 8–20 km from campus. The students were transported to the test sites blindfolded in vans or buses with covered windows.

2.6 Sampling

The sample distributions we have been talking about are created by gathering data. It is no trouble at all, however, to collect worthless data, samples that are biased in some way. If, for instance, you wanted to gather data on female height, you would be asking for trouble if you collected your data outside the dressing rooms of Paris fashion models. One of the largest samples ever collected, the *Literary Digest* poll for the 1936 American presidential election between Franklin Roosevelt and Alfred Landon, was also one of the most biased: the magazine ran its 10-million-questionnaire survey by mail, addressing it to names largely drawn from telephone books. Having telephone service in 1936, the middle of the Great Depression, was too expensive for a large segment of the electorate, so the pollsters were writing mostly to Republicans. And another bias at this stage is likely because only about a quarter of those polled returned the questionnaire.The pollsters concluded on the basis of a sample size of 2.4 million that Landon would win by a large margin (57 : 43), whereas in fact he lost by a landslide (38 : 62). (The Gallup poll, with a sex- and income-weighted sample just 2% as large, predicted that Landon's loss would be 44 : 56.)

Reading serious discussions of possible sources of bias is so depressing that few of the more careful and humble among us would ever dare collect any data at all. (We will look at some particularly amusing examples of covert variables and hidden effects in Chapters 11 and 12.) In general, all that can reasonably be asked of you is that you think first and collect data second; do the best you can to be sure that what you are measuring is really what you think you are measuring; and be sure that the single variable you are seeking to investigate is actually the only variable in play and likely to be a primary cause rather than a secondary effect. To have any meaning, your samples must be randomized and unbiased. You must keep clearly in mind the possible differences between the population you *want* to sample and the one from which you may *actually* be collecting data: an ecologist attempting to sample the population of small mammals in a field using food baits, for instance, is actually looking only at the mammals attracted by the particular bait and not frightened by the trap itself. Often we have had to calibrate the nature of our sample population through an independent experiment just to be sure we weren't fooling ourselves (or that the creatures we intended to study weren't fooling us).

The other important aspect of sampling is simply to be quite sure what sort of information you have. For instance, if you measure the mate preference of a particular female guppy on 50 consecutive days, you learn something quite different from what you discover if you measure the mate preference of 50 different female guppies, one each day for 50 days. In the former case, you have a pretty exact description of one female; in the other, you have a group sample of 50 females. Of course, whether these 50 females represent a particular commercial strain of guppy or the species as a whole is another issue, as your colleagues will hardly hesitate to point out if you fail to mention it first. Similarly, when we measured the heights of 50 female American college students, we were not necessarily collecting data about the heights of American females—indeed, the average height of female American college students is roughly 3 cm higher than the mean height of adult American women.

Many of the examples we use in this book are derived from laboratory studies. They are chosen for their clarity: it's pretty obvious which variables are held constant and which one is changed; the experimental situation is (or ought to be) under the fairly complete control of the researcher; most of the ambiguities alluded to earlier can be reasoned out in advance or detected from anomalies in the data. When working under less controlled conditions outside the lab, however, you may not be able to isolate variables cleanly. For instance, you might want to explore the role of gender in the size of animal home ranges, but controlling for weight or age or reproductive condition may be difficult. These confounding variables are a serious challenge to field studies and limit the interpretations you can reasonably wring from your data. We will describe some approaches (e.g., ANOVA, chi-square for independence, and multiple regression) that can sort out the effects of multiple variables when you have enough data and can identify the crucial factors to be considered. But a strong suspicion that an important confounding variable may be at work must always be maintained: no measure of statistical "significance" has much scientific meaning if a powerful variable has eluded detection.

2.7 Comparing Distributions

The essence of most ***statistical analysis*** lies in comparing two distributions to see how likely it is that they are actually from the same parent distribution. Thus when you use statistics to compare a sample of male heights to the complete (parent) distribution of female heights, you are asking for the precise probability that your particular sample of male heights (or any sample whose distribution is more extreme than what you have observed) could have been obtained by sampling from the parent distribution of female heights instead. If two distributions—in this case, a male sample distribution and a female parent distribution—look quite different, you can guess that the odds are against the sample being from the parent distribution. But what features of the two distributions are you to compare, and how can you calculate the chance that the two distributions are, despite appearances, from the same source?

Parametric distributions get their name from the ability of mathematicians to describe the curve completely with only a few parameters. As we said, for our purposes a "parametric distribution" means a normal curve, describable completely by two parameters: mean and variance (or, more often, standard deviation); for statistical analysis, you will also need to know the sample size. All such parametric curves are identical once they are displaced and scaled on the x axis according to their mean and variance (Figure 2-9; if the y axis is calibrated in terms of number of data points rather than proportion of the data, the curves will have to be scaled for sample size as well). There are also other useful parameters that help reduce a mass of data to its statistical essentials, and even nonparametric curves have some helpful parameters. (If you want to be picky, it is possible to derive a formula with enough variables—call them "parameters" if you like—to approximate any continuous or discrete

Statistical analysis usually calculates the chances that two sample distributions are drawn from the same parent distribution.

2 Data

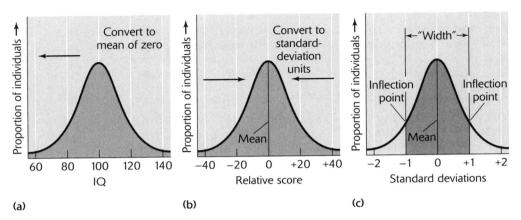

(a) (b) (c)

2-9 Any parametric curve can be converted into any other parametric curve by scaling its parameters: "average" (mean) and "width" (standard deviation; see definition on page 33); sample size may also need to be scaled. For convenience, the standard parametric curve used for statistical analysis has a mean of zero and a width of two standard deviations, and is plotted as a frequency distribution (that is, with an effective sample size of one). Here the ordinary parametric distribution of IQ is converted, step by step **(a, b, c)**, into the standard curve; the sequence begins after the y axis (number of individuals) has already been converted to frequency by dividing all y-axis values by the sample size.

Part **(c)** shows the "width" of a normal curve, the distance between the two points on the curve left and right of the mean where the curve "inflects"—that is, where it switches from accelerating to decelerating. Left of the mean, the inflection point is the point where the curve switches from accelerating up from the x axis to decelerating toward the maximum y-axis value; right of the mean, it is the point where the curve switches from accelerating toward the x axis to decelerating toward the limiting value of zero.

distribution; and certain distributions, including the binomial and Poisson distributions discussed in Chapter 3, have a small number of defining parameters, though they are not mean and standard deviation. To repeat for emphasis, our use of "parametric" to refer *only* to normal distributions conforms to common research usage, but would be laughed to scorn in an advanced probability course.)

Any continuous or discrete distribution has a mean, a median, and a mode. The ***mean*** is simply the arithmetic average of all the data points (Figure 2-10). This value is used in a variety of parametric comparisons. It is the appropriate measure there because a parametric curve is symmetrical about its mean.

The ***median***, on the other hand, is the value of the middle datum when all the values are arranged in order of magnitude. If there are nine points, therefore, the median is the value of the fifth largest (Figure 2-11a). If there are ten points, however, the median lies between the values of the fifth and sixth points (Figure 2-11b); most often the median value is taken to be the average of these two points. The median is a very useful value for describing the average value of a set of nonparametric data. Many researchers, however, use the mean instead of the median for no better reason than force of habit; but the mean is usually a misleading way to characterize nonparametric data. (The mean and median are the same for parametric and other symmetrical distributions.)

The ***mode*** is the most common value in the distribution. While this parameter is rarely used in statistical analysis, it does

The **mean** of a distribution is the arithmetic average of all the data values.

The **median** of a distribution is the value of the middle datum when the data are arranged in the order of their values.

The **mode** of a distribution is the most common data value.

2 Data

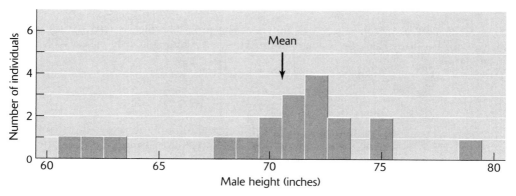

2-10 Heights (in inches) of males in a biology class. The mean is 70.55, about the national average in the United States for male college students despite the overrepresentation in the left tail of the distribution.

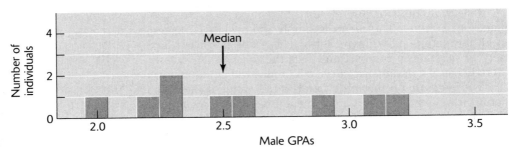

(a)

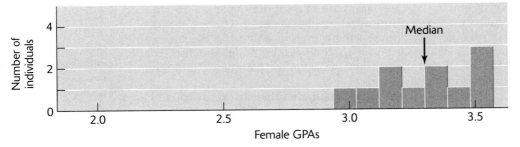

(b)

2-11 Median GPAs of biology students. **(a)** For 9 males, the fifth highest GPA (2.5) is the median. **(b)** For 10 females, the value midway between the fifth and sixth highest GPA (3.5) is the median.

capture the location of the peak of a highly skewed distribution; the mean and median, in contrast, are displaced toward the end, or the *tail*, of the distribution (though the median is least affected). Researchers sometimes distinguish the two peaks in a bimodal distribution as the major and minor modes; the existence and position of multiple peaks can provide useful descriptions of bimodal populations or phenomena. No tests in this book make use of modes.

All data sets are characterized by a ***sample size***. Because the effect of chance errors—which occur despite precautions to randomize your sampling (such as trying to avoid interviewing only basketball players while gathering height data)—decreases with sample size, this parameter is a critical factor in computing statistical values. Sample size is important in both parametric and nonparametric analyses, and is generally referred to as n. Sample size is not, strictly speaking, a parameter, though you can be forgiven for thinking so since it is inevitably asked for along with the mean and variance in most parametric analyses.

The "width" of a parametric distribution is visible in a graph as the distance between the two *points of inflection*—that is, the points where the curve reverses its direction of curvature (see Figure 2-9c). This width is two "standard deviations," as we will show. In practice, the width of a parametric distribution is not measured by analyzing the graph of the distribution—indeed, it is a difficult point to judge by eye. Instead, the width is computed from the data in the distribution. This width can be described either by its variance or its standard deviation.

The ***sample variance*** (code-named s^2, and which we will simply call "variance") is obtained by squaring the difference between each point's value and the mean, summing all these differences, and dividing by $n - 1$, where n is the sample size (as we will illustrate presently). The variance will play an important, explicit role in some of the discussion in this book, particularly in Chapter 7; it will appear again and again in the math associated with many statistical tests. Put intuitively, the variance is something like the square of the average deviation between a sample in the data set and the mean of the set. Most students (like most scientists), however, are more comfortable thinking of distribution width in terms of the standard deviation.

The ***standard deviation*** (***s*** or ***SD***), like the variance, is obtained by first squaring the difference between each point's value and the mean, summing all the differences, and dividing by $n - 1$; this, of course, is the variance. Then, however, we take the square root of this value. Thus SD is something like the average difference between a datum and the mean. In fact, an SD is larger than the average difference (about 37% larger)

The **sample size** of a distribution is the number of independent data values it contains.

The **sample variance** (s^2) of a distribution is a weighted average squared difference between the mean ($\overline{X}$) of a distribution and the n data values (x_i) in the distribution $s^2 = \Sigma(\overline{X} - x_i)^2/(n - 1)$.

The **standard deviation** (**s** or **SD**) of a distribution is a weighted average difference between the mean ($\overline{X}$) of a distribution and the n data values (x_i) in the distribution $s = [\Sigma(\overline{X} - x_i)^2/(n - 1)]^{1/2}$.

because the squaring step weights large differences from the mean more than small ones: a difference of 2 gets a value of 4 when squared, while a difference twice as large (4) gets a value of 16 in the analysis. (This technique of taking the square root of the average of the squared values is known as the root-mean-square, or RMS, method. It turns up in many surprising places—it's the way the power output of stereo amplifiers is computed, for instance.) We'll work through an actual computation of SD in just a moment.

So here, at last, is an actual example of how you might extract the salient parameters from a sample distribution. This is a list of the seven heights (H, in inches) of the male students in a statistics seminar, arranged in order of height. Since height is parametrically distributed, you can legitimately derive all the relevant parameters (the ones used most often are shown in boldface):

	Student	H (height)	H − mean	$(H$ − mean$)^2$
	1	66	66–71 = −5	25
	2	67	67–71 = −4	16
	3	71	71–71 = 0	0
median →	4	72	72–71 = 1	1
	5	72	72–71 = 1	1
	6	74	74–71 = 3	9
	7	75	75–71 = 4	16

$n = 7$; mean = 497/7   variance = 68/6 = 11.3

mean = **71**   SD = $\sqrt{(68/6)}$ = **3.4**

Comparing means of two distributions asks how likely it is that a data set with one mean could have been drawn from the parent distribution of the other data set by chance.

For what it's worth, this very limited data set yields a sample mean within half an inch of the parent (true) mean for college-age males; its standard deviation is somewhat larger than the typical SD for male height.

So now we have the parameters typically used to compare parametric distributions. As we have said, the most common comparison is of means: comparing the means of male and female shoe sizes is a clear example (see Figure 2-6). The tests that make these comparisons nevertheless need the standard deviations in order to take into account whether the two distributions are wide (and may thus overlap greatly), or narrow (and may thus be quite distinct). And they need the sample size in order to allow for the magnitude of chance sampling bias. Another kind of comparison is between the widths of two or more distributions regardless of means—that is, the comparison of variances or SDs. A third kind of comparison is between the individual data points in two distributions, which is typical of paired-parametric and paired-nonparametric analyses. The last major kind of comparison is between two or more variables: the scatterplot of mSAT scores versus vSAT scores is one such example (see Figure 2-5).

We hope you have borne with us through these preliminaries without yielding to despair. We know from personal experience that statistics is never a case of one-trial learning. Thus all the terms used here will be reintroduced in context as they arise in later chapters. (If you ever want a quick summary of this material, look at the first few pages of Chapter 14, where we provide a compact once-over of terms and their meanings.) We turn now to the simplest kind of distribution you are likely to encounter: the binomial distribution.

Comparing widths of two distributions asks how likely it is that a data set with one variance could have been drawn from the parent distribution of the other data set by chance.

Comparing individual data values of two distributions asks how likely it is that pairs of data values could have been drawn from the same parent distribution by chance.

Comparing two variables of two distributions asks how likely it is that the values of both members of various pairs of data are unrelated to each other.

2 Data

Points to remember

✔ The null hypothesis is usually that the data sets being compared are drawn from the same large complete set of data (a parent distribution).

✔ Statistical analysis can never tell you if two data sets are drawn from different parent distributions; it can only estimate the probability that the differences arose from chance alone.

✔ Statistical analysis cannot tell you if the causal factor you are investigating is actually the basis for any effect (difference between two data sets) you observe; there could always be another cause you have overlooked.

✔ Categorical data exist as mutually exclusive types.

✔ Continuous data are measured in terms of a parameter that is infinitely variable over the range in question.

✔ Discrete data are measured in terms of a parameter that can have only certain specific values.

✔ A parent distribution contains all the possible instances of whatever is being measured; a sample distribution contains some subset of these data.

✔ A null distribution is an idealized predicted distribution.

✔ Statistical analysis assumes that data samples are unbiased and independent.

✔ A parametric distribution has a characteristic bell-curve shape; nonparametric distributions are those with other shapes.

✔ A parametric distribution can be completely described by its mean and variance (or standard deviation).

✔ The mean is the sum of all the data values divided by the sample size.

✔ The variance is the sum of the squared differences between the mean and each data value, divided by one fewer than the sample size; the standard deviation is the square root of the variance.

Exercises

For each of the data sets, determine the kind of data (categorical, discrete, or continuous). Plot them in an appropriate way depending on the kind of data. Again depending on the kind of data, compute appropriate summary statistics (mean and SD of the first 10 entries if the data appear to be parametric, median if they do not, and neither if the data are categorical).

1. The admissions committees at many universities attempt to reduce all the information they have available on a student to a number. At Freeman University (an unknown Ivy League school that has the same data distributions as Princeton but different means) there is a 1-to-5 "A" (academic) score and a similar "P" (personal) rating, where 1 is the highest value. Evaluate the A-rank values for the 10 individuals in Table A— that is, determine the kind of data, plot them, and then compute the appropriate summary statistics (if any).

2. Evaluate the GPA data in Table A (4.0 = A).

3. The A-rank is supposed to predict GPA; plot a scatter diagram with A-rank on one axis and GPA on the other. Is there any evident

Table A

Student #	GPA	A-rank	P-rank	vSAT
1	3.7	1	3	570
2	3.2	3	4	470
3	2.9	2	3	650
4	3.7	2	3	600
5	3.3	2	3	540
6	3.7	2	3	550
7	3.2	2	3	550
8	2.5	4	3	490
9	2.7	4	1	400
10	1.7	3	4	480
11	2.3	4	3	530
12	3.7	1	3	550
13	3.1	2	1	520
14	3.0	5	2	350
15	3.5	2	4	640
16	2.3	3	4	560
17	3.1	3	4	580
18	2.5	4	3	510
19	2.9	4	1	450
20	2.6	3	4	410

relationship? Is P-rank any better or worse?

4. Evaluate the verbal SAT scores in Table A.

5. Evaluate the pet-preference data in Table B.

6. Evaluate the student-height–versus–same-sex-parent-height data in Table B.

7. Evaluate the shoe-size data in Table B.

8. Evaluate the student-height data in Table B.

Table B

Student #	Shoe size*	Height	Height vs same-sex parent	Mother vs father	Prefers cat or dog
1	12	5-11	shorter	shorter	C
2	9	5-01	taller	taller	D
3	5.5	5-03	taller	shorter	D
4	6	5-07	shorter	shorter	D
5	6.5	5-08	taller	shorter	C
6	5	5-02	taller	shorter	D
7	11.5	5-02	taller	shorter	D
8	8	5-09	taller	shorter	D
9	9.5	5-09	taller	shorter	D
10	10	5-11	shorter	shorter	D
11	7.5	5-07	taller	shorter	D
12	11	6-00	taller	shorter	D
13	10	6-00	taller	shorter	D
14	8	5-10	taller	shorter	C
15	12	6-01	taller	shorter	D
16	11	5-10	shorter	shorter	D
17	8	5-09	taller	shorter	D
18	9	5-08	shorter	shorter	D
19	8	5-10	shorter	shorter	D
20	11	6-01	taller	shorter	D

*Converted to female size.

9. Categorize the student-height data in Table B into taller half and shorter half; any individual on the tall–short boundary should be assigned as half a tall value and half a short value. Now categorize the shoe-size data in the same way into large versus small. Plot the four category groups: small–short, small–tall, large–short, and large–tall. If there were no relationship between shoe size and height, how should the entries in these four groups be distributed?

10. Look at the actual data provided in Table B by plotting a scatter diagram. Is there any evident relationship between shoe size and height?

Tests Covered in This Chapter

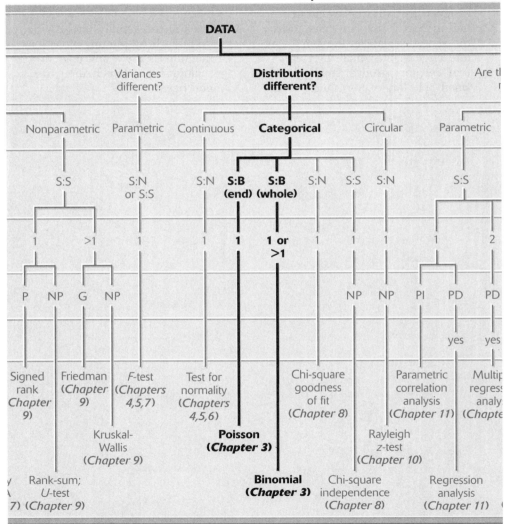

Binomial Distributions [3

The simplest (and thus, it often seems, the rarest) kind of data is binomial—literally, one of two numbers. In this chapter we will see how grappling with this easiest of data types reveals the underlying logic of much of statistics in general.

3.1 What Kinds of Measurements Yield Binomial Distributions?

Binomial distributions are generated when a measurement can have one of two mutually exclusive outcomes. For instance, were you to gather data on the sex of students in biochemistry classes, each datum would have one of two possible states, male or female. A coin flip can have one of two mutually exclusive outcomes: heads or tails. Similarly, when we play alternative mating calls from each of the two ends of a T-maze and score female crickets according to which end of the maze they reached, the data consist of some number of left-hand choices and the remainder are right-hand choices (Figure 3-1). Or when we test honey bee shape-learning by offering a pair of alternative targets, one square and the other triangular, each choice (landing) is either scored as on the square or on the triangle.

Binomial data are both categorical and discrete: there are only two possible categories, and even when they are combined in sets, only discrete values are possible (like four heads and six tails, or seven heads and three tails). We will show in Chapter 8 how to deal with generic categorical data; we are bringing up

> **Binomial distributions** consist of data that can have one of only two possible values or states.

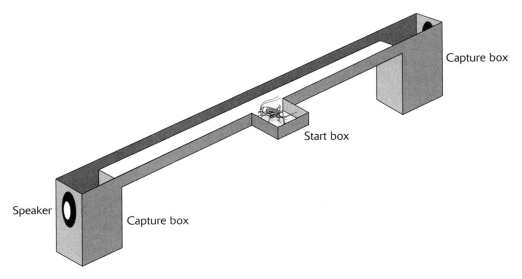

3-1 A T-maze used with crickets. Unmated female crickets are placed in the start box and different sounds are played from each of the two ends of the dark maze. Females turn either left or right out of the start box and end up in one of the capture boxes at the ends.

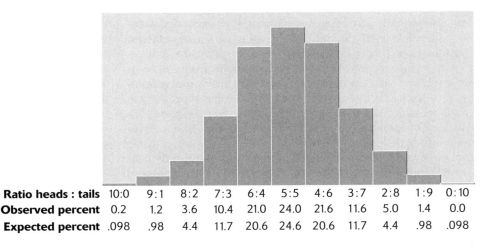

Ratio heads : tails	10:0	9:1	8:2	7:3	6:4	5:5	4:6	3:7	2:8	1:9	0:10
Observed percent	0.2	1.2	3.6	10.4	21.0	24.0	21.6	11.6	5.0	1.4	0.0
Expected percent	.098	.98	4.4	11.7	20.6	24.6	20.6	11.7	4.4	.98	.098

3-2 The distribution of 500 sets of 10 coin flips, with the observed proportions listed along with the theoretical expectation from an infinite number of 10-coin-flip sets.

Box 3.1

BioStats Basics Online | A User's Guide

BioStats Basics Online is an interactive tutorial for probability and statistics that was written to accompany this text. You will find the software, instructions for downloading it, and a troubleshooting guide at www.whfreeman.com/gould. *BioStats Basics Online* provides a concise summary of the topics covered in the book, a variety of user-controlled simulations, interactive cards that allow you to enter and test data, opportunities to save or import data sets, a glossary, and a set of multiple-choice self-test questions. Throughout the chapters in this text, you will see marginal icons that indicate which topics and tests are included in *BioStats Basics Online,* and how you can use the interactive material to better understand how the tests work.

The instructions for using the buttons common to each card are found on the opening screen; the instructions

for saving and importing data are on the following card. The third card is the Table of Contents; clicking on any topic takes you to the card for that topic. For example, the first reference to *BioStats Basics Online* is a mention of the ten-flip demonstration on the Chance and Random Distributions card (card I. B-1). Clicking on that topic on the Topics card takes you directly to the demonstration and accompanying discussion. The card numbers are found in the upper right-hand corner, and correspond to the section and topic hierarchy on the Topics card.

One of the most useful entries on the Topics card is Overview of Tests (III. A-1). Clicking on this topic takes you to a card with a functional overview of the most useful tests in *BioStats Basics Online.* Clicking on any test name takes you to that test.

binomial data now because their discrete nature helps introduce parametric distributions, and because binomial analysis is more precise than the statistical approach to the various other species of categorical data.

Binomial analysis works both for sets of many individual two-choice measurements and sets of binomial tests. As an example of sets, consider an experiment on insect learning: each honey bee is allowed 10 target choices, and the data from 25 individual bees are recorded. Statistics classes regularly generate binomial data by having each student flip a coin 10 times and then analyze the collection of 10-flip data sets (Figure 3-2).

There is a nice demonstration of the coin-flipping case on the Chance and Random Distributions card in *BioStats Basics On-line*. (See the box on page 43.) You can see how long it takes for repeated sampling to produce a distribution that resembles the actual results of an infinite number of 10-flip data sets. (The simulation uses a random-number generator in the software to generate the "flips." For ease of navigation, this card is indexed as "I. B-1" in the upper-right-hand corner; all index numbers in *BioStats Basics Online* correspond to the hierarchy used on the Topics card.)

3.2 The Product Law and the Importance of Independence

The distribution of outcomes of binomial tests in which each alternative has a 50% chance of occurring on any given occasion—clearly the case with a coin flip—was first described by the French philosopher Blaise Pascal in 1654. He reasoned that the chances of obtaining, say, three heads in three flips is the product of the chance of obtaining each outcome on each test (0.5 for heads on each flip; $0.5 \times 0.5 \times 0.5$, or 0.125 for three flips) *times* the number of different orders in which the outcome could occur (only one in this case: heads, heads, heads). Thus the probability of obtaining two heads and a tail in three flips is 0.125 (the odds of obtaining tails also being 0.5) times the three orders that yield this outcome (heads, heads, tails; heads, tails, heads; and tails, heads, heads). Thus the odds are 3×0.125, or 0.375.

The **product law** asserts that the chance of a given set of independent outcomes occurring together is the product of the probabilities of each occurring separately (p_i) times the number of possible orders in which they could occur (N')— usually $P = (p_1 \times p_2 \times \cdots \times p_n)(N')$.

The principle used in this calculation is known as the ***product law***, which assumes that the individual events are independent of one another. This seems like a trivial assumption, but consider what would happen if you were performing a similar binomial test using a deck of cards and scoring red versus black as your two categories. Clearly, the odds of drawing a red card at the outset is 0.5 (26/52, ignoring jokers). But after the first card is recorded the situation changes: unless the card is replaced and the deck thoroughly shuffled, the odds for the next card are 25/51 for the color observed on the first card, and 26/51 for the other color. The reason is obvious: there are only 51 cards left, and only 25 have the already-observed color; the second draw is not independent of the first because one of the 52 possible sample outcomes is now missing. There is an informative demonstration of the effects of two kinds of dependence on the Variance and Independence card (I. H-1) in *BioStats Basics Online*. In one case each outcome tends to repeat the previous outcome ("following"); in the other each outcome tends to avoid the previous one ("alternation").

The product law can be used to compute the odds of any number of concatenations of events so long as the individual probabilities are known and the events are categorical and independent. Thus the odds of rolling a seven with a pair of dice, for instance, can be computed as follows: seven can be obtained from a six on the first die and a one on the second, from a one on the first and a six on the second, from a five on the first and a two on the second, from a two on the first and a five on the second, from a four on the first and a three on the second, from a three on the first and a four on the second. Thus there are six ways to get a total of seven when rolling a pair of dice. The odds of rolling any particular number on a die is 1/6; the odds of rolling any particular combination of numbers with a pair is thus 1/6 × 1/6, or 1/36. Since there are six ways to obtain a total of seven from a pair of dice, the odds are 6 × 1/36, or 1/6 (0.1667).

The product law can also be used to compute the number of unique combinations that are possible from independent events. For instance, when our reproductive cells create gametes, they select at random one member of each of the 23 pairs of chromosomes we inherited from our parents. Thus the number of alternative configurations is 2 × 2 × 2 × 2 × 2 × 2 × 2 × 2 × 2 × 2 × 2 × 2 × 2 × 2 × 2 × 2 × 2 × 2 × 2 × 2 × 2 × 2 × 2 (that is, 2^{23}), or 8,388,608. (Actually, because chromosome pairs first undergo a process of crossing over, which creates many unique "hybrid" chromosomes, the actual number is far, far larger even than this.)

When order does matter, the **product law** says that the chance of a given set of independent outcomes occurring together is the product of the probabilities of each occurring separately: $P = (p_1 \times p_2 \times \cdots \times p_n)$.

3 Binomial Distributions

Pascal's triangle
shows all the possible
outcomes for any
given number of
binomial trials. When
the two outcomes
are equally likely,
then the value, C, of
the nth number in
the Xth row of the
triangle (reading
from either side) is
$_nC_X = n!/[X!(n - X)!]$.

Pascal summarized the frequent case of two equally likely alternatives (e.g., coin flips) as what is now known as ***Pascal's triangle*** (Figure 3-3). Here is the first part:

(a)   1   1
(b)   1   2   1
(c)   1   3   3   1
(d)   1   4   6   4   1

Line (a) corresponds to the possible outcomes of a single flip of a coin: two options, heads or tails. Line (b) represents the possible outcomes of two coin flips: three options, two heads, a head and a tail, and two tails. Because there are two orders for obtaining a head and a tail (head-then-tail, and tail-then-head), the center value in this line is 2. The sum of the three numbers $(1 + 2 + 1)$ is 4, which is the number of different ways the two-flip sequence can come out (head-head, head-tail, tail-head, and tail-tail).

The next line—line (c)—summarizes the possible outcomes of three coin flips. There are four numbers because there are four net outcomes: all heads, two heads and a tail, a head and two tails, and three tails. Each value corresponds to the number of different orders that can yield the particular outcome,

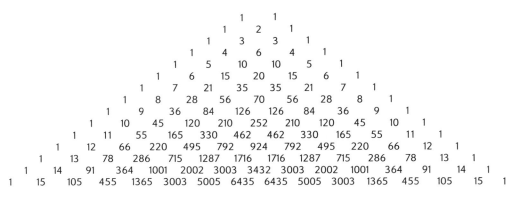

3-3 The first 15 lines of Pascal's triangle.

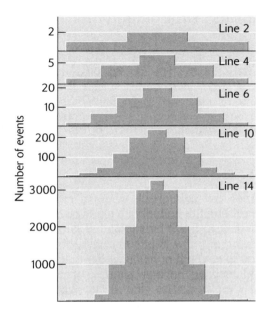

3-4 As the number of independent events increases, the binomial expansion approximates a normal curve more and more closely.

and the total in the row is the sum of all possible orders. Line (d) extends this to the four-flip situation.

If you look closely at the triangle, you may notice that each number is the sum of the two in the row above it, and so the triangle can generate itself indefinitely. The second number on any line is also the number of flips, a useful index to where we are in the triangle. As we move down the triangle, the pattern of numbers across the row—the number of orders for each possible outcome beginning with all heads and ending with all tails—more and more closely approximates a bell curve (Figure 3-4; for a precise mathematical definition of the bell curve, see page 79 in Chapter 4). Thus what we do with a high-order binomial distribution will be analogous to what you will be doing with parametric distributions in the next four chapters, though the treatment of binomials is easier to appreciate intuitively.

3 Binomial Distributions

3.3 Comparing Distributions

As we have said, statistical analysis is most often used to compare two distributions. We stated that a sample distribution, obtained by measuring something, can be compared with one of three types of other distributions: a *null distribution* (a theoretical distribution); a *parent distribution*, composed of all the relevant data from all instances of what is being measured; or another *sample distribution*. In the case of **binomial analysis**, you will almost always be comparing a sample distribution to a null distribution. There is a useful demonstration of how a skewed sample distribution begins to diverge from the null distribution on the Normal and Skewed Distributions card (I. D-1) in *BioStats Basics Online*.

The distributions represented by each row of Pascal's triangle are null distributions: they are the theoretical prediction of the distribution of outcomes that would be expected from an infinite number of repetitions of sets of unbiased and independent coin flips (or any other two-alternative event whose individual probabilities are 0.5). Binomial analysis is the comparison of an actual measured distribution to a null distribution. For example, we can compare the choice behavior of our trained bees (which were being tested to discover if they have learned anything, and if so, how well) to the null distribution (what we would observe if an infinite number of naïve bees were run through the choice tests).

> **Binomial analysis usually compares a sample distribution with a hypothetical null distribution.**

3.4 Probability: Was the Sample Drawn from the Null Distribution?

Let's look at some actual data from a bee-learning experiment. We have 20 bees, each of which has made 10 choices between two alternative targets (Figure 3-5; to control for any shape bias, 10 were trained on a triangle as the "correct" target, while another 10 were trained on a square; all were then offered a choice of triangle or square). The two distributions look quite different: a clueless bee ought on average to select the correct shape 5 times after 10 choices; instead, judging by the histograms in Figure 3-5, the average trained bee is selecting the training

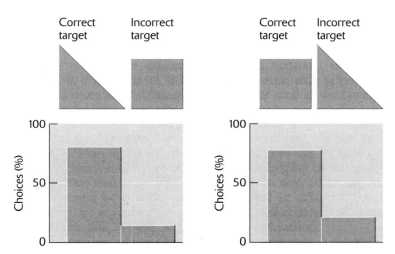

3-5 Twenty bees were trained individually to find food on a target with either a triangular or square shape. The alternative target was present during training but offered only water. (Animals generally learn better when there is an alternative, incorrect choice. The left–right positioning of the targets was reversed frequently to prevent the bees from simply learning *where* the correct target was to be found.) After 10 visits, each bee was allowed to choose five times between new versions of the targets (new so that they did not have any of the odors regularly left on feeders by visiting bees). The percent of correct choices is shown in the left bar of each histogram.

shape about 8 out of 10 times. It certainly looks like the bees have learned to distinguish the training shape from the alternative, but maybe they just got lucky. How can you quantify the chances that this sample distribution is different from the null distribution?

The question we are really asking is this: what are the odds of getting this result, or a result even more different from the null prediction, by sampling the null distribution for 10-choice binomial alternatives 20 times? In short, how unlikely is this result to arise by chance if no learning is actually occurring? Let's

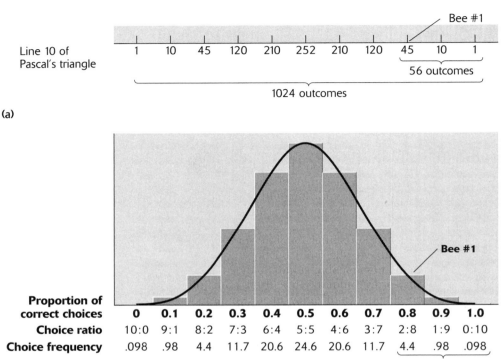

(a)

(b)

3-6 (a) The probability of obtaining by chance the 8-of-10 behavior of the first bee can be computed by using line 10 of Pascal's triangle: sum the appropriate three entries on the far right (45 + 10 + 1) corresponding to 8/10, 9/10, and 10/10 outcomes, and divide by the total number of entries in the row (1024). **(b)** The same result can be obtained by consulting a graph of the null distribution for ten two-choice events with equal probabilities; the behavior of Bee #1 is indicated. This animal chose the training shape on 8 of 10 tries; as the bottom row of figures indicates, this will happen by chance about 4.4% of the time. A result at least this extreme will occur somewhat more often; by summing the odds of all three outcomes (as listed in the bottom line), the chance probability adds to about 5.5%.

start by looking at the first bee to be tested (Figure 3-6a). This individual selected the training shape on 8 out of 10 tests. If you look at the 10-event null-hypothesis binomial curve, you will see that an 8-of-10 outcome should occur by chance 45 times in 1024 (the third entry in the 10-event line compared to the sum of all the possibilities in the row).

But we are asking not only how likely an 8-of-10 result is, but also how likely is a result that is this much different from the null expectation (5-of-10); thus you must also include the 9-of-10 and 10-of-10 alternatives. When you add the three numbers on the 10-event row that take us through what we actually observed, you get a total of 56 (1 + 10 + 45); this means that 56 times in 1024—about 5.5% of the time—an untrained bee should select the correct alternative 8 or more times. The same thing can be done graphically by plotting the binomial distribution for the appropriate number of events and summing the areas from the observed ratio to the end of the curve (Figure 3-6b). Easiest of all, you can go to the Perfect Normal Distribution Test (II. D-1) in *BioStats Basics Online*, use the left side of the card (Standard Two-Choice Data) to select the Sample Mean option, enter 10 for the sample-set size, 1 for n, and 8 for mean of observed data; choose the High (one-tail) option and click on Compute. As we will see, this is the most common form of the binomial test.

The chance *probability* of the 8-of-10 result can be summarized in typical statistical format as $P = 0.05469$, or 5.469%. Similarly, the odds of a 9-of-10 result (the "score" of the second bee) occurring by chance is $(1 + 10)/1024$: $P = 0.01074$, or about 1%. The odds of a 7-of-10 result (observed for the third bee) by chance is $(1 + 10 + 45 + 120)/1024$: $P = 0.17188$, about 17%. The chance odds of a 6-of-10 outcome is about 38%.

A result that has a 38% probability of occurring by chance does not, on its own, tempt most of us to reject the null hypothesis that bees do not learn shape; even 17% is pretty common—this probability will show up roughly 1 time in 6. But what about 5.5%—about 1 in 18? Is this convincing, or is this a result whose probability of turning up by chance is too high to inspire confidence? Even a 10-of-10 result (observed for the fourth bee) has a chance probability of about 0.01%; if you run the test a thousand times with untrained bees you are quite likely to see one or two individuals get the "right" answer 10 out of 10 times.

The probability value (or ***P-value***) cannot tell you if bees learn shapes—that is, that the sample distribution is absolutely for sure not drawn at random from the null distribution. Instead, the P-value tells you only how *unlikely* the outcome is if there is no real phenomenon at work. A 9-of-10 result is unlikely, but it's not impossible. You must decide how small a P-value to accept as indicating that a phenomenon is real; there is no absolute stone-tablet threshold. In most areas of natural science, a value of $P < 0.01$ is taken as convincing; in behavioral research, where large sample sizes are often very hard to obtain, a threshold of $P < 0.05$ is often accepted. In brief three-hour undergraduate science labs, we allow our students to consider a value of $P < 0.1$ to be nonrandom.

Most often, you will be interested in the probability that two distributions are different; at the same time you must avoid the error of concluding that two data sets—drawn from the same distribution—are different when they are in fact *not* different. (In other words, you'll want to avoid being fooled into concluding from your data that there's something interesting going on when in fact there *isn't*.) But as we will sometimes see, the context of the test can be crucial: with issues of safety or health (e.g., is a treatment effective given that the alternative is certain death?), probabilities of 0.2 or even 0.5 might be cause for action. The reason is that in these cases your interest is in avoiding the reverse error: concluding that two distributions are the same when in actuality they are not (that is, thinking that your data reflect chance only, when in fact there *is* some phenomenon at work). Throughout most of this book, our focus will be on avoiding the first class of error (***false positives***); keep in mind that false negatives are usually the major worry in many safety-first contexts. (Just to make things difficult, it would seem, statistical analyses often replace the easily interpreted term "false positive" with meaning-challenged phrase "Type I error." Similarly, the meaning of ***"false negative"*** is made obscure under the rubric of "Type II error.")

The ***P*-value** (probability value) of a result is the chance of that result occurring plus the odds of all possible other outcomes *even more* different from the prediction of the null hypothesis.

A **false positive** (or Type I error) occurs when you mistakenly conclude that your data reflect some causal effect when in fact only chance is involved.

A **false negative** (or Type II error) occurs when you mistakenly conclude that your data reflect only chance effects when in fact a causal process is at work.

The *P*-value of a threshold is usually referred to as a "level of significance," or more often just *significance*. When we say that a result is **significant** at 0.05, or 5%, we are saying that our computation of its probability came out at $P < 0.05$; thus, assuming we didn't do something stupid in designing our experiment or in selecting the animals to be tested, the chances of obtaining a result this different from the null distribution is less than 1 in 20.

Going back to our bees, the first four bees tested had four different scores, corresponding to four different *P*-values. Which one is correct? Maybe we should average the four? In fact, if the individual choices of a bee are independent of one another (which we can show to be the case), we can combine the results from these four bees ($8 + 9 + 7 + 10$ correct responses in 40 total trials, or 34/40) and look in line 40 of Pascal's triangle. Unfortunately, line 40 is too large to fit on a page; indeed, the total number of 40-event orders exceeds a trillion. But we can find the same thing graphically or by simple (if tedious) computation. If we opt for the graphical alternative, we plot the 40-event binomial distribution, mark the 34-of-40 point, and add the number of order combinations for 34/40, 35/40, 36/40, and so on out to the end of the distribution.

These numbers sum to about 4,145,150; fortunately, these values can be computed effortlessly by statistical programs, including *BioStats Basics Online*, which use one of several forms of the **binomial test**. (You should be aware, however, that some statistical packages omit the binomial test, relying instead on a somewhat less powerful technique described in Chapter 8.) So if there are about 4 million ways to get a result as extreme as 34 of 40, and a trillion ways to arrange 40 binomial events, it follows that the chance of observing the 34-of-40 result by chance is about 0.000377%, or $P < 0.000004$. Pretty unlikely. And when we consider all 20 bees, which accumulated a stunning 178 correct choices in 200 tries, the odds of this distribution being drawn by chance from the null (dumb-bee) distribution are vanishingly small. (The binomial test is on card II. E-1.)

There is a crucial lesson here: a statistically wise experimenter would never have wasted energy testing 20 bees when fewer would have done as well. Instead, we should have run a

---

A **significant** result is one for which the *P*-value is lower that the threshold you have set for nonrandom deviations from chance expectation.

A **binomial test** computes the probability that the sample distribution could have been drawn from the null distribution by chance.

Box 3.2

BioStats Basics Online   Computing equal-probability binomials

BioStats Basics Online has a special test for binomial distributions in which the two alternatives are equally likely: $a = b = 0.5$. The Binomial Test, which is on the Perfect Normal Distributions card (II. D-1), has two options: you can enter (or import) the actual data, or you can enter the essential information about the data. The test card will look a little odd: it has two tests on it (only the one on the left is relevant here), and the test we are discussing can also be used with a special case of continuous data, and thus has some features we have not yet discussed. The advantage of this test is that it is fast: by using a couple of approximations, the test yields quick and accurate estimates of *P*-values when they are significant (i.e., <0.05). The (minor) disadvantage is that the test can be off by as much as 4% when the *P*-value is around 0.50.

The test is for standard two-choice data—the number of heads versus tails, for example. If you select the Sample Data option, a field appears into which you can enter or import your data. You must also enter the sample-set size in the field box provided. Thus, if all the measurements are single coin tosses, the number to enter is 1, and there will be only two possible values of the data, typically 0 and 1 (the number of heads, say, in each trial); if instead you are using sets of 10 tosses, you would enter 10 in the sample-set field box, and your data entries will all lie between 0 and 10.

Next you must decide if your test is one-tailed or two-tailed; this issue is covered later in the chapter, so for now you should assume a two-tailed test and skip the rest of this paragraph. If it is one-tailed, you must indicate whether your expectation is that the data will be skewed toward the high or the low end. If you select the high-end option, therefore, you expect that the average will be greater than 0.5 of the sample-set-size value (that is, >0.5 for a sample set of 1, >5.0 for a sample set of 10, and so on).

When you click on the (**Compute**) button, *BioStats Basics Online* calculates and displays the mean number of events per trial (i.e., the average number of heads) and the *P*-value (the odds that

preliminary test to judge how strong the trend was, and then used that information to decide on the optimal sample size for the threshold of significance we consider sufficient. Let's look at how you can make this important experimental call.

the observed distribution could be drawn from the null distribution by chance). *BioStats Basics Online* also graphs the observed distribution (arbitrarily divided into 11 segments) and inserts arrows indicating the observed and expected means. The graphing can look a bit unusual when the sample-set size is small.

The other way to use this test is to choose the Sample Mean option and enter some summary information in the boxed fields that appear. You will need the sample-set size (again), number of sets in the data (*n*), and the mean of the observed data (which will lie somewhere between zero and your sample-set size). Once again you must select the appropriate tails option. Clicking on **Compute** yields the same information as before: the standard error (SE) and standard deviation (which you can ignore) and the *P*-value. Because there are no actual data, *BioStats Basics Online* cannot draw a graph.

As discussed in Chapter 4, the value for SD is actually a second-order calculation of the variance in the means

of sets of binomial data of the sample size you choose. The meaning of SE is discussed in Chapter 4.

Try using this card with some actual data from an undergraduate biology lab. Groups of 10 female crickets were run though the two-choice maze with the sound of one cricket calling at each end, but with the mid-range volume of the experimental sound twice as loud as that of the control. Four lab groups each ran the test twice with the following results (expressed as experimental-end : control-end females): 8:2, 5:5, 4:6, 8:2, 7:3, 7:3, 5:5, 8:2. To analyze this data you would enter a sample-set size of 10 (10 females per test). To use the Sample Mean option you would enter an *n* of 8 (eight tests) and the average number of experimental-end females per test (8 + 5 + 4 + 8 + 7 + 7 + 5 + 8 in eight tests, or 52/8 = 6.5). Then click on **Compute**. Alternatively, you could select the Sample Data option and just enter the results of the eight tests into the blank column on the card; clicking on **Compute** will yield the same *P*-value and a graph.

3.5 The Role of Sample Size

We've just shown that increasing sample size reduces the *P*-value, in this case to an absurd degree. The effect of sample size on *P*-values is well illustrated on two cards in *BioStats Basics Online*. The first, Probability and Sample Size (I. J-1), allows you to set the degree of bias on a coin and compare the results of 10-, 20-, 50-, and 100-flip sets. It is sobering to see the occasional false positive that turns up even with an unbiased coin, and the number of false negatives that result even when the coin is heavily weighted. The other card, Choosing an Optimal Sample Size (I. K-1), takes this point further by illustrating the continuous change in P-value for coins with different degrees of bias as the number of samples increases indefinitely.

Having too much data, and thus a stunningly small P, is a rare problem because most experiments are preceded by a trial run that is used to judge the size of the trend in the data. Say you tested four animals in a preliminary experiment and obtained a P-value of 0.02; this is close to the $P < 0.01$ threshold most researchers usually set, but it's not good enough. There seems to be a trend, and you do have a rough measure of its strength. The next time you do this test, assuming the trend is real and the responses representative of what you will see when the experiment is repeated, how many animals should you test?

This common quandary has a fairly simple rough-and-ready answer: the typical ***chance error*** (the difference between the parental-distribution mean and the sample mean) decreases roughly *in proportion to the square root* of the sample size. (In nearly all parametric cases, and for binomials with large sample sizes, this is an excellent approximation.) Because the P-value associated with a particular difference between sample and parental means varies with the typical chance error (the so-called standard error, as we will see in Chapter 5), it should decline with increasing sample size in about the same way. Thus if $n = 4$ yielded a $P < 0.02$, n of about 16 ought to produce $P < 0.01$, assuming the animals continue to behave exactly as before. Here is the computation: the ratio of the observed P-value to the threshold P-value is $2:1$; the ratio of their squares is $4:1$; the ratio of the sample sizes is thus $4:1$, or 16 animals to 4 animals. Since no two small samples of anything will ever behave exactly the same (chance being always with us and having, we believe, a deeply spiteful nature), and given the approximate nature of the square-root rule, it is best to assume that the original observed trend is stronger than the trend you will encounter when you repeat the test; thus you should increase the sample size still more to allow for this all too common contingency.

Need more than this? See **More Than the Basics:** *Error and Significance.*

Chance error results from normal sampling variability; it tends to decrease in magnitude with sample size. Thus, if the chance error is e_1 for a sample size of n_1, the chance error for a sample size of n_2 should be about $e_2 = e_1(n_1/n_2)^{1/2}$.

We can see that for detecting slight differences the necessary sample size can increase exponentially. Thus a small-scale trial test is absolutely essential: you must be able to judge how big a sample you will need before embarking on a project that may require millions of samples and several decades to complete. There is a button on the Choosing an Optimal Sample Size card (I. K-1) in *BioStats Basics Online* that takes you to the Retrospective card (III. H-1), which allows you to make the calculation just described: you enter the trend you have observed with your original sample size, the *P*-value threshold you are using, and it will supply the approximate necessary sample size for data with the same trend strength.

3.6 The Issue of Tails

We had a clear expectation of what shape-savvy bees should do in our learning experiment: choose the food-reinforced training shape. Foragers unable to learn should choose at random. We know from previous tests that bees land more often on objects of the *colors* on which they have found food previously, so we feel justified in ignoring the possibility that these intelligent insects find food aversive and might learn to *avoid* the training shape. But consider the situation in which we offered two alternative mating calls to female crickets from each end of a maze. One is, say, a normal male's call, while the other consists of a chorus of males. The null hypothesis is that the females will break 50:50, showing no preference. (To control for end bias, we always run the test a second time with each call switched to the other end; this sounds silly, but in fact there are small but consistent end biases whose origins have eluded our best efforts to track them down. We now suspect a conspiracy on the part of female crickets.) What is our expectation? Maybe females are programmed to approach crowds of males, which might increase the odds of finding a mate or opportunities to choose the best male among the callers; or it could be that females are programmed to avoid crowds, which attract parasitic flies.

Given that there is no convincing a priori reason to expect females will prefer one sound over the other, it follows that a 2-of-10 result will be just as interesting as an 8-of-10 result.

A **two-tailed test**
looks for *any*
deviation from the
null hypothesis; a
one-tailed test looks
at deviations in only
one a priori direction.

Since either end, or "tail," of the distribution is interesting, the cricket experiment is a ***two-tailed test***, while the bee-learning experiment was a ***one-tailed test***.

The number of tails greatly alters the magnitude of the P-value, though not the logic and mechanics for computing it. If we observe an 8-of-10 result in our cricket test, we must again count up all the ways of obtaining a result at least this extreme. But this time we must include the 2-of-10, 1-of-10, and 0-of-10 combinations as well. Thus, looking at the 10-event line of Pascal's triangle, we will sum $1 + 10 + 45 + 45 + 10 + 1$. Our P-value is therefore 112/1024, or 0.10938; this is exactly double the 0.05469 value we calculated for the 8-of-10 result in the one-tailed learning test with bees (Figure 3-7). Because Pascal's

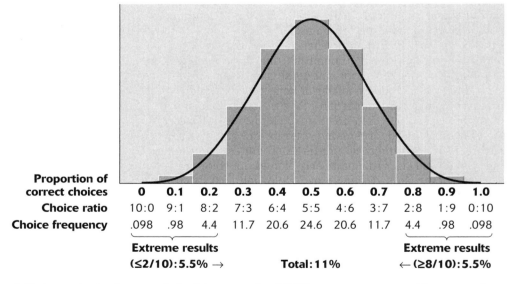

Proportion of correct choices	0	0.1	0.2	0.3	0.4	0.5	0.6	0.7	0.8	0.9	1.0
Choice ratio	10:0	9:1	8:2	7:3	6:4	5:5	4:6	3:7	2:8	1:9	0:10
Choice frequency	.098	.98	4.4	11.7	20.6	24.6	20.6	11.7	4.4	.98	.098

Extreme results
(≤2/10): 5.5% →   **Total: 11%**   ← (≥8/10): 5.5%

3-7 Approximate chances of obtaining a result of 8/10 or more extreme without regard to whether the extreme is high or low. As a result, 0/10 is as unusual as 10/10, 1/10 as unusual as 9/10, and 2/10 as unusual as 8/10. To evaluate this case, the probabilities of all six classes of events must be added together.

triangle is symmetrical, it makes sense that all two-tailed probabilities should be double their corresponding one-tailed values.

3.7 Post Hoc Analysis

Given that two-tailed tests yield much less favorable *P*-values, and that sample sizes must increase exponentially to compensate, the temptation frequently arises to reformulate a hypothesis *after* the data are collected. This can seem very reasonable: researchers often become fully sensible, after the fact, of the powerful logic for expecting what they have now observed. But consider these two examples:

1. You step outside and observe the first license plate you see—YUB 386, let's say. What are the odds of seeing that particular number? Using the product law, with 26 letters and 10 numbers you might quickly compute that it is $26 \times 26 \times 26 \times 10 \times 10 \times 10$, which is 17,576,000; thus the odds are 1/17,576,000 against it (ignoring the minor effect of "forbidden" letter combinations).

2. You wonder what the odds of observing the plate number YUB 386 on the first car you see outdoors must be, and then compute it as above. Only then do you step outside and look.

In the first case you formulated a ***post hoc hypothesis***: the chance of spotting that particular license plate number might seem quite small, but in fact the odds of seeing it were 1 in 1 once you actually *had* seen it. In the second case you used an ***a priori hypothesis***—you found out how likely it would be to spot that particular license plate prior to looking for it. Most of the wonderful coincidences you read about are the consequence of computing post hoc odds—the chance of two American presidents involved in drafting the Declaration of Independence (Thomas Jefferson and John Adams) both dying on the Fourth of July—in fact, on the *same* Independence Day (1826, which, wonder of wonders, is the 50th anniversary of the signing). This sort of thing may be harmless enough in tabloids and popular writing, but in academia it is both dangerous and dishonest.

A **post hoc hypothesis** is formulated *after* looking at the data; an **a priori hypothesis** is formulated *before* looking at the data.

The same applies to the one-tail versus two-tail decision, which must always be made before collecting the data to be tested (though doing so after a preliminary test, the data from which will not be part of the ultimate analysis, is both fair and prudent). Given that your colleagues are bound to be unreasonably skeptical, any argument to support a one-tailed analysis had better be *very* strong—after all, it takes a deviation from chance only half as large to achieve significance if you can use a one-tailed test, and the suspicion naturally arises that the "prediction" might have been post hoc.

3.8 The Uses of Post Hoc Probabilities: Bayesian Logic

Having warned you against the trap of post hoc hypotheses, now we'll look more closely at the cases in which prior knowledge can still be used. We've already indicated (and we'll emphasize this again and again) that preliminary experiments are very useful in formulating hypotheses, designing tests, and setting sample sizes. But why can't you just add these preliminary data to the remainder of the data you collect? Or, to put it another way, why shouldn't you be able to use the product law to combine the preliminary results with later ones? Say the preliminary test yielded a P-value of 0.2, and the follow-up test gave a P-value of 0.06—close to your 0.05 threshold, but not quite there. But if the two tests are independent, then shouldn't the probability of obtaining those two results together be $0.2 \times 0.06 = 0.012$?

First, it should be clear that you cannot use the first test to create one-tailed conditions for the second test *and* then combine the probabilities in this way for a joint one-tailed analysis: to decide to analyze the first experiment with a one-tailed test after you have the data is post hoc analysis. Second, there is the problem that you probably wouldn't have run the second test if the first hadn't shown some sort of a trend. In some disturbing way, the tests are not really independent. Most researchers are deeply suspicious of combining separate experiments; though objectively this might be viewed as merely increasing sample size, it's a post hoc manipulation. At the same time, it is difficult to maintain that there isn't more information from two tests than there is from one.

The problem of dealing with two-step (or multiple-step) probability analysis was first tackled by the English clergyman Thomas Bayes in the mid-1700s. His concern was how prior knowledge of probability should be used in analyzing new information. For instance, if we know that 95% of armed robbers are male, then the report that a particular robber was six feet tall adds very little extra information about the sex of the robber in question since nearly all individuals of that height are males. On the other hand, if an eyewitness reports that the perpetrator was five feet tall, the chance that the criminal is a male is now much smaller because almost no adult males are that short—the probability is about 0.006. Perhaps the robber was one of the rare breed of female bandits. The difference between this scenario and the earlier one involving experimental replication is that the probability before the new "test" (the eyewitness observation) was well known. We can say with confidence that the two observations—there was a robber, the robber was short—have independent and well-understood a priori probabilities with regard to gender. (Bayesian analysis is quite controversial when it is applied to poorly known or subjective probabilities.)

So let's look at how Bayesian analysis can be applied to this example. Throughout this book we will talk mostly in terms of *probabilities*—0.05, for instance. In Bayesian analysis, it is more convenient to speak of the *odds*: 1 in 20, for instance, which is also written $1:20$, or 0.0526, and is equivalent to $P = 0.05$. Where did we get 0.0526? Odds are computed as $P/(1 - P)$, and 0.0526 is simply $1/19$. Conversely, probability is calculated as odds/$(1 + $ odds$)$.

We still need one additional term: the *likelihood ratio*. This ratio is the probability of obtaining a true positive divided by the probability of a false positive. In most cases, this is simply the *P*-value of the next test: in a typical test yielding $P = 0.01$, it would be $1:99$. (We'll look presently at cases in which this naïve computation does not apply.) So, let's say you've performed one test and got suggestive but not significant results;

The **probability (P)** of an event is the fraction of trials in which the event occurs; the **odds** of an event is the probability of it occurring divided by the probability of it not occurring:
odds = $P/(1 - P)$;
P = odds/$(1 + $ odds$)$.

The **likelihood ratio (LR)** is the probability of obtaining a true positive (P_t) divided by the probability of obtaining a false positive (P_f): LR = P_t/P_f.

3 Binomial Distributions

now you perform an independent second test. In our robbery example, the first piece of data—there was a robber—allowed us to presume a P of 0.95 that the culprit was a male. This may be good enough for some biologists, but it isn't satisfactory for police or a jury. The second piece of information yielded a P of 0.006 (1/150) for the robber being male. How can we combine these values?

Bayes' theorem maintains that the odds of a hypothesis being true (O_{new}) after a new test is performed depends on the pre-test odds (O_{old}) times the likelihood ratio (LR) for the new test:

$$O_{new} = (O_{old}) \cdot \text{LR}.$$

In its standard formulation ***Bayes' theorem*** states that post-test odds = pre-test odds × likelihood ratio. Now in many cases this formula is a trivial restatement of the product law: the likelihood ratio for the second test is its P-value expressed as odds. In the case of the robber, the pre-test odds were $.95/(1 - .95) = 19$ that the suspect was a male; the likelihood ratio for the second test (height of five feet) is 1/150 that the suspect is male. Hence, the post-test odds are $19 \times (1/150) = 19/150$ that the suspect is male; as a probability, $P = 0.112$. Later evidence might alter this value again.

But there are cases in which the likelihood ratio is more problematic. Say you are attempting to infer the sex of the students at MIT who are using the exercise facilities. You know that the sex ratio at the institution is $2 : 1$ male to female; these are your pre-test odds that a random user is a male. The only data you have available about equipment use come from the height settings on the exercise bikes. For these data you know that a "tall" setting is used by males 80% of the time, and that 30% of the female users employ the "tall" setting. Your likelihood ratio that a "tall" setting implies a male user is the probability that the setting yields a true positive (0.80) divided by the probability that it yields a false positive (0.30), or $0.80/0.30 = 2.67$. Let's combine this value with the pre-test odds $(2 : 1 = 0.67/[1 - .67] = 2.0)$ to see how likely the "tall" setting is to indicate a male user: post-test odds $= 2.0 \times 2.67 = 5.33$. Converting back to probability you get $5.33/(1 + 5.33) = 0.84$. Thus a single measurement of a "tall" exercise-bike setting identifies a male with $P = 0.84$. Without Bayesian analysis, you might have assumed the value was 0.80 because that is the proportion of males that use the setting. For cases with higher pre-test odds, the difference between ordinary and Bayesian analysis can be enormous; an all too realistic example in the exercises at the end of this chapter yields a roughly 50-fold discrepancy.

3.9 Binomials with Unequal Probabilities

Except for a brief interlude with dice, we have been dealing with binomials in which the two mutually exclusive events have had equal probabilities, 0.5. But binomial analysis works with any pair: 0.7 and 0.3, for instance (Figure 3-8). Clearly, the two alternative probabilities must add to 1.0, and the numbers in Pascal's triangle, while they still indicate the number of possible combinations, no longer directly reveal their probabilities, which are computed according to the **binomial theorem**. The general form of the equation that is used in binomial analysis is $(aA + bB)^n$, where a is the probability of one alternative event (A) and b is the other (B), and n is the number of events (the line in Pascal's triangle). The statistical programs that analyze binomial distributions do not care what values you assign to a and b as long as $a + b = 1.0$.

The **binomial theorem** computes the probability of different combinations of alternative outcomes (A versus B, where a is the probability of A occurring, and b is the probability of B occurring. Put another way, $b = 1 - a$, since all outcomes are either A or B. The probability of any particular number of A/B outcomes in n events is given by the expansion of $(aA + bB)^n$. The probability of i occurrences of outcome A is given by the number that is multiplied by the $A^i B^{(n-i)}$ term.

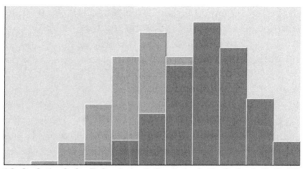

10:0 9:1 8:2 7:3 6:4 5:5 4:6 3:7 2:8 1:9 0:10 **Event ratio**

3-8 When the two alternatives in a binomial distribution have different individual probabilities, the distribution is skewed toward the more likely alternative. Here is the distribution for 10-sample runs with event probabilities of 0.3 and 0.7. (Shown in gray in the background is the typical 0.5-to-0.5 distribution observed when the alternatives are equally likely.)

Box 3.3

BioStats Basics Online Computing binomial probabilities when the alternative events are *not* equally likely

The Binomial Distribution card (II. E-1), which deals with binomial data when *a* and *b* have different probabilities, is similar to the Equal-Probability card in most respects. This explanation assumes you have read the discussion of the equal-probability binomial test earlier in the chapter; if not, what follows may not be entirely clear.

For the unequal-probability binomial test you have the option of entering (or importing) the observed data, or of entering summary information. As with the previous test, you must enter the sample-set size in its boxed field. Unlike the previous test, however, you must indicate the probabilities of *a* and *b* in their boxed fields (and they must sum to 1.0). As before, you must select the appropriate tails option. If you select the

Sample Mean option, you must enter that value (which must lie between 0 and the number in the sample-set-size box) along with the number of sample sets (*n*) in your data. When you click on the **Compute** button you get the *P*-value, but there is no graph.

Because this is strictly a binomial test, there is no extraneous information about standard deviations or standard errors. To prevent *BioStats Basics Online* from computing on forever when the necessary binomial expansion is enormous, the test is limited to cases in which the product of the sample-set size times the number of sets is no larger than 1000. For larger sample sizes with equal-probability alternatives, use the Two-Choice Test (II. D-1).

The demonstration on the card Normal and Skewed Distributions in *BioStats Basics Online* (I. D-1) uses $a = 0.3$ and $b = 0.7$. (Because the $50:50$ ratio of coin flips and two-way choices is so common, many programs, including *BioStats Basics Online*, have a special test optimized for the case $a = b$; it's on card II. D-1.) Let's see how the binomial equation provides the probabilities for the simple case of $a = 0.3$ and $n = 2$. The formula $(a\text{A} + b\text{B})^n$ becomes $(0.3\text{A} + 0.7\text{B})^2$, which is $(0.3\text{A} + 0.7\text{B})$ $(0.3\text{A} + 0.7\text{B})$. Multiplying this out we get $0.09\text{A}^2 + 0.21\text{AB} + 0.21 \quad \text{BA} + 0.49\text{B}^2$. Combining the middle terms, this is $0.09\text{A}^2 + 0.42\text{AB} + 0.49\text{B}^2$. Note that the decimal values sum to 1.0. When we insert the mathematically unnecessary exponents, we have $0.09A^2B^0 + 0.42A^1B^1 + 0.49A^0B^2$. Thus 9% of the outcomes involve two A results and no Bs (i.e., A^2B^0), 42% of the outcomes involve one A and one B result (A^1B^1: 21% in the order A → B, and another 21% as B → A), and 49% of the outcomes involve two B results (A^0B^2). (In *BioStats Basics Online* the test for binomial data with unequal probabilities is on card II. E-1.)

3.10 Binomials with Estimated Probabilities

We have been assuming that you are dealing with a null distribution inferred from an exact knowledge of the values of a and b. Sometimes, however, the two binomial probabilities must be estimated from your data. This introduces some degree of uncertainty in computing P-values: your numbers for a and b are only estimates. When these binomial probabilities are in the range 0.1–0.9 and the sample size is large (>1000), there is probably no real difference between the estimate and a true null distribution. When either of these conditions is not met, you will need to use a test introduced in Chapter 8 (chi-square for goodness of fit). This application of binomial analysis is discussed there.

3.11 Looking at the End: The Poisson Distribution

We have demonstrated how enormous and unwieldy binomial distributions become by line 40, and it gets pretty bad long before that. In 1830 the French physicist Siméon Poisson discovered a way of circumventing the then-impossible mathematical undertaking of analyzing large-event binomial distributions for a special category of cases that interested him: rare events.

A rare-event binomial has a small value for a and a correspondingly large one for b—0.05 and 0.95, for instance—where the value of a is the chance of the event in question (A) occurring, whereas the value of b is the chance of it not occurring (B). The resulting binomial distribution is highly skewed

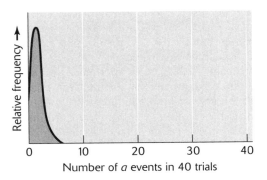

3-9 When one of two alternative events is rare, the distribution of the number of occurrences of the unlikely event is highly skewed in a characteristic manner. Deviations from this skewed distribution indicate a lack of independence in the events contributing to it.

toward the mostly-doesn't-happen end (Figure 3-9). This is the null distribution for these values of a and b, assuming A and B are independent. In the examples up to now we have been asking if the sample distribution is different from the null distribution with an eye for spotting cases in which the actual values for a and b in the sample differ from the null-hypothesis values of a and b that generate the null distribution. Poisson's interest was different: he wanted to know if A and B were actually independent. Naïvely, the chance of observing two As in a row is a^2, because one occurrence of A does not alter the chance that the next event will be A again; if A and B are independent, then the binomial predicts the frequency of never seeing the event occur, seeing it occur once only, seeing it occur twice, and so on. If they aren't independent, there is an interesting interaction between A and B.

Consider the oft-cited folk wisdom that suicides come in waves, one person's act of self-destruction prompting emulation by others who may have been contemplating the same. An alternative hypothesis, of course, is that one well-publicized suicide causes newspaper editors to report subsequent less newsworthy attempts that would otherwise have gone unnoticed by the press. For any limited region and interval of time, the number of suicides is usually zero. One suicide is rare, two even less common, and so on; thus this is a perfect opportunity for ***Poisson analysis***.

Poisson discovered that he could approximate a rare-tail distribution, regardless of the line of Pascal's triangle, simply by knowing the mean number of times the rare event occurred in any sample unit. As long as the product of the probability of the rare event per sample unit times the number of samples is less than 5, the Poisson approximation works well. If we call this probability μ, then the number of times we expect no rare events in the sample interval is $1/1!e^{\mu}$ (where e is 2.718)—if, that is, the events are independent. (The symbol ! indicates a factorial, a number that is the product of a given number and each integer smaller than it down to 1.) We expect to observe one event in $\mu^1/1!e^{\mu}$ cases, two events in $\mu^2/2!e^{\mu}$ cases, and so on. Fortunately you will never need to compute this distribution, because *BioStats Basics Online* (II. F-1) and most other statistical programs can do this for you.

The computation yields an expected rare-end–tail distribution to compare with the observed distribution. Next you must compute a ***coefficient of dispersion (CD)***, a value that quantifies the degree to which the distributions agree; this is done by dividing the variance (the sum of the squared differences between each value and the mean, $\overline{X}$ divided by the sample size, n) by the mean. Again, this value is automatically calculated by *BioStats Basics Online*. The final step is to calculate the probability that what you have observed is just a chance deviation from the prediction of the null hypothesis, which is that the events are independent. (This final step in the analysis uses a method we will discuss in a later chapter, the chi-square test for goodness of fit; for now we will ask you to take the validity of this step on faith.) In the case of suicides, after putting the data through these steps, there is usually a small but clear deviation from independence. We will encounter the CD measure again in Chapter 5 when we use it to see if events or individuals are clumped in time or space.

The Poisson approach can be applied to a surprising number of issues. For example, consider the explosion of the space shuttle *Challenger* in 1986. The disaster occurred when both of the O-rings that joined a pair of segments in one of the two

Poisson analysis is used to determine if two alternative binomial events, one of them rare, are actually independent.

The **coefficient of dispersion (CD)** is the ratio of the variance to the mean: CD = $[\Sigma\,(\overline{X} - x_i)^2 n]/\overline{X}$.

3 Binomial Distributions

Box 3.4

BioStats Basics Online Applying the Poisson analysis
for rare binomial events

This test is on the Poisson Distribution card (II. F-1). To use the Poisson analysis, the product of the mean probability of the rare event (per sample interval) times the number of sample intervals must be less than 5; *BioStats Basics Online* will check this for you *after* you enter the data if you don't want to determine this for yourself first. If your data pass this test, then enter (or import) the observed distribution of samples with 0, 1, 2, . . . 9, and 10 or more events into the boxed fields provided. When you click on (Compute) *BioStats Basics Online* will compute and display the mean probability of the rare event times the number of samples (which must be less than 5), generate the Poisson approximation of the expected rare-tail distribution, and compute a coefficient of dispersion (CD), which quantifies the similarities between the two distributions.

If CD is not close to 1.0, then you will want to perform a test that calculates the probability of the observed distribution being obtained by chance

from the expected distribution. To do this, you must use a test we will discuss in Chapter 8, the chi-square test for goodness of fit. You may wish to save your data first to allow easy importation into the goodness-of-fit test.

You can get to the Goodness-of-Fit card by clicking on the Topics button and then selecting the Categorical Distributions. This takes you to an explanation of the goodness-of-fit test (III. E-1); the next card (III. E-2) is the test itself. Enter (or import) the observed and expected values from the Poisson card. For the goodness-of-fit test to work properly, you need to combine the rarer categories of *expected* values so that no category has an entry less than 5. Click on the (Compute) button and *BioStats Basics Online* will provide a chi-square value and a degrees-of-freedom index (v). Go to the appropriate v row and read across from left to right until the value in the row *exceeds* the chi-square value. Look at the heading of that column and the one immediately to the left. The *P*-value lies between those two numbers.

solid-fuel booster rockets failed. Now failure or other evident damage to O-rings had been observed after several previous launches. (The boosters are recovered and disassembled in preparation for reuse after each launching.) There are 20 O-rings involved in each launch (10 pairs, 5 in each booster connecting the 6 segments), and the frequency of one binomial event—damage—was very low compared with the alternative (surviving intact); the observed rate was 10 in 24 launches, or 10 out of 480 O-rings at risk. Failure of a single O-ring is not fatal: both O-rings in the *same* segment must fail early in the launch to create a problem.

The launch agency, NASA, naturally worried about the possibility of a double failure. If O-ring failures are independent, then the product law can be used to compute the odds of a double failure in the same segment: if failures occurred in 10 out of 480 O-rings in 24 launches, then considering all the ways these failures could occur in the same segment, we can calculate that the odds of a double failure in any single launch are $1:2198$, or 0.00046. These odds represent a $10/480$ chance of failure in a given ring times 20 possible rings per launch, and that intermediate result times the $10/480$ chance of a second failure times—the $1/19$ chance that the second failure will be in the one and only other ring in the same segment. The odds are actually slightly worse than this because we are not considering the three or more failure possibilities that include two rings in the same segment; these increase the probability to about 0.00047. And of course we have been equating "damage" with "failure" just to be safe, whereas in fact damaged rings may not actually represent complete failures.

Not a very likely scenario, it would seem. Perhaps as shuttle safety officers we should shift our attention to other possible problems, since the theoretical chance of a double same-segment—O-ring failure is tiny. But what if the failures are *not* independent? In that case the product law is useless. This is a real worry: what if the failure of the first ring in a segment leads to damage in the second? In that case the odds of a double failure would be much higher.

You can use the Poisson formula to create an expected tail-end distribution for the number of O-ring failures per launch

*Need more than this? See **More Than the Basics:** From Binomial to Poisson.*

using the observed net failure rate before to the accident; then you can compare that curve to the observed distribution of failures per launch. If the two patterns agree, then NASA cannot be blamed for assuming independence and using the product law in their risk calculations. If they do not agree, the agency made a grave error. Figure 3-10 compares the two distributions using the data for both failed and damaged rings. The statistical test for judging the degree of similarity between these two patterns shows no significant difference.

So were NASA and the crew of the *Challenger* just very unlucky? The problem was not independence, but rather that the μ for the failure rate of O-rings was valid only for the range of launch temperatures previously encountered—typically 15° to 25°C (60° to 80°F)—whereas the *Challenger* launch took place with the O-rings at 1°C (34°F). We will look at another way the O-ring data could have been analyzed, to more effect, in Chapter 11. This is also a good place to remind you that we are almost certainly asking the wrong question here: as we mentioned in Chapter 2, you will normally be trying to avoid false positives by computing the odds that two distributions are the same;

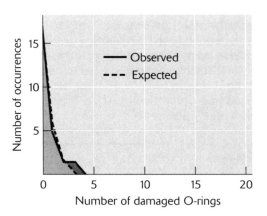

3-10 Observed versus expected distribution of O-ring failures. The two distributions are plotted as line graphs for ease of comparison; technically they should be plotted as bar graphs.

where safety and health are concerned, the goal is (or usually should be) to avoid false negatives.

Checking for independence in binomial distributions in which one alternative is not rare is difficult except for one special but fairly common case: the alternative events have equal probabilities, and the number of events is fairly large (more than about 20). In this case, as we will show in Chapter 5, you can approximate the null distribution with a "normal" curve and apply the F-test.

Points to remember

✔ Binomial data have one of two possible states or values, A and B; the probability of each outcome is a for A and b for B, where $a + b = 1.0$.

✔ The product law states that the probability of a particular set of outcomes is the product of the probability of each outcome times the number of possible outcome orders.

✔ The binomial test compares the mean of a sample distribution with the mean of a null distribution; this is equivalent to testing whether the observed values of a and b correspond to the values used in the null distribution.

✔ The P-value generated by the binomial test (and statistical tests in general) is the chance of drawing this set of data at random from the other distribution (the null distribution in this case), or one more different; this is the probability of a false positive.

✔ A result is "significant" if the P-value—the chance of a false positive— is smaller than the threshold you set.

✔ When there is a real trend in the data, larger sample sizes generally yield smaller P-values because chance deviations are less likely; in such cases the P-value declines roughly with the square root of the sample size.

✔ Most tests are two-tailed: any deviation of the data from the null-hypothesis prediction, whether positive or negative, is equally important; when there are excellent a priori reasons to look for only one kind of deviation, the test is one-tailed; for any given deviation, the P-value generated by a one-tailed test is half that of the equivalent two-tailed test.

✔ A hypothesis formulated after examining the data is valid only for analyzing subsequent data; it cannot

3 Binomial Distributions

be used to test the data already examined (the sin of post hoc analysis).

✔ Odds and probabilities are different ways of representing chance; probabilities are expressed as the proportional chance of an event or outcome (in the range 0–1.0), whereas odds are the chance of the event occurring divided by the chance that the event will not occur.

✔ Bayesian analysis can sometimes be used to combine separate measurements of the odds of an event or outcome by multiplying the odds of all measurements.

✔ Poisson analysis can be used to see if two binomial events or outcomes are actually independent, provided one is rare compared to the other (i.e., $a \gg b$ or $b \gg a$).

Exercises

The data for some of the exercises in this and later chapters can be found in the Data folder in *BioStats Basics Online*. For large data sets, only the first few lines are reproduced in the text (to give you a sense of what the data look like and how they are organized); for shorter sets the data are both on the disk and fully set out in tables in the Exercises section. The file names differ between the Macintosh and Windows versions because older versions of Windows limit users to only eight characters in naming files; the Windows names are easy to recognize—they all end with ".dat."

1. Use the product law to compute the odds that two individuals will have the same birthday (you can ignore February 29). Use the product law to compute the chances that two individuals in a group of three will share a birthday. (The second part of this problem is not as easy as it sounds.)

2. Use the product law to calculate the chances of rolling a twelve with a pair of dice. What are the odds of rolling a six? A seven? What are the odds of being dealt a royal flush (the five highest cards in a suit) in a five-card deal?

3. Table A gives the results of some tests using a T-maze with female crickets. The females could turn either left or right, perhaps choosing randomly as they attempted to flee from one another, or perhaps choosing a maze direction based on the calls played from each end. All tests were done twice, reversing the calls between ends to control for

Table A

Test #	Nature of the alternative call	Number of females moving toward	
		Alternative call	Normal call
1	Silence	702	1131
2	Enhanced treble	103	83
3	Reduced treble	44	16
4	Chorus of males	274	246
5	Heterospecific call	128	194
6	Reversed call	51	47
7	Treble down/bass up	38	64
8	Treble up/bass down	25	10
9	Higher pitch	82	139
10	Lower pitch	106	157
11	Louder call	400	263
12	Scrambled call	50	57
13	More frequent call	190	167
14	Less frequent call	48	72
15	George Bush speaking	85	135
16	Jessie Jackson speaking	100	159

3 Binomial Distributions

any maze bias. Some of the results are given in Table A. The null hypothesis is that there is no effect of playing the calls. For the test involving the normal call against silence, the expectation is that the females will prefer the species' call. All other tests compared the normal call against some variant; for these there is no particular a priori expectation. For Test 1 and three others of your choosing, use the Standard Two-Choice Data form of the binomial test (on the Perfect Normal Distributions card; this version is necessary for Test 1 because the sample size is greater than 1000) or the Binomial Test (on the Binomial Distribution card) to compute the probability that the null hypothesis is correct; justify your choice of tails.

4. For Test 14 in Table A, how much do you need to scale up the sample size to achieve a P-value below 0.01? How much do you need to scale it down to drive the

P-value above 0.1? How much do you need to scale up Test 6 to find significance at $P < 0.05$?

5. You are testing a die to see if it is "fair." You choose to do this by comparing the frequency with which the four-spot side ends face up against the null prediction of 1-in-6. After 100 rolls, you find 12 four-up results; after 200, you find 24; after 400, you find 48; after 600, you find 72; after 800, you find 96; after 1000, you find 120. Which are significant at $P < 0.05$? At $P < 0.01$? Justify your use of one or two tails.

6. Most organisms produce about as many males as females. Red deer females high in the social hierarchy, however, seem to produce more sons than daughters, while low-ranking females tend to produce an abundance of daughters. What are the odds of 61 of 99 offspring born to high-ranking females being sons if there is no ability in this species for females to bias the sex of their offspring?

7. Individual rats in a learning experiment were allowed to explore a featureless eight-arm radial maze with food in each arm. Since each arm extends radially, a rat must return to the center hub before selecting another arm to visit. Through the screened roof of the hub, the rats could see the ceiling of the test room. Each rat was removed after it had visited seven arms and was kept for several hours in isolation. The maze itself was rebaited and rotated 180° so that the position of the unvisited arm was opposite its former direction. The rat was then reintroduced. The researchers wondered if the rat would choose an arm to visit first at random (the null hypothesis), or visit the one in the direction not previously explored (which would imply that the ceiling features were being used for orientation), or select the correct arm (indicating that some sort of odor or other cue within the maze can be used). Of 25 rats tested, about half (11) chose the arm in the unvisited direction. To what degree does this result prove or disprove the null hypothesis? Justify your choice of tails.

8. A new test for colon cancer is about 95% accurate: it yields only 5% false positives and 5% false negatives. What is the probability that you have colon cancer if the test is positive? Suppose that in the general college-age population the chance of having colon cancer is 0.1%. Using Bayesian analysis, find the chance that your positive test result means you have colon cancer.

9. One of the classic examples of the early use of Poisson analysis was its application to the question of whether there were "lax" versus "taut" cavalry corps among the 10 units in the Prussian army. (A lax

Table B

Number of men killed by horse kicks per year per army corps	Observed frequency of occurrence
0	109
1	65
2	22
3	4
4	1
5 or more	0
total:	200

corps would have many accidental injuries, for instance, whereas a taut corps would have few.) Table B gives the data on men per corps per year killed by a kick from a horse. It is possible that if there are lax units that are accident-prone, there could be within-unit clustering of horse-kick fatalities. The null hypothesis in this case is that the distribution of deaths is independent. Check this hypothesis with Poisson analysis. By inspection (that is, without using the goodness-of-fit test referred to on the Poisson Distribution card, which we will not discuss until Chapter 8), which hypothesis is more consistent with the data? (The data are also found in file 3—Prussian Army Deaths or 3-PRUS.dat.)

10. Honey bees approaching artificial feeders are reluctant to land. Some researchers "bait" the feeders with

dark bee-sized objects as decoys. In an experiment using artificial feeders with automatic arrival-time recorders, bees seemed to arrive in clusters, as though when one brave (or foolish) forager landed, her hesitant colleagues were emboldened and landed as well. Table C lists the arrival patterns per minute at feeders. Check this with Poisson analysis. By inspection, is there any evidence of clustering? (The data are also found in file 3—Forager Landings or 3-BEES.dat.)

Table C

Number of bees arriving per minute at feeder	Observed frequency of occurrence
0	422
1	281
2	119
3	84
4	21
5	4
6	1
7 or more	0
total:	932

11. Creosote bushes, which are found in certain arid regions, are almost never found clumped and rarely have other plants growing near them. This observation suggested that the bushes might be producing some toxic substance (toxic to other species, that is) which is released into the soil to act as a competition-reducing herbicide and to act on other creosote bushes as a signal or a seed-germination inhibitor. Table D lists the observed densities of these plants in one study. Check this with Poisson analysis. By inspection, is there any evidence of repulsion, or are the bushes distributed independently? (The data are also found in file 3—Creosote Bushes or 3-CREO.dat.)

Table D

Number of bushes in 10-m^2 patches	Observed frequency of occurrence
0	389
1	122
2	2
3	0
4 or more	0
total:	513

More Than the Basics

Error and significance

The gathering of data is, from a statistician's point of view, the collection of a random sampling of points from a parent distribution, where the parent distribution is the set of all possible results. Statistics is the attempt to answer questions about the parent distribution from the sampled values—rather like attempting to feel one's way around while blindfolded using only a long pole, a job that most academic statisticians regularly reserve for undergraduate research assistants. Obviously, the more data we have, the finer the detail with which we can see the parent distribution, but since we have only our finite sample, we can never really know every detail of the parent distribution.

This also means that we can never answer definitely what the distribution is—it could have some yet-undetermined difference in its fine structure—but we can certainly say what it isn't. Statisticians therefore compose their questions in the following manner: "How likely is it that the parent distribution is no different in some property from some test distribution?" This is called the null hypothesis, or hypothesis of no difference. The property may be any number of things, but it is most often mean or median. The test distribution may be any distribution, but it is most often a well-defined null distribution or the parent distribution of another sample.

A statistical result that does not reflect reality is called an error; the rejection of a true hypothesis is *Type I error*, sometimes called a false positive. Failure to reject a false hypothesis is *Type II er-*

ror, or false negative. It is often possible to reject a hypothesis without much chance of Type I error, but it is never possible to accept a hypothesis without risking Type II error.

The predetermined *P*-value threshold used in a test is the probability that by rejecting the null hypothesis, we would be committing a Type I error.

From binomial to Poisson

The binomial $a = b = 0.5$ (Pascal's triangle):

$$_nC_X = \frac{n!}{X!(n - X)!}$$

The binomial $a = (1 - b)$:

$$C = \frac{n!a^X b^{(n-X)}}{X!(n - X)!}(2^n)$$

Here, the symbol $_nC_X$ represents the xth number in the nth row of Pascal's triangle; the symbol μ represents the mean.

The Poisson distribution:

$$c = \frac{\mu^X}{X!e^\mu} \text{ or } c_0 = e^{-\mu}; \; c_{X+1} = c_X \frac{\mu}{X}$$

Here, the symbol c represents the frequency of samples containing precisely X results.

In fact, the Poisson distribution can be derived from the binomial distribution without great difficulty. If $a \approx 0$ and $b \approx 1$, it is clear that for large X (generally assumed to be $X > 10$), $c \approx 0$; for $X < 10$ and n large, the following algebra may be done:

$$C = \frac{n!a^X b^{(n-X)}}{X!(n - X)!}(2^n) \underset{\substack{b \approx 1 \\ X > 0}}{\Rightarrow} C = \frac{n!a^X}{X!(n - X)!}(2^n) \underset{\substack{c = C/2^n \\ \mu = a}}{\Rightarrow} c = \frac{n!a^X}{X!(n - X)!}(2^n) \underset{X > 0}{\Rightarrow}$$

$$c_{X+1} = c_X(n - X)\frac{a}{X} \underset{X \ll n}{\Rightarrow} c_{X+1} = c_X n \frac{a}{X} \underset{\mu = an}{\Rightarrow} c_{X+1} = c_X \frac{\mu}{X}$$

where c_0 is the probability of the event never occurring. This probability may be

determined by the product law:

$$c_0 = b^n = b^{\frac{\mu}{a}} \underset{a \approx 0}{\Rightarrow} \left(\lim_{a \to 0}\left((1 - a)^{\frac{1}{a}}\right)\right)^\mu = \left(\frac{1}{e}\right)^\mu = e^{-\mu}$$

This yields the Poisson distribution.

Tests Covered in This Chapter

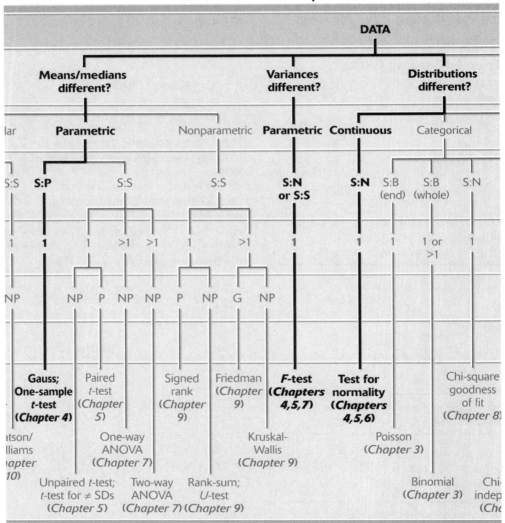

DATA

Means/medians different? | Variances different? | Distributions different?

Parametric | Nonparametric | Parametric | Continuous | Categorical

S:S | **S:P** | S:S | S:S | **S:N or S:S** | **S:N** | S:B (end) | S:B (whole) | S:N

1 | **1** | 1 | >1 | >1 | 1 | >1 | **1** | **1** | 1 | 1 or >1 | 1

NP | NP | P | NP | NP | P | NP | G | NP

Gauss; One-sample t-test (Chapter 4)

Paired t-test (Chapter 5)

Signed rank (Chapter 9)

Friedman (Chapter 9)

F-test (Chapters 4,5,7)

Test for normality (Chapters 4,5,6)

Chi-square goodness of fit (Chapter 8)

atson/ lliams apter 10)

One-way ANOVA (Chapter 7)

Kruskal-Wallis (Chapter 9)

Poisson (Chapter 3)

Unpaired t-test; t-test for ≠ SDs (Chapter 5)

Two-way ANOVA (Chapter 7)

Rank-sum; U-test (Chapter 9)

Binomial (Chapter 3)

Chi-indep (Cho

Continuous Parametric Distributions: I

4

Our discussion of the binomial distribution was focused on the question of whether a sample distribution differed from a null distribution. The binomial test compared the expected values of the alternative event probabilities (*a* and *b*) with the observed values; the Poisson method showed whether, in the case of one event being rare ($P < 0.1$), the two events were independent. In either case the data were discrete—that is, only certain values were possible.

In this chapter we will deal with **continuous parametric distributions**—data that are continuous because any intermediate value is possible. We will ask whether two distributions have different means. This may sound like a very different question for a distinctly different kind of data, but in fact there are strong similarities. These similarities help explain the logic of parametric analysis in terms you already understand.

4.1 What Is "Parametric"?

"Parametric" as we and other scientists use it ("normal" to licensed statisticians) means simply that a distribution is of a special shape that can be described exactly by a particular pair of parameters: its mean value and its width (usually measured as variance or standard deviation). This special shape is the **bell curve**. All bell curves are graphs of the same function, first described by Carl Friedrich Gauss in 1809:

$$y = \frac{1}{\sigma\sqrt{2\pi}} \, e^{\frac{-(x-\mu)^2}{2\sigma^2}}$$

Continuous parametric distributions consist of data that vary in a continuously variable parameter and are distributed in the characteristic bell curve.

The most common test performed on parametric data compares the means of two sets of data.

Parametric data are distributed as a **bell curve** defined by the Gaussian distribution.

In this formula μ (mu) is the mean, e is 2.71828 (rounded), and π (pi) is 3.14159 (also rounded); σ (sigma) is the standard deviation. Thus two parameters, μ and σ, specify the y-x graph of any parametric distribution. The other variable of interest in the analysis of parametric data is the sample size. With these three numbers an entire set of data of any size can be completely summarized and a seemingly endless stream of statistical interpretations wrung from it.

4.2 What Sorts of Measurements Yield Parametric Distributions?

It is truly wonderful that an entire data set can be reduced to two or three numbers if it corresponds to the Gaussian curve. But that any data should happen to distribute themselves according to the strictures of the Gauss formula seems even more amazing. And yet, a tremendous number of distributions *do* have this shape; indeed, to judge from some statistics texts and courses, no other kind of distribution exists. So common is this bell-curve distribution that it has acquired the dubious and misleading distinction of being called the "normal" curve.

Parametric data are distributed randomly about a mean. The randomness can arise from **measurement error** and/or the separate contributions of many independent effects.

Parametric distributions arise from randomness. This randomness has two common origins, which may be mixed together in any particular distribution. The first is **measurement error**. No measurement is either absolutely accurate or precise: if we look for the most sensitive scale or most highly subdivided ruler and independently measure the same sample again and again, the values will still not exactly agree. Even the U.S. Bureau of Standards cannot obtain a consistent value for the weight of its 10-gram standard; the independent weighings typically vary 0.000006 gram above and below the mean. This variation probably results from small imperfections in the scale, variation in the placement of the standard on the scale, and dust on the scale or weight.

Accuracy is the degree to which the data approximate the true value of the thing being measured, whereas **precision** is the degree to which the data agree with one another.

We implied above that there is a distinction between **accuracy** and **precision**. Happily, this difference is not usually important, but when it is you will need to remember that precision is the degree to which repeated measurements agree with one another, whereas accuracy is the degree to which a measurement, or the mean of several measurements, corresponds to the true value of the variable. This distinction is illustrated by the (we hope) now-irrelevant controversy over the accuracy of ballistic missiles: the military claimed an accuracy of about 0.5 km at a distance of 9000 km. However, this value was based on repeated test firings at a particular atoll in the Pacific, and even then the guidance systems had to be repeatedly corrected to place the warheads on this particular target. Accuracy would be the deviation between the impacts and a novel target; the 0.5-km figure represents the precision—the "clumping" of the impacts—made possible by post hoc biasing of the guidance systems that generated the data. Given that other targets (e.g., Moscow) lie at different distances and in different directions over our asymmetrical planet, the actual accuracy of these missiles remains, we are happy to note, unmeasured.

For a variety of reasons, then, no measurement is infinitely precise. More important, measurement errors distribute themselves around the mean in a way that closely approximates the Gaussian curve. Figure 4-1 shows the results of 100 weighings of the same cricket on a digital scale; to assure independence,

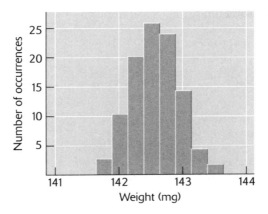

4-1 One hundred independent weighings of a cricket by biology students yielded this distribution. Like most sets of measurement error, this distribution is parametric (normal).

4 Continuous Parametric Distributions: I

*Need more than this?
See **More Than
the Basics:** From
Binomial to Bell.*

each weighing was performed by a different student. The distribution is distinctly bell shaped. The major source of error here was probably air currents, but that does not matter: if we had disposed of that problem we would have observed a narrower bell curve generated by some more subtle source of variation.

Parametric distributions are also generated by measurements that quantify values that have many independent processes contributing to them. Height is a good example: an individual's height is the product of many genes acting together, combined with a variety of environmental influences. The distributions of male and female heights vary in a highly "normal" fashion (Figure 4-2); the influence of genetic and environmental influences swamps the contribution of the equally parametric contribution of measurement error. Instruments routinely used in molecular biology produce some of the most perfect parametric curves around. A scintillation counter, for instance, measures flashes produced indirectly by the decays of radioisotopes. Each unstable atom is independent, and thus the number of scintillation flashes per minute varies in a parametric fashion about some mean value that depends on the number of labeled atoms in the sample; during any given interval of time, therefore, the observed count is some approximation of the mean—sometimes higher, sometimes lower, very rarely exactly on.

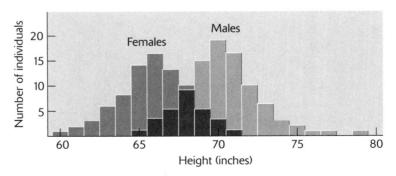

4-2 Height distributions of male and female college students at Freeman University. Both curves are parametric (normal).

4.3 Computing the Parameters of a Parametric Distribution

We have said that any parametric distribution can be completely summarized by its mean (μ or $\overline{X}$) and standard deviation (σ or s); its sample size (N or n) is also critical for statistical analysis. The reason we give two alternative symbols for each parameter is that you will sometimes need to distinguish between the parameters and sample size of a parent or null distribution (μ, σ, and N) and the parameters and sample size of a sample distribution ($\overline{X}$, s, and n). You can summarize any parametric data set with these two parameters because you can take any particular distribution and fit it to a standard version of the normal distribution. Let's see how this is done to some ordinary data (Figure 4-3a), because virtually every statistical test that deals with parametric data does this as a first step. We begin by subtracting the mean from each datum; as a result the curve is now centered on $x = 0$ (Figure 4-3b). If the y axis of the curve is expressed as number of measurements, we will need the sample size as well: we scale down the height by $1/N$. (In our example we are using a proportion-of-the-sample y axis, and so we can skip this step.) Next we scale the dispersion of the

*Need more than this? See **More Than the Basics:** Why Two Sets of Symbols?*

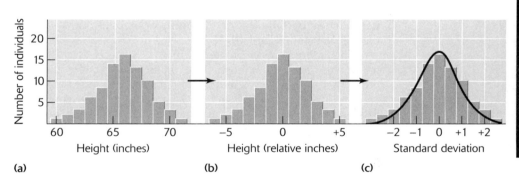

(a)　　　　　　　(b)　　　　　　　(c)

4-3 (a) Transformation of the female-height curve into standard units begins with the untransformed data. **(b)** The mean is subtracted from all values, leaving a graph centered on zero. **(c)** All values are then divided by the standard deviation, producing a distribution whose SD is 1.0.

The **mean** ($\overline{X}$) of a distribution is the average of the n data values (x_i) in the distribution $\overline{X} = [\Sigma(x_i)]/n$.

The **standard deviation** (s or SD) of a distribution is a weighted average difference between the mean ($\overline{X}$) of a distribution and the n data values (x_i) in the distribution $s = [\Sigma(\overline{X} - x_i)^2/(n-1)]^{1/2}$.

A **standard deviation** encompasses 68.26% of the data in the center of the distribution.

Need more than this? See **More Than the Basics:** *The Parameters.*

distribution by dividing each x-value by the standard deviation (Figure 4-3c). The result is a standard curve with a mean of 0.0 and a standard deviation of 1.0.

You've already seen (in Chapter 2) how to compute the *mean*: you add all the measured values and divide by the sample size. You also saw there how the standard deviation is computed, but this bears repeating. The ***standard deviation*** (s, or, more informally, SD) is obtained by taking the square of the difference of each measurement from the mean, summing all the squares, dividing by one fewer than the sample size ($n - 1$), and then taking the square root of the result. The value $n - 1$ is used rather than n because it works better for small sample sizes; for larger samples the two values are so similar it doesn't matter; for the parent distribution, n is exactly right.

Computers calculate the standard deviation in a somewhat different way, one that permits them to forget the individual data values (which they would otherwise need to remember until all the data were entered and a mean could be taken). The computer formula is given on the Mean and Standard Deviation card in *BioStats Basics Online*; this card (I. C-1) also has a nice demonstration of the steps involved in computing both values using data you enter or import. The demonstration runs slowly so you can see what is happening; if you need a mean and standard deviation for some other reason, you can get it much faster by entering or importing your data onto the Testing for Normality card (II. B-1).

The interpretation of the mean is obvious: this is the center of the distribution, so half the values are smaller than the mean, half are larger. The deeper meaning of the standard deviation is less obvious. If we look again at male heights with the SDs plotted (Figure 4-4), the compulsively observant may notice that the parametric curve has an inflection at $+1.0$ and -1.0 SDs—that is, at these points the curve changes from declining more with each step away from the mean to declining less. In quantitative terms, the area under the curve between $+1.0$ and -1.0 SDs is 68.26% of the total area under the curve; the other 31.74% lies outside this region. Moving an equal distance out from the mean on the x axis encompasses two SDs; 95.44% of the data lie between -2.0 and $+2.0$ SDs, while the

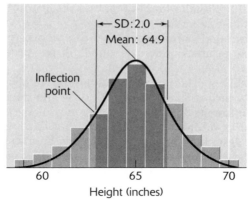

(a)

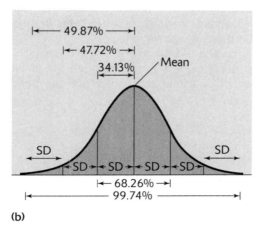

(b)

4-4 The inflection point in a parametric (normal) curve is at $+1.0$ and -1.0 SDs; the area between the $+1.0$ and -1.0 SD bounds contains about 68.26% of the data. **(a)** A sample set of height data fitted to a normal curve. **(b)** A normal curve with SD intervals and the corresponding areas under the curve indicated.

remaining 4.56% lie farther out on the tails. For three SDs, only 0.26% of the data lie farther from the mean. We will see many uses of shorthand phrases like "two SDs above the mean." A complete list of areas under the standardized curve is given in Table 4-1.

One final important comment about the standard deviation: the SD is a property of the variability in the data; increasing your sample size will increase the accuracy of your estimate of the SD, but it will not change its value. But if, as was often the case in the last chapter, you are collecting data as *sets* of measurements, increasing the sample-set size will change the apparent standard deviation when you are using the means of the sets as your data (Figure 4-5). This occurs for the same reason that higher-order binomials (i.e., those that involve more events and thus are represented by a wider line in Pascal's triangle) have a proportionally tighter distribution: the odds of getting many unusually high (or low) measurements in any set declines as you make more measurements.

But in fact, if you think about it, the so-called SD of the binomial sets in Figure 4-5 is not really an SD of the individual data—that is, of two-state measurements like heads and tails.

Table 4-1

SD units from mean	Area under normal curve from mean	SD	Area	SD	Area	SD	Area	SD	Area
0.0	.0000	1.0	.3413	2.0	.4772	3.0	.4987	4.0	.499968
0.1	.0398	1.1	.3643	2.1	.4821	3.1	.4990	4.1	.499979
0.2	.0793	1.2	.3849	2.2	.4861	3.2	.4993	4.2	.499987
0.3	.1179	1.3	.4032	2.3	.4893	3.3	.4995	4.3	.499991
0.4	.1554	1.4	.4192	2.4	.4918	3.4	.4997	4.4	.499995
0.5	.1915	1.5	.4332	2.5	.4939	3.5	.49977	4.5	.499997
0.6	.2257	1.6	.4452	2.6	.4953	3.6	.49984	4.6	.499998
0.7	.2580	1.7	.4554	2.7	.4965	3.7	.49989	4.7	.499999
0.8	.2881	1.8	.4641	2.8	.4974	3.8	.49993	4.8	.499999
0.9	.3159	1.9	.4713	2.9	.4981	3.9	.49995	4.9	.500000

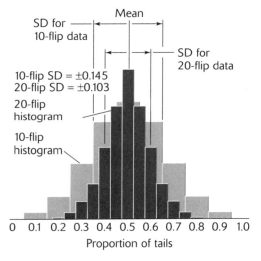

4-5 The apparent standard deviation of sets of binomial data changes as the number of elements in each set changes. This figure compares the binomial distributions for 10-set and 20-set coin-flip data, expressing the apparent SD in terms of proportion of heads to tails. We use the term "apparent" because the "SD" of binomial distributions is actually a second-order estimate of variance, based on the distribution of the means of multiple measurements of a given sample size. The SD of other types of data is first-order, based on the raw data rather than means of sets; as a result, the SD measures variance in the data, a property that does not change with sample size.

How could it? How could you encompass 68.26% of the outcomes of such data where all data have values of either, say, 0 or 1? Instead, what we call the SD of a binomial distribution is actually the variance in the mean values of equal-size sets of binomial data. And so in the special case of binomials, the "SD" is really what we will call a second-order SD—that is, it is removed one step from actual data. Any **second-order statistic** is characterized by the second use of some important summary information about the data. In this case, sample size is used twice: first to define the set size and number of sets and then to compute the binomial SD.

A **second-order statistic** is any value computed using the same piece of information twice.

The distinction between the true first-order standard deviation, which is a measure of the unchanging intrinsic variance in the data, and the distribution of various measurements of the mean of sets should be clear: SD doesn't change with sample size, but the variance in measured (sample) *mean* values (centered on the true but unknown parent mean) ought to decline with increasing sample size. After all, the larger the sample size, the better your estimate of the true mean should be. (The more times you flip a coin, the more likely you are to find the proportion of heads to be close to 0.5.) As we will show in Chapter 5, this property will allow you to judge the accuracy of your sample-based estimate of the true (parent) mean; the term we will develop there is known as the standard error of the mean (SE, or SEM, see page 126), a key component in the analysis of whether two distribution means are significantly different.

4.4 How Do We Know If a Distribution Is Parametric?

Given that parametric curves have many useful properties and can be analyzed to death by powerful statistical tests, how do you know if a sample distribution you have collected is parametric? This question is often answered, either explicitly or implicitly, by the convenient dispensation, "Just assume it." If you are dealing with data for which the parent distribution—the set of all possible data on the measurement of interest, like the IQs of all Americans for example—is known to be parametric, this is a satisfactory way to proceed. It also works when the theoretical null distribution is known to be parametric. But if you've collected data on, say, approach times by female mollies (a common aquarium fish) toward male fish models in a two-choice test (Figure 4-6a), there is no plausible basis for assuming the data will be parametric. Indeed, a glance with your on-board statistical-analysis system, the eye and brain, tells you that the approach times for females of a closely related fish, the platy, are clearly *not* parametric (Figure 4-6b). On the other hand, it would be foolish to reject out of hand the possibility that your data may be parametric just because they are novel—if you did so you would forgo much statistical weaponry and convenience.

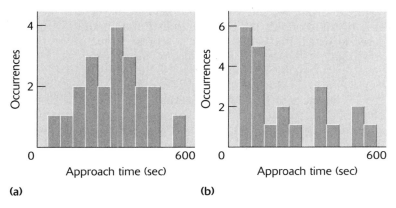

(a)   (b)

4-6 **(a)** Approach-time data from an experiment in which female mollies were offered a choice between two model males, one positioned at each end of the tank. These data may be parametric. Approach time for each model was measured as time spent at an end actively swimming toward that model. Most of the females in these tests showed a lively interest in the two models, and frequently swam back and forth, apparently comparing the models at the two ends of the tank. **(b)** Approach-time data from a similar experiment using platy females. Our eyes tell us these data are not parametric; the results are dominated by females that showed little interest in either model.

So how do you determine whether novel data sets are parametric? This is not a trivial problem. Here is a data set: 1.2 cm, 1.5 cm, and 1.7 cm. Is it from a parametric parent distribution? Clearly, there is no way to tell with so few values—the parent set could be skewed *or* parametric. With a larger set, however, departures from normality become increasingly unlikely if the parent distribution is in fact parametric. There is a complication, however: we don't know the mean and standard deviation of the parent distribution. We must infer those values from the sample data, thus adding to the overall uncertainty.

A good test for normality, therefore, will compute the odds that a given set of data are not from a parametric distribution

Parametric sample distributions have three kinds of **uncertainty**: potential errors (relative to the true values) in the sample mean, the sample SD, and the shape of the sample distribution.

while allowing for three kinds of **uncertainty**: (1) the mean of the distribution, (2) the SD of the distribution, and (3) the degree to which the sample data are representative of the parent distribution. *BioStats Basics Online* provides a very conservative test for normality on the Testing for Normality card (II. B-2); we will also need this test in Chapter 6 when we deal with the tricky business of transforming certain nonparametric distributions into ones that are suitably parametric. But though this test provides a satisfyingly quantitative *P*-value for normality, we encourage you to "eyeball" the distribution; if it looks parametric, it probably is—or rather, it's probably close enough that parametric analysis will work just fine on it since most parametric tests are robust enough to handle slightly skewed data. (This remarkable ability of parametric tests to digest less than perfect data is the reason the *P*-values used in the test tables for normality are 0.5 and 0.2 rather than the more usual 0.05 and 0.01.)

There is an easy mistake you can make when testing for normality (it must be easy, since we keep making it): you can enter *summary* data rather than the *original* data for testing (both in the Test for Normality card and on other cards in other tests). This pitfall is especially tempting because data are so often presented as summaries. For instance, when we were evaluating the distribution of scores in a departmental evaluation, we had 5 cases of a score of 1, 9 of 2, 21 of 3, 11 of 4, and 6 cases of score 5. This looks pretty parametric, but if we enter 5, 9, 21, 11, and 6 as the data set, the distribution is a surprise: after being told our sample size is small (5, in fact, rather than 52), the distribution plots two events in the left-most column, then two cases of one event each, and then way off to the right another event. In fact, we entered only *five* bits of data—5, 9, 21, 11, and 6—and that was what was analyzed, whereas we meant to enter 52 bits of data: 1, 1, 1, 1, 1, 2, 2, 2, 2, 2, 2, 2, 2, 2, 3, 3, 3, 3, 3, 3, 3, 3, 3, 3, 3, 3, 3, 3, 3, 3, 3, 3, 3, 3, 3, 3, 4, 4, 4, 4, 4, 4, 4, 4, 4, 4, 4, 5, 5, 5, 5, 5, and 5. The distribution in this case is the expected bell curve. To make it easy for you to convert from summaries to raw data, there is a (**Redistribute data**) button on the Testing for Normality card (II. B-2) that re-creates the original distribution for you automatically. Just follow the instructions that appear.

Box 4.1

BioStats Basics Online | Testing for normality

The Testing for Normality card (II. B-2) provides a way for computing the odds that a sample distribution is *not* drawn from a normal distribution. The test uses the mean and standard deviation of your sample as estimates of the true mean and SD; the sample size is its guide to the degree of inevitable imprecision in these estimates. It computes the difference between the sample distribution and a parametric distribution that has the same apparent mean and SD. From this it derives a measure of difference which is then compared to a list of values we obtained by the Monte Carlo method (described in Chapter 13). This permits *BioStats Basics Online* to determine how often a particular number of data values (your sample size) drawn from a real parametric distribution would produce a measure of difference as small as the one calculated from your actual data. (Be sure to use the original data, not a summary; if you only have the summary data, click on (**Distribute data**)

and follow the instructions to re-create the original distribution. This card will come in handy whenever you need to "deconstruct" summary data.)

To use the test, enter or import your data. Click on (**Compute**). *BioStats Basics Online* will generate an estimate of the mean and standard deviation, a degrees-of-freedom index (v), plus an index of difference (D); for now, don't worry about what is meant by "degrees of freedom." The table of critical values on the card scrolls—that is, by clicking on the shaded vertical bar at the right of the table, or dragging the slider, you can look at hidden parts of the table. Scroll to the line with the appropriate v-value and read across to the first value (the $P < 0.20$ column). If your D-index value is smaller than this number, you can treat your data as parametric. The test is not very useful for sample sizes smaller than 20; the larger the sample size, the more likely an actual set of parametric data will be judged to be parametric by this test.

4 Continuous Parametric Distributions: I

4.5 What Do Binomial and Parametric Distributions Have in Common?

We said at the outset of this chapter that though parametric distributions are continuous whereas binomial distributions are discrete, they have much in common. This similarity is most obvious when we set about plotting a continuous parametric distribution: researchers almost inevitably subdivide the x axis into discrete regions, combine the data within that region, and use that data count to plot a *y*-axis value. This does little violence to the phenomenon being described and greatly improves the clarity of the graphs. It also makes the point that distinguishing graphs of discrete data (with many possible *x*-axis values) from graphs of continuous data is often impossible; indeed, the distinction is lost on most research workers because they automatically lump continuous data into discrete bins simply by rounding the *observed* values.

When you are dealing with a high-order binomial (line 20 or beyond in Pascal's triangle) in which the alternatives have equal probabilities (that is, $a = b = 0.5$), the null distribution is an excellent approximation of the Gaussian curve (Figure 4-7). The reason that the curves have the same basic shape is that each is derived from a large number of independent contributions—many genes and environmental influences in the case of height, many coin flips in the case of a typical binomial.

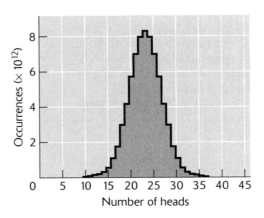

4-7 The binomial distribution for a sample-set size of 46. As sample-set size (the number of independent events) increases, the distribution is a better and better approximation of a parametric (normal) curve.

4.6 Comparing a Parent and a Sample Distribution

Because the shape and logic of parametric and equal-alternative binomial distributions are the same, it is easy to understand why testing a sample parametric distribution against a parent distribution has much in common with testing a sample binomial distribution against its null distribution. In both cases, you are asking the same question: what are the odds of obtaining the observed distribution, or one more extreme, by sampling from the parent/null distribution? As usual, the null hypothesis is that there will be no difference. And the first steps in the analysis are the same: you determine whether a one-tailed or a two-tailed test is appropriate, find the average (mean) value of the sample, measure the corresponding area from that value to the end of the parent/null distribution's tail, and then double the number if you are using a two-tailed test. This number is the probability of obtaining a *single* sample from the parent/null distribution as extreme as the value of the *mean* of your sample distribution. Figure 4-8 shows the comparison graphically for contrasting a sample of heights drawn from 25 American college males against a parent distribution of adult American females. So far, so good.

The differences in the two analyses turn on two points:

1. In binomial tests, you must count actual discrete combinations of outcomes (or approximate this process mathematically), and the combinations are different for every line of Pascal's triangle. In parametric analysis, however, you simply compute an area (that is, integrate along the x axis) using the well-understood Gauss formula to get the single-sample P-value.

2. In binomial analysis, the effect of sample size is dealt with by the choice of the line in Pascal's triangle to use. In parametric tests the effect of sample size must be calculated; the simple analysis already described yields the P-value for obtaining *one* datum with the observed mean value by drawing one value at random from the parent distribution. Clearly, the odds of getting, say, a sample set

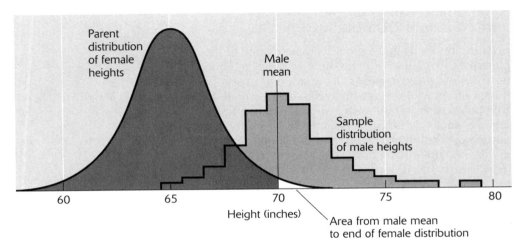

4-8 A sample distribution of male heights plotted on the same axis as a female parent distribution. The mean of the male distribution is indicated, and the area under the female curve to its nearest tail is shaded. The unshaded proportion is the one-tailed probability that an individual of mean male height was selected at random from the female distribution, using the experimental hypothesis that males are taller than females. The probability of obtaining a sample mean with the sample size of 92 used here is far less than the chance of finding a single female of this height or taller—about 10.4% of the already-small chance of a single randomly selected individual of 70.8 inches in height being a female.

If an observed trend in a parametric sample distribution is real and consistent, the P-value in a parametric test (P_a) will roughly vary inversely with the square root of the sample size (n_a):
$P_a/P_b = (n_b/n_a)^{1/2}$.

of 100 measurements with a mean equally far from the parent mean must be lower—in this case, about 10 times lower. So parametric tests, like most statistical tests, have an added step that factors in sample size. To a first approximation, as we said in Chapter 3, the P-value associated with a particular difference in means often declines at least roughly (and often almost exactly) with the square root of the sample size.

Here is an example, using a parent distribution represented by 100 points, and a sample mean carefully chosen to make the necessary computational steps simple; *BioStats Basics Online*, of course, will do this for you automatically:

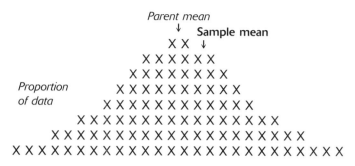

Using this observed sample mean, you must begin by integrating to the nearest tail, that is, determining the proportion of the values under the curve from the mean to the end. Since this graph just happens to have exactly 100 entries, each represents 1% of the area under the curve. Counting from the mean to the right end of the curve, you will find 35 points, corresponding to 35% of the area under the curve. Assuming this is a two-tailed situation, this means that 70% of the possible outcomes for a single sample from this parent distribution will be at least as extreme as the one corresponding to the sample mean; thus, in this case, $P = 0.70$. (For a one-tailed situation, $P = 0.35$.) (A trivial complication here: because for an infinite sample size the tails of the distribution never drop completely to zero, tests actually integrate from the sample mean back to the parent mean and then subtract this value from 0.50—the area from the mean to the end of the tail—to get the area under the tail.)

But this computation is correct *only* if the sample size is one. As we have said, the *P*-value declines with the square root of the sample size (in consequence of the greater certainty that our observed sample mean is an accurate measure of the true mean of our sample, which we can determine precisely only by collecting vast amounts of data). Thus if the sample size were four, we would divide the area (or, what is the same thing, the

P-value) by 2 (the square root of 4). For $n = 4$, then, $P = 0.35$ for two tails. For $n = 25$ (whose square root is 5), the two-tailed $P = 0.14$ (i.e., $0.70/5$); for $n = 100$, $P = 0.07$; for $n = 400$, $P = 0.035$. This is exactly what the software in *BioStats Basics Online* is doing when you ask it to compare a sample mean to a parametric parent distribution (II. D-1, when you know the SD of the parent distribution, and III. C-2, when you do not). If you know the SD of the parent distribution, you can do this by hand using Table 4-1 and correcting for sample size.

The situation is somewhat more complex when you are comparing two sample distributions, each with its own internal er-

---

Box 4.2

BioStats Basics Online   Testing two-choice distributions

There is a special test in *BioStats Basics Online* for this most basic of parametric comparisons. It is found on the left side of the Perfect Normal Distributions card (II. D-1). It allows you two options: entering (or importing) the sample data or entering the parameters of the sample distribution. This test is a shortcut for dealing with equal-alternative two-choice data, where the alternatives are scored as either 1 or 0 and have equal null probabilities of occurring; the parent mean is therefore one half of the sample-set size (e.g., 0.5 when each "set" consists of a single measurement).

To use this Standard Two-Choice Data Test, enter the sample-set size in the boxed field and select either the Sample Mean or Sample Data option. If you choose **Sample Data**, you must enter or import the data into the boxed field that will appear. Next choose

between the two-tail, one-tail-high, and one-tail-low options. When you click on **Compute**, *BioStats Basics Online* will display the sample size, the standard deviation, the standard error (which you should ignore for now), the P-value for the comparison, and then will graph the two distributions and indicate their means with arrows.

If you choose **Sample Mean**, you must enter the sample size (the number of sample sets in the data) and the observed mean. After selecting the appropriate option for tails, click on **Compute**. *BioStats Basics Online* will calculate a P-value. Since the original data were not available, there is no SD or graph displayed.

Because this test uses the standard deviation of the parent distribution, it is more precise than the one-sample t-test described later in this chapter.

rors and unreliability. This will be the topic of Chapter 5, and we will not be able to present a simple, intuitive, visual analogue to reflect the underlying mathematical manipulations necessary to deal with these multiple uncertainties. Rest assured, however, that except for the correction factors necessary for two-sample tests, the underlying logic of the computation of P-values is the same. For now, though, we are assuming you have a perfectly known parent distribution as a benchmark.

The simplest use of parametric analysis, as we have indicated, compares the mean of a sample distribution to the mean of a parent distribution. The most trivial form of this analysis deals with high-order binomial data similar to some you looked at in the last chapter. *BioStats Basics Online* provides a shortcut for testing standard two-choice data where the two choices have equal null probabilities (II. D-1). As with all testing of parametric sample data against a parametric parent (or, in this case, null) distribution, you must know the parameters of the parent distribution—the mean and standard deviation. For the equal-alternative case, these are defined automatically by the sample-set size.

If you use either of these tests, or almost any that you will encounter later, you will see the term ***degrees of freedom (df)***. Almost all tests generate a test statistic, the meaning of which depends on the sample size. Degrees of freedom is generally a measure of sample size, but not quite; in most cases the value is one fewer than the number of samples. The logic is that if you have a sample mean, then once you have looked at all but one of the data values, you can infer the last one: it's the value necessary to yield the mean when averaged with the other data. For other tests, however, this uncertainty is resolved earlier or in a different way, and thus the degrees-of-freedom value is $n - 2$ or smaller. Still other tests yield a pair of df values. The same tables are often used by very different tests, each with its own degrees-of-freedom logic; in fact, one test generates a fractional value of degrees of freedom! Happily, *BioStats Basics Online* automatically supplies the appropriate degrees-of-freedom number.

As with the previous test, the more general form of simple sample-to-parent parametric analysis again compares the means

The **degrees of freedom (df)** in a statistical test describes the number of independent values used in the test; in most (but not all) cases, df is one fewer than the sample size: $df = n - 1$.

4 Continuous Parametric Distributions: I

The **Gauss test** compares a parametric sample distribution to a parent distribution whose mean and SD are exactly known.

of two distributions. We call this version of the sample-to-parent comparison the ***Gauss test*** to distinguish it from the one-sample *t*-test, to be discussed presently. To use the Gauss test, you must know both parameters of the parent distribution—that is, you must be able to enter a parent mean and standard deviation. (The sample size is irrelevant since a parent distribution is assumed to be exact and free of errors.) To compare the sample distribution to it, you need either to know the apparent mean of the sample, or have the data available so that your computer can calculate it. *BioStats Basics Online* provides both options (II. D-1). The test standardizes the parent distribution (mean of zero, SD of 1.0) and integrates the area under this normal curve from the transformed sample mean to the tail; then it corrects this value to account for sample size and the number of tails. When we apply this parametric approach to

Box 4.3

BioStats Basics Online   The Gauss test

BioStats Basics Online provides an exact test for this basic parametric comparison in which the standard deviation of the parent distribution is known. This analysis, which we call the Gauss test, is found on the right side of the Perfect Normal Distributions card (II. D-1). It allows you two options: entering (or importing) the sample data or entering the parameters of the sample distribution. In either case, you must enter the mean and standard deviation of the parent distribution in the boxed fields.

If you choose (**Sample Data**), you must enter or import the data into the boxed field that will appear. Next choose between the two-tail, one-tail-high, and one-tail-low options. When you click on (**Compute**), *BioStats Basics Online* will display the sample size, the standard deviation, the standard error (which you should ignore for now), the *P*-value for the comparison, and then will graph the two distributions and indicate their means with arrows.

If you choose (**Sample Mean**), you must enter the sample size (the number of sample sets in the data) and the observed mean. After selecting the appropriate option for tails, click on (**Compute**). *BioStats Basics Online* will compute a *P*-value. Since the original data were not available, there is no SD or graph displayed.

Because this test uses the standard deviation of the parent distribution, it is more precise than the one-sample *t*-test described later in this chapter.

our height data, we find that the two-tailed chance of obtaining the observed male mean by sampling females is very small: $P < 0.001$. We can conclude with confidence that the means of the two distributions are (probably) different.

Finally, there is an even more useful form of the sample-to-parent comparison that does not require you to know the standard deviation of the parent distribution; the mean (often all we know) is enough. This clever analysis is one of four forms of the so-called *t-test*. The first *t*-test was developed in 1908 by W. S. Gosset, who wrote under the pseudonym of "Student." His anonymity successfully protected his employer, the Guinness Brewery of Dublin, from the risk of having its competitors see the value of the *t*-test approach to judging product quality from a few small samples of large batches. Previously, it was possible only to compare a sample to a completely characterized parent distribution.

The essential brilliance of the sample-to-parent (one-sample) *t*-test (III. C-2) is that it uses the sample data themselves to

The one-sample **t-test** compares a parametric sample distribution to a parent distribution whose mean is known but whose SD is unknown.

Box 4.4

BioStats Basics Online The one-sample *t*-test

The One-Sample *t*-Test is found on the second (III. C-2) of five *t*-test cards (linked to Testing Imperfect Normal Distributions on the Topics card). It compares a parametric sample distribution to a parametric parent distribution. It presupposes that the standard deviations of the sample and parent distribution are the same; it does not require that you know the parent SD. To use the test, you enter or import your data and enter the mean of the parent distribution in the boxed field. When you click on **Compute**, *BioStats Basics Online* displays the sample size and computes the *t*-statistic, the sample mean,

and a standard error (which you can ignore). It also graphs the distribution.

To evaluate the *t*-statistic, click on the **Table** button in the lower right-hand corner of the card. This takes you to a card with a table of critical *t*-values. The *t*-statistic is automatically transferred to this card, and the number of degrees of freedom (v) is displayed. Find the appropriate v row and read across from left to right until you find a *t*-value larger than your *t*-statistic. Look at the top of the column and the one to the left; the odds of your data having been drawn by chance from the parent distribution lie between these two values.

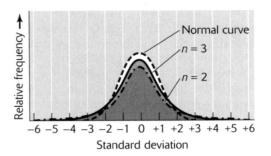

4-9 A comparison of two *t*-test curves, for sample sizes 2 and 3, and the normal curve (which is identical to the *t*-curve for very large sample sizes). The *t*-curves are broader because of the uncertainty in the size of the standard deviation in the sample: the curve is lower and broader than the normal curve, and thus has more area from any given value to the nearest tail than does the normal curve. [R. R. Sokal and F. J. Rohlf, *Biometry* (New York: W. H. Freeman, 1981), Figure 7.6, p. 146.]

estimate the standard deviation of the parent distribution. Recall that the width of the Gaussian curve of a parametric distribution is determined by its SD. When the SD can only be estimated from the data, it stands to reason that the curve can be drawn only approximately. The *t*-test factors in the uncertainty of that SD estimate in calculating probabilities. The basic trick is to create a new Gaussian curve, widened just enough to account for the uncertainty about the SD of the parent distribution given the sample size in question (Figure 4-9); in effect, the test integrates the area under this synthetic curve from the sample mean to the tail, making the usual corrections for sample size and number of tails in the test. Because the *t*-test curve is wider than the corresponding Gaussian curve, the *P*-values will necessarily be larger.

Need more than this? See **More Than the Basics:** *The One-Sample t-Test.*

In Chapter 5, we will tackle the problem of comparing two sample distributions with each other, using Gosset's *t*-test approach as the key to unlocking this doubly difficult problem. Then we will see how the *t*-test logic has been extended to the even more perplexing problem of comparing multiple samples to one another—a technical impossibility until R. A. Fisher saw how to apply Gosset's approach from a different direction, and thereby spawned the largest collection of special-purpose statistical tests to be found in most texts (and which we will cover with a merciful degree of selectivity in Chapter 7).

Points to remember

✔ Parametric distributions are bell curves that correspond to the Gauss formula; they typically arise from measurement error and the contribution of several independent effects.

✔ Accuracy is a measure of how closely data correspond to the true mean of the distribution being sampled; precision is a measure of how closely the data correspond to one another.

✔ Parametric distributions are completely described by two parameters: the mean and variance (or standard deviation, SD); statistical analysis also requires the sample size.

✔ One SD encompasses 68.26% of the data centered on the mean; two SD encompass an additional 68.26% of the data centered on the mean lying outside the one-SD area—95.44% of the data in all.

✔ The probability (P-value) of obtaining given measurement from a parent distribution is the chance of drawing that particular value or any result more extreme; in a one-tailed situation, this is the area under the

unit parametric distribution from that value to the end of the predicted tail; in a two-tailed situation, it is twice this value (to take into account outcomes equally distant from the mean in the opposite direction).

✔ The P-value of obtaining a particular sample mean value from a parent distribution is the probability of drawing one result that extreme divided by the square root of the sample size contributing to the sample mean.

✔ Degrees of freedom (df) is a measure of the number of independent measurements used in statistical analysis; usually df is one fewer than the sample size.

✔ The Gauss test compares the mean of sample distribution to that of a parent distribution whose mean and SD are perfectly known.

✔ The one-sample t-test compares the mean of sample distribution to that of a parent distribution whose mean (but not SD) is perfectly known; the SD of the parent distribution is estimated from the sample data.

4 Continuous Parametric Distributions: I

Exercises

1. Partial data on the batting averages for long-term pitchers and fielders in the major leagues are given in Table A, which lists the first 20 individuals from each full list. (The full data sets are on the *BioStats Basics Online* disk as file 4—Pitcher Batting Averages and 4—Fielder Batting Averages, or 4-PBA.dat and 4-FBA.dat.) Is either distribution normal? (Use the test for normality.)

Table A   Partial list of batting averages

Pitcher batting averages				Fielder batting averages			
.127	.150	.132	.169	.242	.325	.271	.277
.111	.138	.119	.131	.288	.244	.267	.288
.153	.162	.181	.215	.260	.262	.252	.294
.174	.215	.143	.167	.269	.256	.273	.291
.145	.165	.156	.168	.257	.291	.251	.256

2. Partial data on the heights and weights of long-term pitchers in the major leagues are given in Table B, which lists the first 20 individuals from each full list. (The full data sets are on the *BioStats Basics Online* disk as file 4—Pitcher Heights and 4—Pitcher Weights, or 4-PHTS.dat and 4-PWTS.dat.) Is either distribution normal? (Use the test for normality.)

Table B   Partial list of heights and weights

Pitcher heights				Pitcher weights			
5-11	6-01	6-00	6-01	185	185	177	190
6-04	6-03	6-02	6-04	210	205	210	215
6-01	6-05	6-00	6-06	195	225	175	230
6-03	6-03	6-01	6-05	203	190	185	215
6-03	6-05	6-00	6-08	205	210	200	230

3. The parent distribution for adult male height has a mean of 69.5 inches and an SD of 2.4 inches. Athletes, and particularly baseball pitchers, are not a random slice of ordinary mortals. Is the sample distribution of fielder height (Table C; file 4—Fielder Heights or 4-FHTS.dat) significantly different from the parent distribution? If so, at what P-value? (Use the Gauss test.)

Table C   Partial list of fielder heights

Fielder heights			
5-11	6-02	6-02	6-03
6-02	6-02	6-05	6-02
6-01	6-01	6-02	6-00
6-02	6-05	6-00	6-02
6-01	5-08	5-10	5-11

4. A slow generation-by-generation increase in adult height has been observed for almost 200 years. Given an adult female mean height of 63.5 inches and an SD of 2.4 inches, and remembering that immigration may be shifting the mean as well, can you detect this generational shift from a small sample of female biology students (Table D; file 4—Female Bio Student Heights or 4-FBHTS.dat), who represent a group about 20 years younger than the mean age of the women in the (literally) parent distribution? If so, at what P-value? (Use the Gauss test.)

Table D   Student heights

5-03	5-07	5-11	5-03
5-07	5-10	5-05	5-09
5-07	5-09	5-09	5-04
5-02	5-10	5-08	5-09

5. The parent distribution of major league batting averages has a mean of .2217 and an SD of .0299. Pitchers are legendary as poor hitters. Is the sample distribution of batting averages of long-term pitchers significantly lower than that for players in general? If so, at what P-value? (Use the Gauss test.)

4 Continuous Parametric Distributions: I

6. Darwin hypothesized that the female birds that returned from migration first were more fit than those that returned later, which would explain why males attempted to court them so feverishly. If this were true, we would expect that pairs that included an early arriving female ought to lay more eggs and fledge more young than ordinary nests. For one species of sparrow the average clutch size is 3.1 eggs per nest; the SD was not measured. The data on egg-laying by birds nesting before April 15 are given in Table E (and file 4—Sparrow Clutch Sizes or 4-SEGGS.dat). Are the early birds laying more eggs? If so, at what P-value? (Use the one-sample t-test.)

Table F   Finch data

Beak depth (mm)	Number in 1978
6.0	2
6.4	0
6.8	3
7.2	0
7.6	4
8.0	2
8.4	6
8.8	11
9.2	13
9.6	15
10.0	16
10.4	19
10.8	14
11.2	8
11.6	4
12.0	0
12.4	0

Table E   Clutch sizes in nests of early returning female sparrows

4	5	1
6	7	4
3	5	2
3	3	4
4	5	2
7	6	5

of the birds and all but one of the nestlings died—the beaks of the survivors were measured again for evidence of evolution as a result of this environmental challenge. The post-drought data are given in Table F (and file 4—1978 Beak Data or 4-78BEAK.dat). Is there a significant change in beak size? If so, at what P-value? (Use the one-sample t-test.)

7. In a typical year, the mean beak depth for finches on a small island in the Galápagos is 9.4 mm. After a severe drought—so severe that most

8. Certain automakers market both sedan and station wagon versions of the same car. Station wagons are inevitably heavier and often have poorer aerodynamics (and thus lose energy to drag). Are these disadvantages reflected in poorer gasoline mileage? The average station wagon marketed in the United States in 1995 got 28.0 MPG. Table G (file 4—Sedan MPG Subset or 4-SEDAN.dat) lists the miles-per-gallon values for the equivalent sedan versions. Is there a significant difference in fuel efficiency? If so, at what P-value? (Use the one-sample t-test.)

Table G   Fuel efficiency of selected sedans

24	32	29	35
29	31	29	29
28	30	26	25
27	30	28	29

9. Roughly 15,000 women are diagnosed annually with cervical cancer in the United States. Because cervical cancer is more common among smokers, one study set out to discover if the potent carcinogens in tobacco smoke (N-nitrosamines, and in particular one known as NNK) actually reach the mucus of the cervix, where they could promote cancer. The data from 15 smokers and 10 nonsmokers matched for age and other important variables are given in Table H. (The data are also in file 4—NNK Levels or 4-NNK.dat.) Are these data suitable for parametric analysis? If so, which test is appropriate? Are the levels of NNK significantly different?

Table H   NNK levels (ng/g NNK in mucus)

Smokers	Nonsmokers
31	31
20	19
85	7
115	6
33	11
49	4
64	18
18	0
43	10
43	23
58	
12	
15	
23	
111	

10. A certain cookie manufacturer claims that there are more than a thousand chocolate chips in each bag of (naturally) chocolate chip cookies. A dedicated group of cadets actually counted the chips in 42 bags of cookies. Their data are shown in Table I (file 4–Chips Ahoy! Count or 4-CHIPS.dat). No bag had fewer than 1000 chips.

But because the ends of a parametric distribution stretch forever, with enough bags there should be one that dips below the advertised number. Compute the mean and SD, then use Table 4-1 to calculate the average number of bags one would have to census to find a chip-challenged bag.

Table I   Number of chocolate chips in bags of cookies

1087	1143	1213	1244	1294	1377	1098
1154	1241	1247	1295	1402	1103	1166
1215	1258	1307	1419	1121	1185	1219
1296	1325	1440	1132	1191	1219	1270
1345	1514	1135	1199	1228	1279	1356
1545	1137	1200	1239	1293	1363	1546

More Than the Basics

From binomial to bell

The Gaussian distribution ("bell curve"):

$$y = \frac{1}{\sigma\sqrt{2\pi}} e^{\frac{-(x-\mu)^2}{2\sigma^2}}$$

It is fairly obvious by now that the combination of a number of random factors should give something roughly like the binomial distribution. But what does this have to do with the bell curve, and why does anything distribute itself in a bell curve anyway? The answer is both surprisingly simple and profound.

Both the binomial and the Poisson distributions approximate the bell curve at their extreme values. The math in either case requires a fair bit of integral calculus, which we will not include here, though in the case of the Poisson distribution it is well within the reach of most students.

The essential fact is this: just as the binomial and Poisson distributions describe precisely the sum of some finite number of chance variations, the sum of a large (ideally infinite) number of chance variations will be distributed in a Gaussian fashion.

Why two sets of symbols?

As we have already mentioned, two sets of symbols keep popping up for the same ideas. Standard deviation can be s or σ, mean $\overline{X}$ or μ, sample size n or N. The distinction is more than just typographic, though it is certainly subtle. The distinction is that between a population and a sample.

A distribution of every possible datum is a population; it is not possible to get any more data into a population, because it already contains them all. A set of data that contains only a few individuals from a population is a sample. As sample size becomes increasingly large, the sample mean and SD become increasingly better estimates of the population mean and SD.

A sample mean is represented as $\overline{X}$, sample SD as s, and sample size as n. A population mean is shown as μ, population SD as σ, and population size as N. The distinction, while not generally necessary for calculations, is important to the understanding of the underlying logic of statistical tests.

The parameters

Sample and population means:

$$\overline{X} = \sum \frac{X_i}{n} \quad \mu = \sum \frac{X_i}{N}$$

Sample and population SDs:

$$s = \sqrt{\frac{\sum X_i^2 - \dfrac{\left(\sum X_i\right)^2}{n}}{n-1}}$$

$$\sigma = \sqrt{\frac{\sum X_i^2 - \dfrac{\left(\sum X_i\right)^2}{N}}{N}}$$

where X_i is the ith datum.

A parametrically distributed population of infinite size may be entirely characterized by the population mean and population SD, the parameters μ and σ. As sample size increases, $\overline{X}$ will approach the value of μ and s will approach σ.

4 Continuous Parametric Distributions: I

The mean   Everyone knows that the mean is the value "in the middle" of the distribution, but to state this in a mathematically precise way is not straightforward. Statistics relies heavily on the property of "least squares" of the mean. One can imagine calculating the sum of the squared differences (SS) between each datum and some number:

$$SS(x) = \sum (X_i - x)^2$$

The mean is the number for which this sum of squares is the smallest.

The variance   This sum of squared deviations, when computed as deviation from the mean, obviously increases with both the number of data and the spread of those data. To correct for the number of data, we divide by the number of data to get an average squared deviation. This is often called the "mean square" deviation, MS, or variance:

$$MS = V = \frac{\sum (X_i - \mu)^2}{N}$$

Unfortunately, this will always be larger than an equivalently calculated sample statistic. Why? In a small sample, the variance calculated by the above formula would be held down by the fact that the mean was calculated from the same data, and therefore the distance of a point from the mean will tend to be artificially shortened by the difference between sample and population means. Because the sample mean has the property of least squares, the SD estimated from it by the uncorrected formula will always be lower than the true SD. Sample variance, therefore, must be calculated with a correction to provide a true estimate of population variance.

$$v = \frac{\sum (X_i - \overline{X})^2}{n - 1}$$

The -1 in the denominator of this formula may be omitted if the population mean rather than the sample mean is used in the numerator—a rare situation, but not unheard of.

The standard deviation   The variance seems at first to be a useless number; it doesn't represent anything measurable on our graphs or in real life. If, for instance, X is measured in inches, the variance is in units of square inches. The solution seems obvious: define a new quantity equal to the square root of the variance. This number, originally known as the root-mean-square deviation, is now called the standard deviation.

$$s = \sqrt{v}; \quad \sigma = \sqrt{V}$$

This statistic describes the width of a distribution and therefore is a useful way to estimate how extreme is any particular result.

The one-sample *t*-test

The formula for a one-sample *t*-test is

$$t = \sqrt{n}\,\frac{|\overline{X} - \mu|}{s}$$

where *n* is the sample size,

$\overline{X}$ is the mean of your sample,

s is the SD of your sample, and

μ is the mean to which you are comparing your sample.

The degrees of freedom is equal to one fewer than your sample size.

The value of *t* computed should then be compared to the table in Selected Statistical Tables. A calculated value of *t* larger than that in the table indicates that the mean of your data is significantly different from the number to which you were comparing it.

Tests Covered in This Chapter

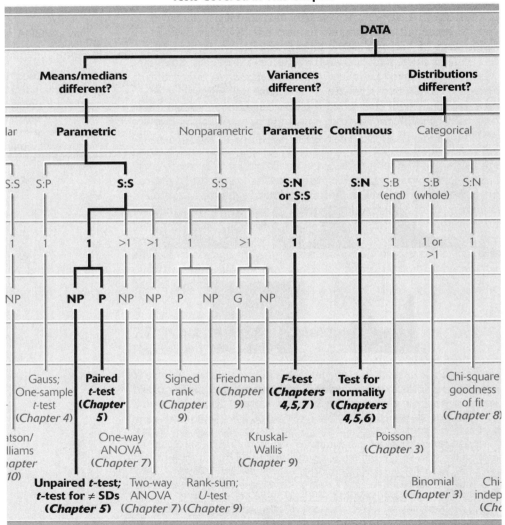

Continuous Parametric Distributions: II

Comparing a sample distribution to a parent distribution to see if they have different means is, at least in theory, very simple for parametric distributions. Because the exact shape of a parametric parent distribution is completely defined by its mean and standard deviation, you have only to compute the proportion of the area under the curve from the sample mean to the end of the parent-distribution tail (doubling the value for two-tailed tests). You then decrease this value to take into account the sample size.

5.1 Sample Size and Certainty

The **sample size** is an indirect measure of our certainty about the value of the mean. The probability that a sample distribution with a particular mean is actually drawn from a parent distribution with a slightly different mean—that is, the distributions *do not* have different means—hinges on this question: how sure are you that your estimate of the mean of the population being sampled is accurate? With an infinite sample size—another parent distribution, essentially—you can be perfectly sure that the two means are what they appear to be; if they have different values, then that is because the two distributions have different means. But when one of the means is based on a limited sample, you have only an estimate; your *P*-value reflects both the magnitude of the difference in means (relative to the standard deviation) *and* the sample size (the certainty of our estimate of the sample mean; Figure 5-1).

The **sample size** of a distribution is the number of independent data values it contains.

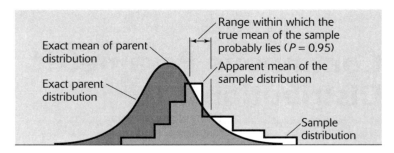

5-1 The mean of a parent distribution is exactly known, whereas the mean of a limited sample ($n = 32$ in this case) must be estimated from the data. The actual mean lies somewhere near (how near depends on the sample size and the SD in the data). Shown here is the range of means that has a 95% probability of including the true mean of this sample. (We will demonstrate later in the chapter how to calculate this range of mean values.) The consequence of using an estimate on determining a P-value should be clear: you integrate the area under the parent curve from the sample mean to the end of the parent distribution and then correct for sample size; but since there is uncertainty in the exact value of the sample mean, there is similar uncertainty in the proportion of area under the curve.

5.2 Comparing Sample Distributions: The Logic of the t-Tests

We saw in Chapter 4 that Gosset developed a way of using the sample distribution itself to estimate the standard deviation of the parent distribution. This is so often necessary that some statistical packages omit the option of inputting the SD of the parent distribution altogether, relying instead in all such cases on the one-sample t-test. But the one-sample t-test is potentially less powerful than the Gauss test. Recall that two parametric distributions are compared by first centering the parent mean on zero (by subtracting the parent mean), then setting the standard deviation of the parent curve to 1.0; then the sample mean gets the same treatment—the parent mean is subtracted and the values scaled with the same SD factor. But in the one-sample t-test, the standard deviation of the parent distribution is unknown; it is estimated from the sample distribution. Thus the

scaling step has an inherent uncertainty analogous to your uncertainty about the true sample mean: if the sample size is large,
the estimate of the SD is probably pretty close, but if it is small,
there is less reason for confidence (Figure 5-2).

So Gosset's one-sample *t*-test method has a price: in order
to analyze data previously beyond the reach of conventional statistics, you must rely on two estimated values, the sample mean
and the parent SD. The situation should become worse when
you extend his powerful approach to compare two samples—for
example, male versus female scores on a reaction-time test—
where there is no parent distribution at all. This is perhaps the
most common single sort of situation encountered in experimental statistics, and without the *t*-test, you would be forced to
use less powerful nonparametric approaches. We will consider
the three alternative forms of the *t*-test that deal with two-
sample data in this chapter, but first we need to look at the
problems that all three face.

As we will show, though things are not necessarily always
quite this approximate, in the extreme, some two-sample *t*-tests

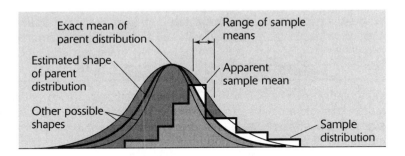

5-2 When the SD of the parent distribution is estimated from the
apparent standard deviation of the sample distribution, there is
uncertainty about the exact shape of the parent distribution. This
adds a new level of ambiguity to the computation of *P*-values: you
must integrate the area under the parent curve from the sample
mean to the end of the parent distribution and then correct for
sample size; not only are you unsure of the exact value of the
sample mean, you also don't know the exact shape (width) of the
parent curve, and thus the exact area to be integrated from any
specific value for the sample mean.

are estimating *four* parameters from the sample data themselves: the mean and standard deviation of the data in both the first and second distributions (Figure 5-3). Therefore there is more uncertainty—and thus, less statistical precision—inherent in the comparison, and the importance of sample size increases even more. And you can probably also guess that if the sample sizes of the two samples are unequal, the smaller sample will limit the precision of the entire analysis.

If you look more closely at the logic of the two-sample *t*-tests, you will see another cause for concern. Let's suppose the test arbitrarily selects the larger sample (which we will call Sample A) to work with first, and treats it like a parent distribution: the mean of Sample A (μ_A) is set to zero, and the standard deviation of the distribution is scaled to 1.0 by computing some cor-

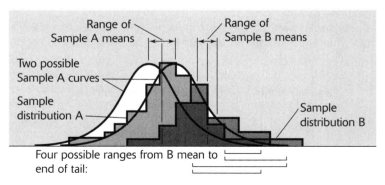

5-3 When both distributions are sample distributions, there is uncertainty in the mean and SD of each. This is illustrated here by reproducing the 95% sample-mean range and its underlying data distribution (here labeled "Sample B") from Figure 5.2 and adding a new sample distribution ("Sample A") with the same apparent mean as the parent distribution in Figures 5-1 and 5-2. The 95% range of the Sample A mean ($n = 51$) is shown, along with two of the infinite number of possible parametric approximations of the parent curve for Sample A. The leftmost of the two assumes a low true mean and SD; the rightmost assumes a high true mean and SD. The ambiguity regarding the area to integrate from the true Sample B mean (whatever it may be) to the end of the true Sample A distribution (whatever it may be) is now greater still.

rection factor, SD_A, which does the job. Now apply the same transformation to Sample B. Then, with Gosset's mathematical magic, you compare the two means with careful attention to the sample sizes of the two distributions (which, converted to a degrees-of-freedom value, direct us to the correct line of the P-value table). But in doing all this, not only have you been relying on parameter estimates, you have assumed that the standard deviations of the distributions from which the two samples are drawn are the same. If the SDs are *not* the same, the SD-scaling step cannot be legitimately performed. What a priori reason do you have to believe that, say, the standard deviation of the distribution of female reaction times will be the same as that for males?

5.3 Checking That Variances Are Equal

Before we can get to the two-sample t-tests, therefore, you have to satisfy yourself (and your colleagues) that the samples being compared have roughly the *same* dispersion. As we said in Chapter 2 (page 33), there are two measures of dispersion: the variance and the standard deviation. They differ only in that the SD has a square root taken in the penultimate step of the calculation, yielding a number that makes some intuitive sense. But tradition (and mathematical convenience in some cases) dictates that we use the word "variance" in this discussion, even though for our purposes "SD" might seem to make more sense.

Comparing the variances of two sample distributions, each with inherent sampling uncertainties and so on, is the task of the **F-test**, developed by the statistician R. A. Fisher around 1925. (The F-test—the "F" is for "Fisher"—has many other remarkable uses, as we will show presently.) Note that you are testing the hypothesis that the two sample distributions to be compared are *not* drawn from parent distributions with the *same* variance—that is, in this unusual case, the null hypothesis is that there *is* a difference; we'll see in a moment why the question is phrased in this way. Now there is no way to know the exact value of those parent variances; you can only estimate them from the variances of your limited samples. Let's call the sample variances SD^2_A and SD^2_B, where SD^2_A is the larger of the two. The F-test simply takes the ratio of the two variances: SD^2_A/SD^2_B. Since SD^2_A is the larger of the two variances, the

The **F-test** compares the variances to two sample distributions; most often it is used to see if they are similar enough to be used in the more common forms of the t-test.

value is never smaller than 1.0; if the two are equal, the ideal state for a parametric test, the ratio would in fact be exactly 1.0. (In *BioStats Basics Online* the test is on card III. B-1.)

The next step is to consult a set of tables listing critical *F*-statistics for each combination of sample sizes. You may notice that the variances can be quite different; again, that's because the *F*-test is *not* asking if you can be 95% certain the variances are the *same*, but rather if you can be sure they are *different*. The reason you can ask the easier-to-pass version of the question is that parametric tests are very robust and work even when the variances are not very similar. So, if your data pass this fairly undemanding test, you can subject them to one of the two *t*-tests to be described shortly. If your data fail, do not despair: a later section in the chapter describes a special version of the *t*-test that estimates the two SDs separately and allows for their different magnitudes. But as you can probably guess, this extra step in estimation further reduces the power of the analysis and leads

---

Box 5.1

BioStats Basics Online   The *F*-test

The *F*-test computes the probability that two sample distributions have the same variance (and thus the same standard deviation). The test is found on the Confirming Distribution Similarity card (III. B-1). It offers you two options: you can select the **Input SDs** alternative, in which case you must enter the SD and sample size of each sample. The other alternative to is select the **Input Data** option and enter (or import) the two sets of sample data.

When you click on Compute, *BioStats Basics Online* will display an *F*-statistic and two values for degrees of freedom (one for each sample); if you chose the **Input Data** option, the computed SDs will also appear and the distributions will be graphed. *BioStats Basics Online* automatically displays the portion of the table of critical *F*-values that has degrees-of-freedom values for the first of the two samples you entered; the row with the second one can be found by scrolling. Some interpolation may be necessary. If your *F*-statistic is larger than the corresponding one in the table, the SDs are significantly different at $P < 0.05$; if your *F*-statistic is smaller, the difference in SDs is not significant and you can use the conventional form of the *t*-test.

to less useful *P*-values. Still, you're not thrown on the mercy of nonparametric analysis, which assumes you don't even know that the data are parametrically distributed.

5.4 Other Reasons to Compare Variances

Opening the way to use the *t*-test is not the only value of the *F*-test; comparing two distributions to see if they have the same variance is often itself interesting. The most obvious application is in looking for ***independence***. Researchers almost always take it for granted that data values are independent, but what if the choice of one cricket in a maze biases the behavior of the next? The bias could consist of a tendency to follow the previous individual's lead, or to avoid it (Figure 5-4).

Data are **independent** when the value of one datum does not depend on the value of any other datum in the sample.

Our favorite use of the *F*-test to check for independence is in examining the common skeptic's hypothesis that the outcomes of sports events depend on nothing but luck: there are no good and bad teams, just lucky and unlucky ones. Now this is a question in which means will us no good. The mean win:loss record under any hypothesis for a sport that does not permit ties (e.g., baseball) is 0.5, because all games are won by one of the two teams playing and lost by the other; the total number of wins equals the total number of losses. What will differ between the two hypotheses, however, is the variances.

For a season of 162 games, line 162 of Pascal's triangle will yield an expected distribution based on chance alone (Figure 5-5, page 119); on the same graph we can plot the win:loss records of Major League Baseball teams over the past few years. (We have generated the null curve from the SD supplied by the Standard Two-Choice Data Test in *BioStats Basics Online*.) It is clear that the variance in win:loss records is much greater than the chance prediction; it is equally clear that it is less than what we would expect if the "best" team (the one with the highest win:loss record) simply won all its games, the next best won all games except the ones played against the best team, and so on. Thus, the win:loss records, like most of life's events, seem to reflect a mixture of talent and luck, of causation and chance; an enterprising student could even attempt to estimate the relative roles of these two competing factors in Major League Baseball.

5 Continuous Parametric Distributions: II

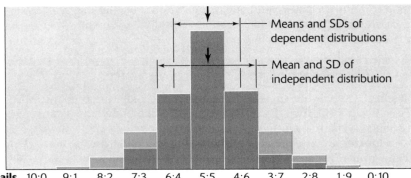

Ratio heads:tails	10:0	9:1	8:2	7:3	6:4	5:5	4:6	3:7	2:8	1:9	0:10
Observed percent	0.0	0.0	0.40	6.80	22.8	41.6	23.6	4.60	0.20	0.0	0.0
Expected percent	.098	.98	4.4	11.7	20.6	24.6	20.6	11.7	4.4	.98	.098

(a)

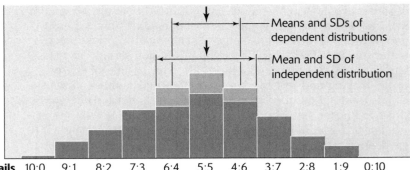

Ratio heads:tails	10:0	9:1	8:2	7:3	6:4	5:5	4:6	3:7	2:8	1:9	0:10
Observed percent	0.8	5.0	8.40	14.0	14.8	19.0	16.4	12.0	6.20	3.4	0.0
Expected percent	.098	.98	4.4	11.7	20.6	24.6	20.6	11.7	4.4	.98	.098

(b)

5-4 All the statistical techniques we will discuss until Chapter 11 assume the data are independent. Here is what happens to a simple binomial distribution when there is a certain degree of dependence of one datum on another. **(a)** The consequences when one datum state (heads versus tails, for instance) tends to match that of the preceding one. The independent distribution is shown in light blue in the background, along with its standard deviations; the SD of the dependent distribution is shown in dark blue (the arrow indicates the mean, which is about the same in each case). **(b)** The consequences when one datum tends to be the opposite of the preceding one.

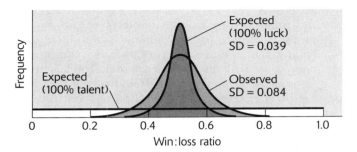

5-5 The observed distribution of win:loss ratios of Major League Baseball teams compared to the null prediction of independence. The actual SD is more than twice the expected SD, indicating that the results are not independent but depend to some extent on the "quality" of the team. If quality were a complete guide, however, so that there is complete statistical dependence, the curve would be a flat line just above the x-axis with an SD of 0.341 (except, of course, as the curve would not be parametric, the SD value would be meaningless).

We will also encounter the *F*-test when we look at how to make multiple-mean comparisons when faced with several sets of parametric sample distributions. Until now, all comparisons have been between two distributions; as we will demonstrate in Chapter 7, there are compelling reasons for treating multiple-sample comparisons differently.

An analogous use of variances is common in analyses of spatial or temporal distributions. Again, the implicit question is whether the distribution is random, and if not, in what way it is biased. The Poisson analysis in Chapter 3 dealt with an important class of special cases and asked whether the observed distribution was random (independent) or showed evidence of clustering or repulsion. The question whether cricket maze choices are independent or influenced either by following or alternation is a less restricted version of the same issue. The most general form of the question of independence in a distribution is best illustrated by a common ecological problem: are individual plants or animals distributed (in two dimensions rather than one, as was the case in earlier examples) independently (randomly), uniformly, or are they clustered (Figure 5-6)?

The most usual way to take data on spatial distributions is to subdivide a habitat into squares ("quadrats") and then count the number of individuals in each subunit. You might expect that a frequency-distribution plot of individuals per quadrat could then be compared to a null distribution constructed with the same mean; a larger sample variance would suggest clustering (more cases of many or very few individuals per quadrat than would occur by chance if the organisms were randomly scattered), while a smaller one would suggest unusually uniform spacing. This does not in fact work because the outcome depends on quadrat size. Look at Figure 5-6c: if you happened to choose a quadrat dimension large enough to include several clusters, the data would appear random even though the individuals are clumped; if the quadrat is even larger, both clumped and random distributions would appear to be uniformly spaced. And if the quadrat is quite small, even a clumped distribution would appear random.

A number of clever ways have been devised to circumvent this scaling problem, but the simplest is known as nearest-neighbor analysis. Individuals are selected at random and the distance to the nearest neighbor is measured. When the data are analyzed, you can extract the mean distance between individuals and the variance in the distribution. If the individuals tend to be clumped, there will be an excess of short (within-clump) and large (between-clump) distances, whereas if the spacing is uniform there will be very little variation in the distance data. Ecologists summarize the distribution patterns by simply taking the ratio of the variance to the mean: clumped distributions have variances much larger than the mean; random (independent distributions) have variances close to their means;

(a)   (b)   (c)

5-6 Three common kinds of spatial distributions seen in nature: (a) uniform, (b) random, and (c) clumped.

uniform distributions have variances smaller than the mean. This is the same measure—the *coefficient of dispersion (CD)*—used in Poisson analysis, as we saw in Chapter 3.

Determining whether a given CD ratio is significantly different from the 1.0 predicted for a random distribution requires generating a number that can be compared to a table of critical values. To use the table of values in *BioStats Basics Online* for the *t*-tests, you need the mean nearest-neighbor distance in the sample (r); the mean you would expect if the distances were randomly distributed (E, which is given by $1/[2p^{1/2}]$, where p is population density—the number of individuals per unit area, the unit being that used in measuring the value of r); and the standard error (SE, discussed later in this chapter) of the expected value of r. The expected value of r is given by $0.26136/[(Np)^{1/2}]$, where N is the sample size. For computational simplicity, the SE is used instead of variance (V); SE is $(V/N)^{1/2}$. You then combine these to obtain the *t*-test statistic as follows: $(r - E)/\text{SE}$. A good rule of thumb is that, for sample sizes larger than 50, a test statistic greater than $+2.0$ means that the distribution is significantly uniform, while one lower than -2.0 implies significant clumping.

> The **coefficient of dispersion (CD)** is the ratio of the variance to the mean: $CD = [\Sigma\ (X - x_i)^2 n]/\overline{X}$.

5.5 Comparing the Means of Paired versus Unpaired Data

Two-sample comparisons, whether of parametric or nonparametric data, exist in two forms: paired or unpaired. Paired data usually come from experiments that measure two variables in the same individual: mSAT versus vSAT scores of Freeman University undergraduates, for instance (Figure 5-7a), keeping track of which student obtained which pair of scores. Unpaired data usually come from experiments that measure the same variable in different individuals: vSAT scores for males versus those obtained by females (Figure 5-7b).

The mechanics of the *paired-sample t-test* are elegant (card III. C-3), and the test is surprisingly powerful. As usual, the null hypothesis is that the distributions being compared are the same; in this case (as in all paired tests), the null hypothesis asserts that the two samples are drawn from the same parent distribution. You take the difference between each paired sample; the

> *Need more than this? See **More Than the Basics:** Two-Sample t-Test.*

> The **paired-sample t-test** tests the null hypothesis that each member of a data pair comes from the same parametric parent distribution.

result is a distribution of differences. You then take the mean of those differences and compare them to a predicted mean of zero. The actual math is the same as in the one-sample t-test. You know the predicted mean exactly—0.0—and thus it takes the place of the parent-distribution mean. You estimate the SD from the sample data, just as in the one-sample t-test. Thus the paired t-test rescues a two-sample comparison, with its multiple ambiguities, through the neat expedient of using the pairing to produce a parental mean. (The same sort of internal comparison goes on in paired nonparametric tests, as we will show, and enhances their ***power*** as a result. By the way, "power"—a term we've used again and again without a statistical definition—in practice basically means how small a sample size you will need to detect a phenomenon: a powerful test requires fewer data than a less powerful test.) A lesson to take to heart here is that

In practice, the **power** of a test is simply how small a sample size you will need to detect a phenomenon of a given magnitude.

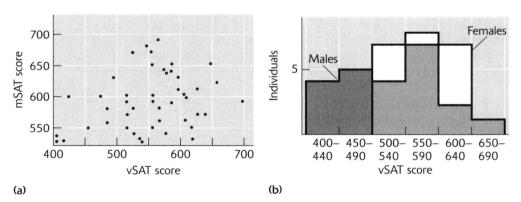

(a)   (b)

5-7 (a) Paired data preserve the relationship between two different measures obtained from the same individual; here each point represents the combination of vSAT and mSAT scores of a number of different biology majors. **(b)** Unpaired data usually compare the same measure from different groups of individuals; here we see the vSAT scores of male versus female biology students. Each of these graphs is taken from the same data set and illustrates quite different phenomena: in the paired comparison we see that there is a rough correlation between increasing vSAT and mSAT scores, mSAT scores are generally higher than vSAT scores, and the mSAT range is much narrower than the vSAT range; in the unpaired comparison we see that male biology majors tend to have lower vSAT scores than their female counterparts, and it is the low male scores that account for the larger vSAT range in the paired graph.

whenever possible, find a way to make your tests depend on paired data. (For the paired SAT data, the paired *t*-test yields a *P*-value below 0.01 even just considering the first 10 students, confirming the well-known truism that, before the great "recentering" of 1995, the mean mSAT was higher than the mean vSAT.)

This trick is not possible when the data are unpaired. In this case, you have the scenario described early in this chapter: two unrelated sample distributions whose means and SD you can only estimate. The ***unpaired t-test*** compares the means using the appropriate allowances for the various uncertainties and produces a *t*-statistic that allows you to determine the probability of obtaining a sample with a mean that extreme by drawing data at random from the other sample distribution. (The test is on card III. C-3.) In our example, the *P*-value for the first 25 students in the male-versus-female vSAT comparison was below 0.01 (against the null hypothesis of no difference between the sexes); in general, females do somewhat better than males on this test. To illustrate the relative power of the paired and unpaired versions of this test, we took the paired vSAT-mSAT and unpaired it—that is, we randomized the order of the scores. Then we analyzed it using the unpaired test until we reached the same *P*-value we found with only 10 students using the paired *t*-test; we needed 23 scores of each type to reach the same level of significance employing the unpaired *t*-test. Again, the moral is to try to collect paired data whenever possible.

The **unpaired *t*-test** tests the null hypothesis that two parametric sample distributions are drawn from the same parent distribution.

5.6 What Do You Do If the Standard Deviations Are Not Similar?

As we have said, the paired and unpaired *t*-tests assume the SDs of the distributions the two samples are taken from are the same parent distribution. You confirm this condition with the *F*-test. If your distributions fail the *F*-test but the data are parametric, your best bet is the ***unpaired t-test for different SDs***. There is no paired version of this test; if your data are paired, you must nevertheless treat them as unpaired. (The test is on card III. C-4.) You would apply this approach, for instance, when testing to see if the mean height of baseball players is the same as that of the general male population. (We must use fielders

The **unpaired *t*-test for different SDs** tests the null hypothesis that the means of the parent distributions (which have different SDs) of each of two sample distributions are the same.

5 Continuous Parametric Distributions: II

Box 5.2

The test for paired data whose variances are similar is found on the left side of the third *t*-Test card (III. C-3). This Two-Group Paired *t*-Test requires you to enter or import your paired data; each pair must be in the same row. Because of simple size limitations on the card, only 33 pairs of data can be entered, though any number can be imported. To enter more than 33 pairs, click on the (Large Data Set) button at the bottom; this will take you to a card that makes it easier to enter larger sets without losing track of which entries are paired.

When your data are in place, click on the (Compute) button. *BioStats Basics Online* will display the sample size,

the *t*-statistic, and a mean for each data set, and graph the two distributions (Figure A). (*BioStats Basics Online* will also display a standard error for each data set.) To interpret the *t*-statistic, click on the (Table) button; this will take you to a card (III. C-5) that will list the *t*-statistic you just obtained, and a degrees-of-freedom value (v). As with previous tables, use the v-value to find the correct row of the table, and then read across from left to right until you find a value that is larger than your *t*-statistic. The *P*-value of your statistic lies between the number at the top of that column and the number at the top of the column to the left.

Box 5.3

The test for unpaired data whose variances are similar is found on the right side of the third *t*-Test card (III. C-3). This Two-Group Unpaired *t*-Test requires you to enter or import your data. When your data are in place, click on the (Compute) button. *BioStats Basics Online* will display the sample sizes, the *t*-statistic, and a mean for each data set, and graph the two distributions. (*BioStats Basics Online* will also display a standard error for each data set.) It will also calculate a df value, which takes the place of the degrees-of-freedom number

in other tests; df is generally the sum of the two sample sizes minus 2.

To interpret the *t*-statistic, click on the (Table) button; this will take you to a card (III. C-5) that will list the *t*-statistic you just obtained, and a df value. Use the df value to find the correct row of the table, and then read across from left to right until you find a value that is larger than your *t*-statistic. The *P*-value of your statistic lies between the number at the top of that column and the number at the top of the column to the left.

rather than pitchers, since only fielder height is parametrically distributed; the pitcher distribution has a truncated short tail, reflecting the reality that height—and therefore long arms—is at a premium.) The range of heights among baseball players (SD = 1.83 inches) is much smaller than that of a sample of male introductory biology students (SD = 2.76; Figure 5-8). The corresponding variances are 3.33 (fielders) and 7.59 (students); this yields an F-ratio of 7.59/3.33, or 2.28. For our sample size the maximum allowable F-ratio is about 1.82; thus the data fail the F-test.

*Need more than this? See **More Than the Basics:** Two-Sample t-Test with Dissimilar SDs.*

This least powerful version of the t-test approach was developed by the statisticians W. V. Beherens and R. A. Fisher to fill this statistical gap. It estimates a separate scaling factor for each sample based on their apparent SDs. A surprising consequence of the mathematical manipulations that are required before a t-statistic can be calculated is that the "df" value—the equivalent of degrees of freedom for unpaired tests—is almost never a whole number. As a result, you will usually need to interpolate between lines in the P-value table. Since the t-statistic is only rarely near the threshold of significance, this is not a major problem.

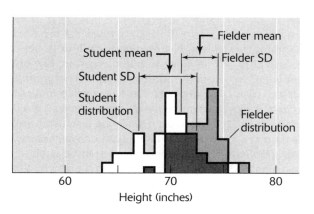

5-8 The standard deviation (and thus the variance) in the heights of Major League Baseball players is much smaller than the SD in the heights of male biology students. As a result, the ratio of the variances is too large to permit the use of ordinary parametric comparisons.

5 Continuous Parametric Distributions: II

5.7 The Standard Error and Confidence Intervals

Recall that we began this chapter worrying about the precision of your estimate of the true mean of your sample: if the sample size is large, your estimate is likely to be fairly precise, but is if it small, you cannot have very much confidence in it.

In statistical analysis, this uncertainty is quantified as the *standard error (SE).* If you were to draw a particular number of measurements from a given parent distribution, you would obtain a mean value for that sampling. If you were to repeat the process, you would obtain another mean value. After a considerable number of such resamplings, you could then plot the various means (Figure 5-9). The resulting distribution is a parametric curve centered on the true mean of the parent distribution. Clearly, any particular sample mean you measure is an approximation of the parent mean. This distribution of sample means, being a parametric distribution, has its own "standard deviation" encompassing 68.26% of all sample means of this

The **standard error (SE)** is the region about the sample mean within which 68.26% of all sample means of sample size n will be found: $SE = SD/(\sqrt{n})$.

Box 5.4

BioStats Basics Online   The unequal SD *t*-test

The test for paired data whose variances are *not* similar is found on the fourth *t*-Test card (III. C-4). This unpaired *t*-test requires you to enter or import your data. When your data are in place, click on the (**Compute**) button. *BioStats Basics Online* will display the sample sizes, the *t*-statistic, and a mean for each data set, and graph the two distributions. (*BioStats Basics Online* will also display a standard error for each data set.) It will also calculate a df value, which will almost always be fractional.

To interpret the *t*-statistic, click on the (**Table**) button; this will take you to a

card (III. C-5) that will list the *t*-statistic you just obtained and a df value. Use the df value to find the correct row of the table, and then read across from left to right until you find a value that is larger than your *t*-statistic. Because the df value will usually be a fraction, you will need to interpolate between entries in the table if your *t*-statistic is close to the significance threshold given by the nearby table values. The *P*-value of your statistic lies between the number at the top of the first column with a value larger than the *t*-statistic and the number at the top of the column to its left.

sample size from this parent distribution. It is this "SD" of sample means that is the standard error.

The standard error has much in common with the standard deviation: it is a measure of the width of a parametric distribution. But recall that the SD does not change with sample size, but is a characteristic of the variance of the data. The SE, on the other hand, clearly gets smaller with increasing sample size. In fact, the simplest way to remember how to compute the SE (though *BioStats Basics Online* will do this for you automatically) is simply to divide the SD by the square root of the sample size. As we said in the last chapter, the SE is, by our definition, a second-order statistic; the reason is that sample size is used both to compute the SD and then (in combination with the SD) to calculate the SE.

So what use is the standard error? In general, whenever parametric means of serious data are graphed, they are decorated

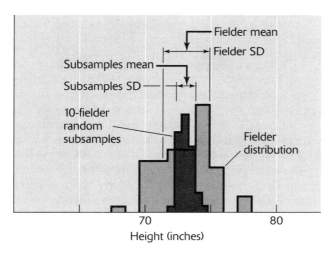

Need more than this? See **More Than the Basics:** *Variance, Additivity, and the Standard Error.*

5-9 When we take random 10-fielder samples from the player height distribution and plot them, the result is a much tighter distribution around the about same mean. The "standard deviation" of this 10-sample distribution is equivalent to the standard error (SE) for a randomly chosen sample of 10 fielders.

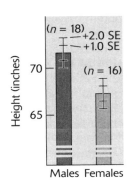

5-10 Heights of male and female biology majors compared in
summary form as a histogram showing the mean of each and the
associated standard errors. When there is no overlap between the
one-SE brackets enclosing two means, the means are usually
different at $P < 0.16$ (two-tailed; $P < 0.08$ one-tailed in the predicted
direction); when, in this case, there is no overlap between the two-
SE brackets, the means are usually different at $P < 0.02$ (two-tailed).
To provide higher resolution of the confidence intervals, most of the
histogram below 60 inches is omitted, as indicated by the horizontal
gaps in the bars.

with *error bars*. These bars typically represent one standard error (Figure 5-10). Now if one SE encloses 68.26% of the means,
31.74% lie outside these bounds, and the SE brackets seem
pretty crude. But if you have two means with their SE brackets
plotted and the bracketed regions do *not* overlap, the probability that the two samples are drawn from the same parent distribution is low. If the brackets almost touch, there is less than a
16% chance that the samples are the same (two-tailed—the figure is about 7.5% for a one-tailed comparison if the means are
different in the predicted direction); if the SE brackets are farther apart, the chance is lower.

Sometimes researchers plot both 1.0 and 2.0 SEs on their
graphs; the corresponding P-value for a nonoverlapping pair of
2-SE brackets is about 0.02 (two-tailed—0.01 one-tailed in the
predicted direction). Thus standard errors are a convenient visual aid, saving the weary reader the trouble of deciphering the
accompanying table. Standard errors are even useful when you

are comparing a sample distribution to a parent distribution; just remember that the SE of any parent distribution is zero, so if the 2-SE bracket of the sample does not enclose the parent mean, the two values are different with a two-tailed probability of about 0.02.

A related concept used by some researchers is the ***confidence interval***. This is the range around the sample mean within in which the parent mean lies (as opposed to SE, which is just a description of the distribution of sample means). This seems like a trivial distinction, and at large sample sizes it is. As a good rule of thumb, the area between the points two SEs above and below the mean of a sample roughly defines the ***95% confidence interval***; the parent mean will be found within this region 95 times out of 100. The actual values are 1.96 SEs for the 95% confidence interval of large samples; the equivalent 98% confidence interval is 2.33 SEs; the analogous 99% confidence interval encompasses about 2.58 SEs. For smaller sample sizes, it should be clear that your estimate of the standard error is not as reliable. As a result, the confidence intervals are necessarily larger. Figure 5-11 shows how the number of SEs

The **confidence interval** is the region about the sample mean within which the true (parent mean) will be found a stated percentage (usually 95%) of the time. One SE is the 68.26% confidence interval.

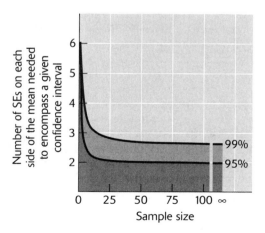

5-11 The number of standard errors encompassed by the 95% and 99% confidence intervals depends on the sample size, as the curves here illustrate.

5 Continuous Parametric Distributions: II

encompassed by the 99% and 95% confidence intervals increases with decreasing sample sizes; if you need to assign confidence intervals to your data, refer to this graph and the computation of SE in *BioStats Basics Online* to calculate the relevant confidence intervals.

Sometimes you will see data with confidence intervals marked without regard to precision. In physics, the convention is to use one SE as the confidence interval—which is a 68% confidence interval. In most other fields, the convention is to use a 95% confidence interval.

The decline of the standard error with sample size is sometimes called the ***central-limit theorem***. The phenomenon is easy to picture in connection with coin flipping: the more coins you flip, the closer your total head : tail ratio is likely to be to 0.50. In Figure 5-12, we have counted sport-utility vehicles trying to determine a ratio of models on the road (which could range from 0.0 to 1.0). The proportion after the first observation is unlikely to be an accurate measure of the true ratio; chance deviations from the true mean are highly likely with small sample

The **central-limit theorem** states that the standard error of the mean (SE) will decline with sample size (*n*); the decline generally goes with the square root of the sample size.

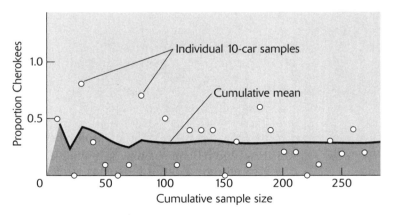

5-12 The points are proportion of Jeep Cherokees among all Jeeps observed in each of a series of samples. The line represents the cumulative proportion of Cherokees. The individual samples display considerable variance, but as the sample size increases the long-term proportion becomes increasingly stable near the true mean.

sizes. But as the sample size increases, the long-term sample mean will tend to approximate the parental mean ever more closely—that is, the sample mean *regresses* toward the true mean. The typical degree of difference between the two means, of course, declines with the square root of the sample size.

Points to remember

✔ The null hypothesis of two-sample *t*-tests is that the parametric data sets (sample distributions) being compared are drawn from the same parent (complete) distribution.

✔ Comparisons in two-sample *t*-tests are made between the means of the sample distributions being compared.

✔ To perform most *t*-tests, the distributions being compared must have roughly the same variance; variances are compared with the *F*-test.

✔ The *F*-test is sometimes used to compare the variances of two distributions without regard to their means in order to judge their independence.

✔ The larger the sample size in a distribution, the more certain we can be that the sample mean is a good estimate of the parent mean.

✔ Statistical power generally refers to how small a sample size a test requires to detect a given difference between two distributions.

✔ The most powerful two-sample *t*-test is the paired *t*-test (which is as powerful as the one-sample *t*-test); the next most powerful *t*-test is the unpaired *t*-test; the least powerful is the unpaired *t*-test for different SDs (variances).

✔ The standard error (SE) is the region about the sample mean within which about 68% of all means with the same sample size drawn from the same parent distribution will be found.

✔ The confidence interval is the region about the sample mean within which the parent mean will be found the stated percentage of the time. One SE marks the 68% confidence interval.

✔ Knowing whether the test is one-tailed or two-tailed, and observing whether the SE brackets of two distributions overlap (or the confidence-interval brackets of two distributions overlap), permits us to judge by sight whether two distributions have significantly different means.

5 Continuous Parametric Distributions: II

✔ The central-limit theorem posits that the standard error will decline with increasing sample size (generally with the square root of the sample size). (The standard deviation, because it measures the variance of the distribution, does not decline with sample size; the *accuracy* with which a sample distribution's SD reflects the true SD of the parent distribution, however, increases with sample size.)

Exercises

For the first eight questions, keep a list of the means and standard errors; this list will be needed for Question 9.

1. If we wanted to compare the mSAT scores of male and female biology majors, we would first need to see if the two data sets have similar variances. Run the *F*-test on these data (a partial list is in Table A; the full set is in file 5—Male vs Female mSATs or 5-MFMSAT.dat). Do they pass or fail?

Table A Partial list of mSAT scores

Male mSATs		Female mSATs	
640	600	590	550
620	560	600	540
680	600	580	570
670	630	610	550
530	530	530	540

2. If we wanted to compare the beak lengths of the finches in the

Table B Partial list of beak lengths

Pre-drought beak lengths		
6.0	7.6	8.0
6.4	7.6	8.0
6.8	8.0	8.4
7.2		

Post-drought beak lengths		
6.0	8.8	8.4
7.6	6.8	8.4
8.0	7.6	8.8
8.4		

Galápagos before and after the drought that killed most of them, we would need to first check the data with the *F*-test. Table B is a partial list of these data (a larger set, but still just a subset of the data, is found in file 5—Partial Finch Beak Lengths or 5-BEAKS.dat). See if they pass. (Because one of the data sets is quite large, the loading and testing may take more time than you expect.)

3. Male guppies perform "sigmoid" courtship displays (they twist their bodies into an S shape and show off the tail) when they approach females. We wondered if males display more frequently to females if another "rival male" (actually a wooden model) is visible. Table C presents the data for each of several males both with and without the rival present. (The data are also in file 5—Guppy Display Rates or 5-GDISP.dat.) Check to see if the data are parametric. If so, see if they pass the *F*-test. Which is the appropriate version of the *t*-test? Run it for these data. Is there a significant difference between male behavior under these two conditions?

4. Chicks that have been reared under certain conditions will "freeze" when

silhouettes pass overhead. Konrad Lorenz and Niko Tinbergen, early students of behavior, believed that chicks innately recognize the shape of predator birds (a short neck relative to tail length) and react more to predators than to the outlines of unthreatening birds (e.g., ducks) with a long neck relative to tail length. Skeptics argued that what Lorenz and Tinbergen observed were chicks that had become inured to common shapes and learned to freeze to silhouettes that elicited alarm calls from the parents. Table D presents data from naïve chicks,

Table C   Guppy display rates (displays/minute)

Guppy	Without rival	With rival
A	2.1	2.0
B	1.7	2.5
C	2.4	2.8
D	2.3	2.2
E	1.9	2.4
F	2.8	3.0
G	2.2	2.6
H	2.5	1.9
I	2.5	2.5
J	2.6	2.7

Table D   Freeze times

Chick	Gull model	Hawk model
A	3	1
B	2	4
C	3	6
D	1	5
E	2	2
F	0	3
G	2	4
H	1	4
I	1	7
J	2	3
K	2	3
L	2	4
M	0	2
N	1	4
O	3	6
P	4	5

5 Continuous Parametric Distributions: II

hatched and reared in indoor cages, that were tested with both hawk and gull silhouettes. (The data are also found in file 5–Freeze Times or 5-FREEZE.dat.) Are the data parametric? If so, do they pass the *F*-test? With this knowledge, apply the proper *t*-test and determine whether there is a significant difference in freeze times in these chicks.

5. You have already determined whether the mSAT data pass the *F*-test, and SAT data are known to be parametrically distributed. Select the appropriate *t*-test and see if there is a significant difference in scores between male and female biology majors on this part of the SAT exam.

6. You've checked to see if the finch data have a sufficiently low *F*-ratio. You may assume the lengths are parametrically distributed. Select the appropriate *t*-test and find out if there is a significant difference between the pre-drought and post-drought means.

7. Pitchers are, as we've noted in an earlier chapter, notoriously poor hitters. Table E provides a partial sample of batting averages of the Major League Baseball pitchers and fielders with the largest number of at-bats. (The full list is in file 5– Fielder vs Pitcher Batting or 5-FPBAT.dat.) See if both data sets are

Table E Partial list of batting averages

Pitcher batting averages	
.127	.111
.153	.174
.145	.150
.138	.162
.215	.165

Fielder batting averages	
.242	.325
.288	.244
.260	.262
.269	.256
.257	.291

parametric, and if so whether they pass the *F*-test. Then apply the appropriate *t*-test to determine if the two groups are significantly different at batting.

8. The shoe-size data we looked at briefly in Chapter 2 are listed in Table F. (They are also in file 5–Male vs Female Shoe Size or 5-MFSHOE.dat.) See if these data are parametric, and if so whether they pass the *F*-test. Apply the appropriate *t*-test to see if these two data sets are significantly different.

9. Plot the data in the preceding questions as simple histograms of the means, with SE brackets. Is there

Table F   Shoe sizes

Male shoe sizes*		Female shoe sizes	
14	11	6.5	7
13.5	11.5	7.5	6
12	13	9	7.5
12	14	8	8
13	11	6.5	10
13	12.5	7.5	9
12	13	7	8
15	17	6	8
12	13.5		
14			

*Converted to female size.

good correlation between the cases in which the SEs overlap and the data have a P-value > 0.05? Is there is good correlation between the cases in which the SEs do not overlap and the data have a P-value < 0.05? For the cases in which $P < 0.01$, do the 2-SE brackets about the means being compared overlap?

More Than the Basics

Two-sample t-test

To perform a two-sample paired t-test, use the formula

$$t = \sqrt{n} \, \frac{\left| \sum (X_i - Y_i) \right|}{s}$$

where X_i and Y_i are the X and Y values of the ith pair.

The degrees of freedom equals one fewer than the total number of pairs.

To perform a two-sample unpaired t-test, use the formula

$$t = \frac{|\bar{X}_1 - \bar{X}_2|}{\sqrt{\left(\dfrac{(n_1 - 1)s_1^2 + (n_2 - 1)s_2^2}{} \right)\left(\dfrac{1}{n_1} + \dfrac{1}{n_2} \right)}}$$

where $\overline{X}_1$ is the mean of the first group, $\overline{X}_2$ is the mean of the second group, n_1 is the sample size of the first group, n_2 is the sample size of the second group,

s_1 is the SD of the first group, and s_2 is the SD of the second group.

The degrees of freedom are $n_1 + n_2 - 2$.

Two-sample *t*-test with dissimilar SDs

To perform a two-sample *t*-test with dissimilar SDs, use the formula

$$t = \frac{|\overline{X}_1 - \overline{X}_2|}{}$$

The degrees of freedom is

$$df = \frac{\left(\dfrac{s_1^2}{n_1} + \dfrac{s_2^2}{n_2}\right)^2}{\dfrac{\left(\dfrac{s_1^2}{n_1}\right)^2}{(n_1 - 1)} + \dfrac{\left(\dfrac{s_2^1}{n_2}\right)^2}{(n_2 - 1)}}$$

Variance, additivity, and the standard error

Why, if the SD is such a convenient and sensible tool, do we bother about the variance at all? As it turns out, from a mathematical perspective, the variance is, if anything, more important. Consider the following situation: if randomly (that is, parametrically) distributed variable X (length of a phone call, for instance) is the sum of randomly distributed and independent variables Y_1 (the time you wait for the other party to pick up the phone), Y_2 (the time you spend talking), and Y_3 (the time required for the switch to disconnect the call), then the variance $V(X)$ of total call time is equal to the sum of the variances of the three variables— $V(Y_1)$, $V(Y_2)$, $V(Y_3)$—that contribute to the value of X. To put it more generally, as

may be derived with a little bit of integral calculus, when normally distributed independent variables are added ($X = Y_1 + Y_2 + Y_3 \ldots$), the variance of the resulting distribution is

$$V = V_1 + V_2 + V_3 \ldots$$

where V_1 is variance of distribution Y_1.

This is a powerful result: many interesting, normally distributed variables can be thought of as the sum of other, roughly normal variables. It suggests that the variance of such an interesting variable could be partitioned into the parts due to its "causative" variables, an idea that will be explored in more detail in Chapter 12.

It also tells us (with a little bit of algebra) that the distribution of the means of fixed-size samples of a population will themselves have a variance of

$$V_{\text{means}} = \frac{V}{n}$$

This is called the error variance (or error mean squares) because it describes the variance of sample means due to the finite sample size. From this, we may take the square root to calculate the error SD, or standard error: the SD of the distribution of means caused by the inaccurate estimation of population means by a small sample. This is why the standard error is often called the standard error of the mean, a term heavily used in physics.

To summarize:

The population variance is

$$\sigma^2 = V = \frac{\sum (X_i - \mu)^2}{N}$$

where N is the size of the parent population.

Sample standard error is

$$SE = \sqrt{\frac{V}{n}}$$

where n is the size of the sample.

Tests Used in This Chapter

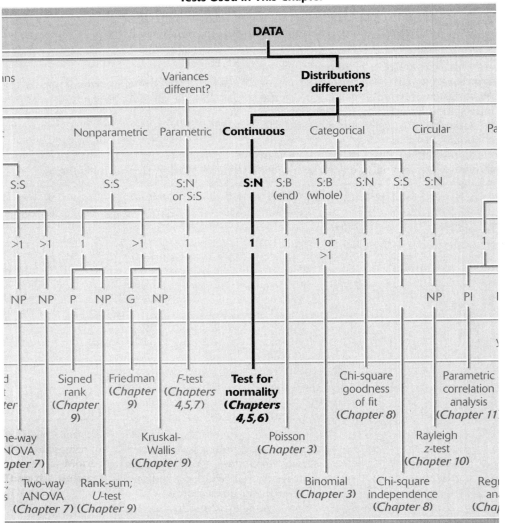

DATA

ins · Variances different? · **Distributions different?**

Nonparametric · Parametric · **Continuous** · Categorical · Circular · Pa

S:S · S:S · S:N or S:S · **S:N** · S:B (end) · S:B (whole) · S:N · S:S · S:N

>1 · >1 · 1 · >1 · 1 · **1** · 1 · 1 or >1 · 1 · 1 · 1 · 1

NP · NP · P · NP · G · NP · NP · PI

Signed rank (*Chapter 9*) · Friedman (*Chapter 9*) · *F*-test (*Chapters 4,5,7*) · **Test for normality (*Chapters 4,5,6*)** · Chi-square goodness of fit (*Chapter 8*) · Parametric correlation analysis (*Chapter 11*)

ne-way NOVA apter 7) · Kruskal-Wallis (*Chapter 9*) · Poisson (*Chapter 3*) · Rayleigh z-test (*Chapter 10*)

Two-way ANOVA (*Chapter 7*) · Rank-sum; *U*-test (*Chapter 9*) · Binomial (*Chapter 3*) · Chi-square independence (*Chapter 8*) · Reg an. (*Cha*

Data Transformations

6

All the previous chapters have been concerned with parametric data, indulging the popular statistical fantasy that all data are really parametric. This is, unfortunately, not true—but the reasons for wishing it were bear consideration.

6.1 Why Parametric Is Better

The shape of a "normal" distribution can be described by only two parameters, the mean and SD. In contrast, no small number of parameters can describe the shape of a genuinely nonparametric distribution: we simply don't know what it actually looks like unless we have a huge sample size, and even then fitting it to an equation will usually be quite difficult. While we may find the median value, this really doesn't tells us very much, since the distribution around that median could be of any shape and any width.

In addition, since a nonparametric distribution does not have the central-limit property of a parametric distribution, you cannot obtain a P-value simply by counting SDs from the mean and doing that relatively simple computation. So while the t-test and other parametric tests can base their math on the distance (in SDs) of a point from the mean, the relative distance of a point from the median of a nonparametric distribution is impossible to describe precisely.

In short, tests analyzing nonparametric data are much less powerful than those analyzing data drawn from a normal distribution, and we should like to avoid them whenever possible.

6.2 Nonparametric May Be Only Skin Deep

It is an unfortunate fact of life that not all data sets are parametric. For one reason or another—perhaps the nature of the distribution, perhaps limitations in the experiment—you may find yourself with nonparametric data. Indeed, in many areas of science (e.g., animal behavior) data are almost never parametric. As you will see in Chapter 9, it is much harder to find significant trends in nonparametric data. Given the power of parametric statistics, it is natural to wonder: Can I make my data parametric?

*Need more than the basics? See **More Than the Basics:** Additivity.*

The answer is a "maybe." To see why, look at the data for weekly wages in the United States (Figure 6-1a). A quick glance reveals that the distribution is not at all normal: the high-income end extends much farther than the low-income end, leaving a distribution that is obviously not symmetrical. But neither

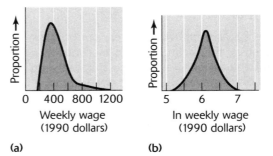

(a)   (b)

6-1 (a) The distribution of median weekly wages in the United States in 1990; this distribution is not parametric because of its long tail to the right (high-wage earners) and the absence of a tail to the left (the minimum-wage floor). **(b)** The distribution of ln(wages)—that is, the natural log of the original distribution. This transformed distribution easily passes the test for normality, and thus can be used in *t*-tests. Keep in mind, however, that if you convert your mean, SD, and SE back into the original units, the SDs and SEs will not be symmetrical—that is, the value of $+1.0$ SD and SE will be larger than the value of -1.0 SD and SE; they were symmetrical, of course, in the transformed units. [Parts **(a)** and **(b)** based on data from U.S. Department of Labor, 1991.]

does the distribution look entirely nonparametric: it still has the characteristic hump and tails that we associate with a bell-shaped curve.

What is to be done with this distribution? Clearly, wages are not parametrically distributed. But if we plot a different variable on the horizontal axis—for instance, the logarithm of the wage—we can get a parametric distribution and analyze it with parametric tests (Figure 6-1b). This method of analyzing a number derived mathematically from a datum rather than the datum itself is called a ***transformation***. This distribution is precisely what we wanted, but it has some drawbacks.

A **transformation** of a distribution involves performing the same mathematical operation on every datum in the distribution.

6.3 The Catch

Analyzing the log-wage rather than wage seems a little bit dishonest. For instance, the mean and SD of the transformed distribution are in units of log-dollars, which doesn't really tell us much about the original distribution. In addition, since the variable being analyzed is not measured in dollars but in log-dollars, any distribution with which this is being compared must also be transformed in this manner, a step that could well make a parametric distribution nonparametric. As you can probably guess, a parametric distribution cannot be analyzed parametrically against a nonparametric distribution.

There are other drawbacks to transformations which will become clear in the chapter on regression (Chapter 11).

6.4 Popular Transformations

One obvious question to ask would be: how did we know to take the log of the distribution above? How, indeed, can you know which of the many possible transformations to apply? A few simple rules of thumb can guide you in this question, and *BioStats Basics Online* lets you try them on your data. Use the Normalization Techniques card (II. I-2) and transfer them automatically to the Testing for Normality card (II. B-2) to see if the transform worked.

1. The distribution in Figure 6-2a, which looks as though it has been given a shove to the left, may often be rescued by taking the square root of each datum; such distributions occur in variables like natural animal weights in which a given distance below the mean is much less likely (healthy in the case of weight) than the same distance above. If this doesn't work, try the "ln" option, which takes the natural log (this transformation is illustrated in the *BioStats Basics Online* box; it is identical to taking the log except for a scaling factor, and computers compute natural logs faster).

2. The distribution in Figure 6-2b is right-skewed and is susceptible to squaring; such distributions occur in

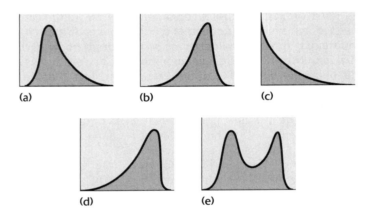

6-2 (a) This sort of left-shifted distribution can often be transformed into a parametric distribution by taking the square root or the natural log of each point. (b) Right-shifted distributions like this one can sometimes be made parametric by squaring each datum. (c) This kind of "decaying" distribution can sometimes be transformed by adding 1.0 to each datum and then taking the natural log of each value; this also is worth trying when the distribution has a far-left peak and then declines steeply for x-axis values near zero. (d) Distributions with a steeply declining right-hand side can in many cases be transformed by the e^x technique—that is, by raising e to the power of the datum for each value in the distribution. (e) Distributions like this one require sampling to create a parametric curve.

Box 6.1

BioStats Basics Online   Transformations

The Normalization Techniques card in *BioStats Basics Online* (II. I-2) allows you to apply transformations to your data. Simply enter your data in the left-hand column (or import it) and click on the appropriate transformation button. The transformed data will appear in the right column, and the mean, SD, and SE of the transformed sample will be displayed on the right. Click the (**Graph**) button to see how the shape of the sample distribution has been changed. This card will also perform the sample-mean transformation, but consult the text for warnings on proper sample sizes. Note that when you save these data, *BioStats Basics Online* saves both the original and transformed data.

Click the (**Go to the test for normality**) button and the transformed data will be imported onto the Test for Normality card (II. B-2).

variables, like tree heights, for which a given distance above the mean is more susceptible to environmental conditions than the same distance below.

3. For the distribution in Figure 6-2c, $\log(x + 1)$ will work; you may expect Poisson-like distributions such as this in variables like time, in which there is a hard minimum on the possible values of the variable.

4. For the distribution in Figure 6-2d, e^x will work; you may expect distributions like this in cases in which a hard maximum of some sort is present.

5. The distribution in Figure 6-2e is hopeless. No simple transformation can help a distribution like this one; this particular distribution is most likely a combination of two separate phenomena (for instance, heights, which are distributed differently for males and females). But don't despair: read the next section on sampling.

Need more than the basics? See **More Than the Basics:** *Guiding Principles.*

6.5 Sample Means: The Ultimate Transformation?

There is another method for making a distribution normal: if you repeatedly take sample sets of a fixed size and use the means

Any distribution can be made parametric by **sampling** the data and plotting the distribution of **sample means**.

of these sample sets as the data, then (provided that the sample-set sizes were large enough) the sample means will be distributed parametrically (Figure 6-3). This may be hard to believe, but it's really no different from what we've been doing all along: a single coin flip gives two widely separated peaks, a distribution that is certainly not parametric. But as we have seen, if you take a sample of means (or sums) of a large number of coin flips, you do get a normal distribution. (There is a nice demonstration of the sample-mean technique in *BioStats Basics Online* on the

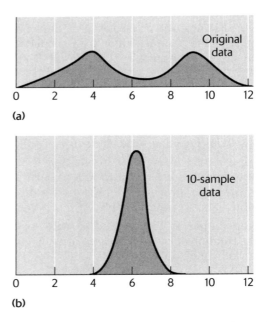

(a)

(b)

6-3 **(a)** This two-peaked distribution cannot be transformed by simple mathematical tricks. **(b)** Sampling from the original data in 10-datum units and plotting the means of these samples does yield a parametric curve. This approach requires lots of sample data (the total number of values going into the 10-sample curve cannot exceed the number of values in the original data set) and yields an SD that depends entirely on the sampling size chosen (10 in this case); thus it can only be compared to other 10-sample-means data.

6 Data Transformations

Transforming Skewed Null Distributions card, II. I-1, which can be reached from the Topics card. The transform itself is an option on the Normalization Techniques card, II. I-2.)

But there is a danger to this method: you seldom have the luxury to gather the vast amounts of data required, and it is tempting simply to take a single data set and take all the sample sets randomly from that. This is a simple though tedious procedure with an obvious pitfall: you could generate an infinite number of sample sets, and thus an infinite quantity of data, from only a small sample simply by repeating this process. Clearly, this cannot be legitimate; you must never take more sample sets than the total size of the original sample divided by the sample-set size. Otherwise you find yourself making broad generalizations from tiny amounts of real data, the statistical equivalent of "All police cars are Fords—at least the one I saw was."

In practice, this means that you must gather huge quantities of nonparametric data in order to get even a small amount of parametric data in this way. However, the parametric data so obtained will have a tight SD as a result of the sampling and may therefore prove significant. If you're really lucky, someone else (perhaps the U.S. Census Bureau, or a German lab in almost any field) has already gathered enough nonparametric data (e.g., family income) to allow liberal sampling of means.

Need more than the basics? See **More Than the Basics:** *Sample Sets and SEs.*

6.6 Comparisons from Sample Means

A set of data gathered as sample means must be treated with care. Because its SD is a consequence both of the data distribution *and* the sample-set size, it will most likely be different from that of any sample distribution not obtained by the same method and same sample-set size. Therefore you cannot use the equal-SD versions of the *t*-test except to test one set of sample means against another taken with the *same sampling method.* Just as with the other transformations, transformed data sets may usually be compared only with other sets transformed in the same way. This SD issue has another consequence: the SD of the sample means has no independent meaning—that is, it reflects no characteristic of the original distribution. Finally, taking sample means from *paired* data removes almost any useful information that the pairing might have given, especially for large sample-set sizes. Paired data subject to sampling may just as well be treated as unpaired for subsequent tests, because the matching of individuals is damped out by the taking of sample means.

On the positive side, however, sampling of the means is the only transformation that permits the comparison of parametric to nonparametric data, thereby providing a powerful tool for analyzing a kind of comparison often thought to be immune to statistics.

Points to remember

✔ The most powerful statistical techniques require parametric data sets.

✔ A distribution that is not parametric can sometimes be transformed into a parametric distribution by the applying the same mathematical operation to all members of the data set.

✔ The transformed distribution can only be compared to other distributions that have undergone the same transformation.

✔ The SE and SD of transformed distributions have no independent meaning but can be used in the usual way in the F-test and in statistical comparisons.

✔ The sample-mean technique cannot involve more data values than were present in the original distribution.

Exercises

For each data set, test it first for normality and record the D-index (index of difference). Then try the most promising transforms including, for large data sets, sampling. Save the transformed values that yield an improved distribution and test them for normality, comparing

their D-index values with that of the untransformed set.

1. Table A is a partial list of annual family income data from the United States in 1970, rounded to the nearest thousand. (The data are also

Table A   Partial list of income (thousands of dollars/year)

2	5	7	8
10	12	14	16
19	23	34	55

found in file 6–1970 Family Income or 6-INCOME.dat.) Which transform works best?

2. Table B is a list of the weights of Freeman University female students (in pounds, rounded). (The data are also found in file 6–Female Weights or 6-FWTS.dat.) Which transform works best?

3. Table C is a partial list of cumulative GPAs of biology majors. (The data are also found in file 6–GPAs or 6-GPAS.dat.) Which transform works best?

4. Table D is a partial list of the adult weights (in grams) of wild-caught mice. (The data are also found in file 6–Mice Weights or 6-MICE.dat.) Which transform works best?

Table B   Female weights

90	100	100	100	110	110
110	110	110	110	110	120
120	120	120	120	120	120
120	120	130	130	130	130
130	130	130	130	130	140
140	140	140	140	140	140
140	150	150	150	150	150
150	160	160	160	160	170
170	170	180	180	190	190
200	210	220	230		

Table C   Partial list of GPAs

3.2	2.9	3.7	3.3
3.7	3.2	2.5	2.7
1.7	2.3	3.7	3.1
3.0	3.5	2.3	3.1

Table D   Partial list of mouse weights

1.0	1.3	1.4	1.6
1.7	1.8	1.9	2.0
2.0	2.0	2.0	2.0
2.1	2.1	2.1	2.2
2.2	2.2	2.2	2.3

More Than the Basics

Additivity

Bell curves are generated by additive phenomena, that is, situations in which the quantity being measured varies with a number of contributing factors, each one adding in its effect to form a grand total. Nonnormal distributions, by contrast, usually come about when the variable being measured depends in a nonadditive way on its various contributing factors.

For example, some factor may not contribute by simple addition but rather by multiplication or in some other way; electric bills, for instance, are not based simply on the sum of all electricity consumption in a household but include nonlinear factors such as taxes and transmission (infrastructure) charges, surcharges for excess summer use, and other complications. In another familiar case, tennis scores are not normally distributed, in part because the first winning shots count 15 points while later ones count 10. Pension values of normalized wages are not normal because the interest earned in one year itself earns interest in subsequent years, with multiplicative rather than additive effects on the accumulation. Thus any variation in nonlinear factors can render a distribution nonnormal.

Additivity will become quite important in the two-way ANOVA presented in Chapter 7, which also provides a statistical test for nonadditivity.

Guiding Principles

Mathematical transformations seem so powerful that one almost suspects that they are "cheating"—that there must be something wrong with them unless there is some underlying assumption that justifies their use. In fact, the principle of additivity (See More Than the Basics: Additivity) gives us just such a justification.

For instance, if a nonnormal distribution is caused by multiplicative factors—a common cause of nonparametric data—taking the logarithm of each data point will restore additivity:

$$X = x_1 x_2 x_3 x_4 \ldots \Rightarrow \log(X) = \log(x_1) + \log(x_2) + \log(x_3) + \log(x_4) + \cdots$$

Each transformation handles a type of nonadditive situation, restoring additivity and thus normality.

This should not, however, be taken as carte blanche to use any transformation that does the job on the assumption that it is restoring additivity. There are restrictions on when we may use transformations and what transformations are legitimate to use: first, the theory behind transformation of data is that transformations are being applied to whole distributions rather than small samples. While a large sample may legitimately be transformed, the choice of a transformation for a small sample may result as much from chance variation as from the parent distribution, and may even border on post hoc analysis. In addition, not just any mathematical function may be used as a transformation. A function must be *monotonic*—always increasing for larger numbers. This means that it is generally impossible to transform a bimodal distribution (see Figure 6-2e).

Sample sets and SEs

In the text we say, "We must never take more sample sets than the total size of the original sample divided by the sample-set size." You may be curious as to how we came up with this seemingly arbitrary dictate. The answer is that it is a simple property of our favorite quantity, the SE.

Pretend for a moment that the sample is from a parametric distribution. Naturally, the parent distribution has no SE, but does have some variance, V. The sample, therefore, has variance and SE:

$$v \approx V; \ SE = \sqrt{\frac{v}{n}}$$

Clearly, if our distribution after sampling has a different SE, we have collected either too many or too few sample sets: sampling should not leave us any more or less certain about the mean. So if we take *a* sample means, each from a sample set of size *b*, the math follows beautifully:

$$v_{\text{sampled}} = \frac{v_{\text{original}}}{b} \Rightarrow$$

$$SE_{\text{sampled}} = \sqrt{\frac{v_{\text{sampled}}}{a}} = \sqrt{\frac{v_{\text{original}}}{ab}}$$

$$SE_{\text{original}} = SE_{\text{sampled}} \Rightarrow \sqrt{\frac{v_{\text{original}}}{n}}$$

$$= \sqrt{\frac{v_{\text{original}}}{ab}} \Rightarrow ab = n$$

And we determine that the number of sample sets times the sample-set size must be equal to the size of the original sample.

Tests Used or Covered in This Chapter

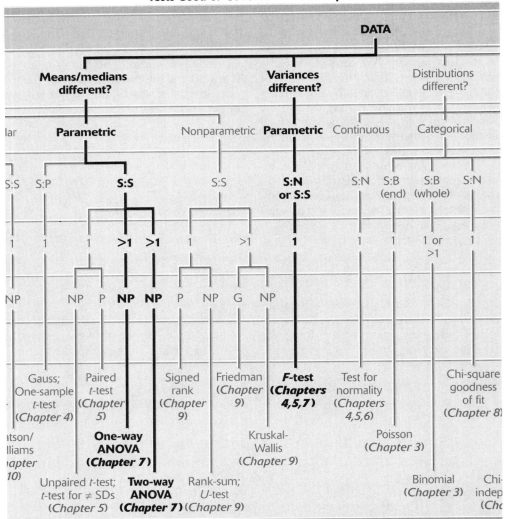

Multiple Parametric Distributions: ANOVA

7

As we showed in the last three chapters, Gosset's approach to parametric data sets provides an enormously powerful tool for comparing two means. Quite often, however, you may have more than two sets of data. In tests to determine which conditions, if any, are favorable to the growth of a common roadside weed (*Erigeron annuus*, daisy fleabane), one study compared sterilized soil from four different sites in each of three separate locations that differed in their soil condition, being either plowed, un-plowed, or a roadside verge; in addition, each sample was ei-ther inoculated with symbiotic root fungi or left alone. Ten pots were prepared for each of the 24 distinct combinations possi-ble (e.g., Site 2, unplowed, uninoculated), and a seed was planted in each. After a season of constant and favorable growth con-ditions in a greenhouse, the 214 surviving plants were weighed. The weights were roughly parametrically distributed. The prob-lem now is to compare these 24 data groups to see what these plants "like" from among the variables in play. (The null hy-pothesis, of course, is that all the distributions are the same.)

7.1 Why It's Illegal to Perform Multiple *t*-Tests within a Data Set

One possibility would be to plot the 24 sample means (with their standard errors) and look for differences (Figure 7-1; for simplicity we've plotted the pooled replicates from only two of the four sites for each combination of soil and fungi variables).

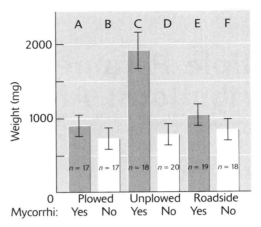

7-1 The mean weights and standard errors for plants grown from seed planted in sterilized soil gathered from one of three sites (at four locations) and either inoculated with symbiotic root fungi (mycorrhizae) or left uninoculated. Data from the first two locations only are combined; sample size is shown. Group C appears to be different from the others, but because there are $5 + 4 + 3 + 2 + 1$ possible pairwise comparisons (15) with six means, the chance of having one mean this far from the others cannot be computed with a simple t-test.

The **P-value** (probability value) of a result is the chance of that result occurring plus the odds of all possible other outcomes *even more* different from the prediction of the null hypothesis.

But which means do you compare with which other means? This seems simple enough: analyze the ones that look most different. If this sounds too easy, it's because it is. Think back to what a **P-value** really means: it is the chance of drawing one sample distribution from another distribution. Thus a $P < 0.05$ tells us that if you find a difference this large, the odds of obtaining it by sampling the parent distribution itself are about 1 in 20; we generally take this to imply that the sample is probably *not* drawn from the parent distribution. But turn this around for a moment: if you were to sample from the same parent distribution 20 times, you are likely to obtain one or more sample means with a P-value of 0.05 or lower; in fact, the product law (see page 44) allows us to calculate a 65% chance of *at least* one false positive. In short, when you compare Group 1 with the other 23 plant-treatment groups in this experiment, there is a very good chance of seeing one or more "significant" differences even if all 24 groups had been identical to begin with.

When we select the two *most* different, the situation is even worse. There are 276 unique comparisons: the first sample can be compared with 23 others; the second can be compared with 22 others (having already been compared to the first); the third can be compared with 21 others (having already been compared with each of the first two); and so on. Thus there are $24 + 23 + 22 + 21 + 20 + \cdots 4 + 3 + 2 + 1$ possible comparisons. A faster way of computing this value is to say: there are 24×24 treatment groups (576) less the 24 cases in which a group is compared with itself (552) divided by 2 (since the order of the comparison doesn't matter; 276). This corresponds to $(n^2 - n)/2$. It would be surprising indeed if 276 independent samples of the same parent distributions did not produce many "significant" differences, and on average you will find two false positives with P-values below even 0.01.

The lesson here is clear: when you have several groups to compare, you cannot apply any version of the one- or two-sample t-tests. Table 7-1 lists the actual sobering odds of obtaining false positives for a variety of multiple pairwise comparisons of means for several different t-test thresholds. (With only three groups, for which there are three possible comparisons—

Table 7-1   Probability of at least one false positive

Samples	Number of paired comparisons	Level of significance				
		0.10	0.05	0.01	0.005	0.001
2	1	0.10	0.05	0.01	0.005	0.001
3	3	0.27	0.14	0.03	0.015	0.003
4	6	0.47	0.26	0.06	0.030	0.006
5	10	0.65	0.40	0.10	0.049	0.010
6	15	0.79	0.54	0.14	0.072	0.015
10	45	0.99	0.90	0.36	0.202	0.044

2 + 1—using three t-tests with a stricter P-value threshold will generally be considered reasonable except by your personal enemies; your close friends might even let you get away with four means—six possible comparisons—and a still-stricter threshold. Beyond that, only your mother will believe you. If you elect this simpler course, use this crude approximation: divide the P-value threshold you are using by the number of possible comparisons, and use the new value as your cutoff, citing the original value as your threshold.)

A crude (i.e., overly conservative) but valid way of compensating for **multiple comparisons** is to divide the P-value by the number of pairwise comparisons possible. The ANOVA test, though more difficult, has more statistical power.

The problem, then, is to find a way of making multiple comparisons of parametric data without risking false positives at every turn. This dilemma can manifest itself in two ways: you may need to compare multiple tests with each other (as in the case just described), or you may have several independent pairwise tests. The ANOVA test described below, though more trouble than just dividing P-values, will generally yield much better results for the first circumstance. The proper approach to the second situation will be discussed presently.

7.2 Comparing Means without Comparing Means

The ANOVA test, or **analysis of variance**, determines if one or more distributions from a set is significantly different from the others.

The problem of how to look for differences in means without actually comparing the means themselves (a process that, as we've just seen, generates false positives at an alarming rate) was solved in principle by R. A. Fisher. Since Fisher's approach depends on looking at variances rather than means, it is called an ***analysis of variance*** (or, more often, ***ANOVA***). (As with Gosset's t-test approach, Fisher's initial solution has stimulated many innovative improvements and elaborations.) The logic is fairly clear, and the math in the simplest cases isn't even too forbidding, but first let's be clear about our assumptions:

1. as always, the samples are independent,
2. the samples are all from parametric distributions, and
3. the samples all have similar SDs.

If there is any doubt about this last point, plot the distributions and then, using the F-test, compare the two whose SDs look most different. *BioStats Basics Online* automatically plots the data in these multiple-mean comparison tests and allows quick transfer of the two most different distributions to the *F*-test card (III. B-1). As far as the parametric criterion is concerned, experience has shown that ANOVA is less sensitive to skewed data than any of the other parametric tests. Finally, keep in mind that variance is SD^2; we've mostly used SD as a measure of the variability in a sample distribution, but ANOVA uses variances because they can be added, as More Than the Basics: Variance, Additivity, and the Standard Error section in Chapter 5 pointed out (we will elaborate on this in Chapter 12).

Need more than the basics? See **More Than the Basics: The F-Distribution.**

Fisher's logic was this: first, the variance in a set of data involving two or more sample groups is, by definition, just the square of the difference between each data value and the mean of the entire set, summed for all data values, and divided by $N - 1$ (where N is the total sample size). But the variance can *also* be described as a combination of two separate variances. Imagine that all the means of the groups (the sample means) are plotted as a distribution all their own. (This sample-mean distribution is parametric since the values should represent sampling errors.) This sample-mean distribution has its own mean and variance; the mean is the same as the set mean, but the variance of the group means should be much smaller than the variance of the data; indeed, when the sample sizes within groups are large, the group means should all be the same, and thus the group-mean variance should be near zero. So much for the variance between group means. Where is the rest of the set variance? That's easy: within each group, the data have a variance about the group mean. Under null-hypothesis conditions (all groups are drawn from the same parent distribution), most or all of the set variance is in the group variances.

So now we have two new variances to deal with: (1) there is the variance *within* each sample group (the sum of the squared differences between each datum and its group mean, divided by $n - 1$, where n is the sample size of the group), (2) *plus* the variance of the sample-mean distribution (the sum of the squared differences between the mean of *all* the data and the mean of each sample group, divided by one fewer than the number of groups). The total variance in all the data is the sum of the variances (scaled for sample size) of these two group-based elements (Figure 7-2).

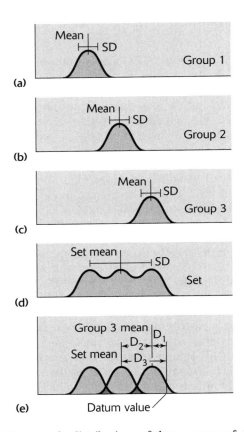

7-2 **(a)–(c)** The sample distributions of three groups of data that
are part of the data set to be analyzed. Each group distribution has
its own mean and variance (shown here as SD; variance is SD^2).
(d) The distribution of the entire data set—that is, of all the groups
combined—has its own mean and variance. The combined variance
is typically larger than the individual group variances because the
groups have different means. **(e)** The difference (d, or deviation)
between any datum and the overall set mean (expressed in terms of
the variance of an individual datum, d^2, so they can be added
without regard to sign) can be described in two steps or one. As two
steps, it is the sum of its deviation from its group mean (D_1) *and* the
deviation of the group mean from the set mean (D_2); as one step, it
is the deviation of the datum value from the set mean (D_3). The
actual analysis uses group and set variances (summing the individual
variances and dividing by $n - 1$); the result is the same.

The second step in Fisher's logic was this: if each group's sample distribution was drawn from the *same* parent distribution (the null hypothesis), then the variances (SD^2) of the sample distributions of the various groups should all be about the same (allowing for sample size, and thus any chance variation, in approximating the variance in each sample). As indicated above, you must assume this in any case as a prerequisite for making this comparison.

It's possible to get confused at this point because the variances we are talking about are all scaled by dividing by $n - 1$, where n is the sample size of the set, the sample size of the group, or the number of groups. As we said in Chapter 2, "variance" technically is the square of the deviation between a single isolated datum and the mean (d^2); "sample variance" (SD^2) is the "average" variance in the sample set. (It's not quite an average, since the division of the sum of d^2s is by $n - 1$ rather than n.) We said then that since we would always be talking about sample variances, we would shorten that term to "variance" (as do all real-world users of statistics outside math departments). Usually the averaged nature of variances doesn't matter to us; it's just part of the math we take for granted. But with three different variances in play in ANOVA analysis, this one time we need to remember that sample variances are "averages" rather than sums, and thus do not automatically increase or decrease with the sample size (or number of groups) we are considering. With this in mind, we can continue with the third step in Fisher's elegant argument (again, using "variance" to mean "sample variance").

Finally, Fisher reasoned, the ("average") variance of all the sample data combined (the "sample set," as we are calling it) should be similar to the typical variance of the individual groups of sample data: since all the data are assumed to be from the same parent distribution, they should therefore have the same mean (as illustrated for the null-hypothesis scenario in Figure 7-3), and thus the group variance should be small or zero. But if one (or more) of the sample groups has a *quite different* mean from that of the set, it will greatly increase the total group variance, even though its own within-group variance will be similar to that of other groups (as was illustrated in Figure 7-2).

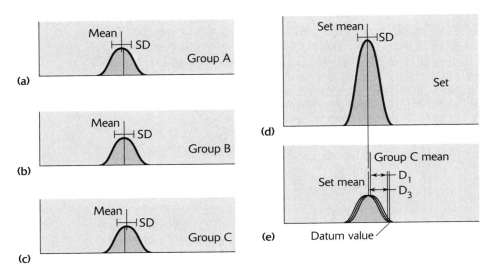

7-3 (a)–(c). The sample distributions of three groups of data that are part of the data set to be analyzed. Each group distribution has its own mean and variance; in this case, for the sake of simplicity, the three group distributions are identical, but their means are slightly different. **(d)** The distribution of the entire set—that is, of all the groups combined—has its own mean and variance. In the unlikely case illustrated here where all groups have the same variance, the set variance is the same as the group variances. **(e)** The deviation of any datum from the overall set mean can be described either as the sum of its deviation from its group mean (D_1) and the deviation of the group mean from the set mean (D_2), or as the deviation of the datum value from the set mean (D_3). When all groups are exactly identical, $D_1 = D_3$ because $D_2 = 0$. (This almost never happens because even if the groups sample the same parent distribution, there will be chance variation between samplings; the usual case when all groups are samples of the same parent distribution, is that the group means and variances are quite close to—but not exactly equal—the set mean and variation.)

So let's look at how this plays out in the plant data. Figure 7-4a shows the distributions for the five sample groups that seemed to have about the same mean in Figure 7-1; Figure 7-4b adds the odd-looking group at the top. In Figure 7-5a (page 160) we've plotted the variances for the five similar sample groups, and the total (set) variance; they are very similar. This is to say that the variance in the sample groups is about the

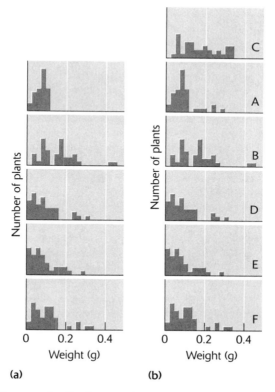

7-4 (a) The sample distributions for groups A, B, D, E, and F are plotted; they are fairly similar. **(b)** The sample distribution for the plants in Group C are plotted at the top; this distribution is shifted to the right.

same as the variance of the set—all the data combined. In Figure 7-5b, we've added the group that looked different. The mean variance in the (now) six groups is not much changed, but the variance in the data taken as a whole is much increased. Figure 7-5c shows why: the sixth group adds data at the right end of the overall set distribution, and this increases the average

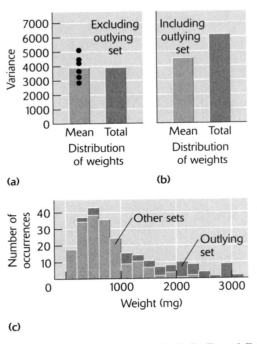

(a)

(b)

(c)

7-5 (a) The mean variance for samples A, B, D, E, and F is indicated by the left bar of the histogram, along with the individual variances plotted as filled circles; the right bar shows the variance of the data from these five sets taken together. The two bars are very similar because the five sample sets are similar: they have about the same means and variances. Thus combining the data from all five sets has about the same effect (that is, none) that you would expect from simply increasing the sample size of one of the sets fivefold. **(b)** When one of the data sets has a distinctly different mean, a different pattern emerges. The total variance of all the data taken together (right bar) is much larger than the mean variance of the six sets of data (left bar). **(c)** This increase reflects the greater spread in data values that Sample C (dark blue portions of the histogram) contributes to the total; because the variance depends on the square of the difference between individual values and the mean, the influence of the many outliers at the right end of the distribution is substantial.

variance of data values from the overall mean. (The increase is so large because we square deviations to obtain variance.)

In short, when the sample group distributions have about the same mean, the average variance *within* sample groups will be only slightly less than the variance of the entire sample set; when the sample-group distributions do not have the same means, however, the variance of the lumped sample set will be substantially greater than the average variance of the sample sets. In the ANOVA test, the actual comparison is between the mean variance *within* sample groups—the ***within-group variance***—and the variance *between* sample-group means—the ***between-group variance***. When the between-group variance is substantial, one or more of the means may be different from the others. The statistic generated by taking the ratio of these two measures of variance is evaluated by the ubiquitous table of *F*-values. Thus the *F*-test provides both the entry and exit points for ANOVAs.

Because of the importance of ANOVA in many, many fields, let's look briefly at the actual computation in the simplest case, the ***one-way ANOVA***. It has much in common with the basic calculations we've already worked through. Remember that the sample variance is the sum of the squared deviations of the measurements from the mean divided by the degrees of freedom (that is, one fewer than the sample size). So, if X is the mean and x_i the ith datum in a set of n, then sample variance of group X is

$$SD_X{}^2 = \left[\sum (\overline{X} - x_i)^2\right] / (n - 1)$$

Since we will be talking about several different groups of data, we have to distinguish among the different groups with a group subscript; thus for the first group of data the variance is simply

$$SD_{X_1}{}^2 = \left[\sum (\overline{X}_1 - x_{1_i})^2\right] / (n - 1)$$

In the simplest case, the groups are of the same size, so we don't need to worry about weighting the several sample variances involved. (The formula in More Than the Basics: One-Way ANOVA Computation, at the end of this chapter, cannot

The ANOVA approach compares variances (SD^2). One variance is the mean **within-group variance**:

$SD_{wg}{}^2 = (SD_1{}^2 + SD_2{}^2 \cdots + SD_n{}^2) / n - 1$, where $SD_1{}^2$ is the variance of group 1. The other variance is the **between-group variance**:

$SD_{bg}{}^2 = [(\overline{X}_T - \overline{x}_1)^2) + (\overline{X}_T - \overline{x}_2)^2 + \cdots + (\overline{X}_T - \overline{x}_k)^2] / (k - 1)]$, where k is the number of groups, and $\overline{X}_T$ and are set and group means, respectively. The statistic used for analysis is the *F*-value: $SD_{bg}{}^2 / SD_{wg}{}^2$.

Need more than the basics? See **More Than the Basics:** *One-Way ANOVA Computation.*

The **one-way ANOVA** determines if one or more distributions in a set is different from the others. It does not specify which distribution(s) is (are) different.

make this convenient assumption.) Thus if there are three groups of equal size, their individual variances can be averaged:

$$\left(\left\{\left[\sum(\overline{X}_1 - x_{1_i})^2\right]/(n-1)\right\} + \left\{\left[\sum(\overline{X}_2 - x_{2_i})^2\right]/(n-1)\right\}\right.$$
$$\left. + \left\{\left[\sum(\overline{X}_3 - x_{3_i})^2\right]/(n-1)\right\}\right)/3$$

This is the mean within-group variance, SD_{wg}^2.

To get the variance of the group relative to the mean of the entire set, we first must either combine all the data to get this combined mean (which we will call $\overline{X}_T$), or simply take the **weighted average** (the average adjusted for the different sample-set sizes of the groups) of the several group means (which gives the same result). Next we must compute the *between-group* variance; this is done by taking the squared deviation of each group mean from the set mean—$(\overline{X}_T - \overline{X}_1)^2$ for Group 1, for instance—and summing them, and finally dividing by one fewer than the number of groups. So, the between-group variance for our three groups of equal size (and which thus require no weighting) is

$$SD_{bg}^2 = [(\overline{X}_T - \bar{x}_1)^2 + (\overline{X}_T - \bar{x}_2)^2 + (\overline{X}_T - \bar{x}_3)^2]/(3-1)$$

The number subjected to the *F*-test via ANOVA is SD_{bg}^2/SD_{wg}^2. If the samples (groups) are drawn from the same distribution, this value will be close to one; if not, it will be larger. The *F*-test tells you if it is significantly different from one; the null hypothesis of no difference is that $SD_{bg}^2/SD_{wg}^2 = 1$. (The ANOVA test we have been describing is found on card III. D-2; the *F*-test value computed by *BioStats Basics Online* for your data is compared to tables on the card.)

An *F*-value of one even when the data in the groups are drawn from the same parent distribution is not plausible except when there are few groups and lots of data in each group; in this case, the sample means are likely to be similar, and thus the between-group variance quite small. The "2-20" curve in Figure 7-6 illustrates this case; this is the distribution of *F*-values for two groups of 20 values each drawn from the same

---

A **weighted average** combines the means of several groups such that the mean of each group contributes to the overall mean ($\overline{X}_T$) in proportion to the sample size in each group, where N is the total number of data values in the set (i.e., the sum of the sample sizes from all groups):

$$\overline{X}_T = \left(\overline{X}_1 \times \frac{n_1}{N}\right)$$
$$+ \left(\overline{X}_2 \times \frac{n_2}{N}\right)$$
$$+ \cdots + \left(\overline{X}_n \times \frac{n_n}{N}\right)$$

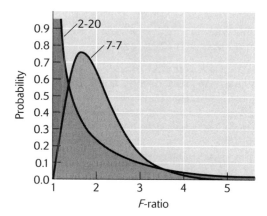

7-6 The null F-ratio distribution depends on the number of groups and the sample size in each group. The "2-20" line represents two groups of 20 data values each; an F-ratio near one is fairly likely. The "7-7" line represents seven groups of seven data values each; for these an F-ratio of one (meaning all seven means, each based on only seven data points, are virtually the same) is extremely unlikely, but a value of about 1.5 (meaning between-group variance is about half as large as within-group variance) is quite common.

parent distribution. A value near one is likely, but F-ratios above 2.0 are evident even by chance. When there are more groups, the chance variation between group means is likely to be much greater, particularly if the sample size in the group is small. The "7-7" curve represents the chance F-ratios for seven groups with seven samples in each.

7.3 But Which One Is Different?

Thus far we have been discussing a one-way ANOVA that looks for a difference in means among a set of parametric samples. We will look at other ways to analyze multiple samples in a moment. For now, we are faced with the problem of what to do if the ANOVA tells us that there is a significant difference in the means of a multisample set, as it does with our 24-condition-

Box 7.1

BioStats Basics Online   One-way ANOVA test

The one-way ANOVA test is found on the second ANOVA card (III. D-2). It allows you to enter or import up to seven data sets. If you click on the (Graph) button, *BioStats Basics Online* will both compute the *F*-statistic for the ratio of between-group to within-group variances and plot the sample distributions; it also displays the number of sample sets (*k*) and the total sample size (*N*). If you click on (Compute), the time-consuming (though often informative) graphing step will be omitted.

The card automatically displays a scrolling field with the appropriate number of degrees of freedom for the sample sets (one fewer than the total number of sets). You will need to scroll down to the appropriate degree-of-freedom line in the $k - 1$ column you are using; this line will be, as usual, one fewer than the value of *N*. If the *F*-statistic computed by *BioStats Basics Online* is larger than the *F*-value listed in this conjunction of column $k - 1$ and row $N - 1$, then at least one of the means is significantly different from the others. If you opted for graphing, your eye can probably tell you which one or ones are different. If you need objective evidence for which mean or means are different, proceed to the next card, on the Tukey-Kramer method.

The **Tukey-Kramer method** usually identifies which distribution is significantly different from the others in the set.

set greenhouse data. For though the ANOVA indicates that there *is* a difference, it does not tell us *which* sample is (or which samples are) significantly different from the others.

No wholly satisfactory test has been developed for answering this question. (Indeed, for many purposes, the mere existence of a significant difference is all that the researcher needs to establish.) The most conservative test—that is, the one least subject to being fooled by special cases—is the *Tukey-Kramer method* (which is available from the first ANOVA test card in *BioStats Basics Online*; it is on card III. D-3.). On the other hand, this test can sometimes fail to isolate a significantly different mean even when the ANOVA demonstrates that there is one; it may even ignore two very different sample means when one is based on a small sample size. The math behind this test involves computing a minimum significant difference of means in terms of standard errors for the data sets, corrected for the mul-

Box 7.2

BioStats Basics Online **Figuring out which sample mean is different**

If the one-way ANOVA test indicates that one or more means are different from the others, and it is important to know which one or ones are responsible for the difference, then you must use the Tukey-Kramer test. This test is found on the third ANOVA card (III. D-3). To use it, begin by entering or importing your data. Select a P-value threshold of 0.05 or 0.01 using the buttons at the center right. When you click on (**Compute**), *BioStats Basics Online* will calculate the minimum significant differences in means for each pair of samples and list them in the upper half of the table on the card and compute the actual observed differences in the lower half. It will display the

significant observed differences (for the significance threshold you chose at the outset) in boldface. You may want to repeat the process using the other threshold. If one mean is significantly different from the others, you will see boldfaced differences for all (or nearly all) entries involving a comparison with that mean. If two means are significantly different *and* in the same sense (i.e., both high), you may see two sets of boldfaced differences but without significance for the comparison between the two means that differ from the other samples. If there are two significantly different means with opposite senses, then their comparison should also be significant.

tiple comparisons; it is not pretty, but it does the job in most cases. In the example we have been tracing here, it demonstrates that all sample sets differ from the one case involving plants in unplowed, inoculated soils.

Need more than the basics? See **More Than the Basics:** *The Tukey-Kramer Method.*

7.4 Two-Way ANOVAs

If you think back about our 24-treatment plant data, you might wonder whether it was the unplowed soil or the inoculation with symbiotic root fungi that was the critical factor. Perhaps instead it was the fortuitous conjunction of two slightly favorable factors (fungi and unplowed soil) that come together only in this one sample. Or perhaps it was a synergistic effect: neither of the variables (soil condition nor the presence of root fungi) on its own enhances growth at all, but together they create

7 Multiple Parametric Distributions: ANOVA

The **two-way**
ANOVA compares
the effects of two
variables on parametric
distributions to see if
one or more is
associated with
distributions
significantly different
from the set as a
whole.

mutually ideal circumstances. If there were systematic effects of
soil condition or inoculation, you ought to be able to combine
the data in some way to bring this out. If the effect arose from
a synergism or the combination of several subtle sources of en-
hancement, lumping the data in other ways might show something. But how do you go about combining the data?

The *two-way ANOVA* approach can compare any number
of "groupings" against any number of "experimental conditions";
these terms simply to two sets of variables and provide convenient ways to keep track of which parameter values are varied in
the rows versus the columns that hold the two-way data. (The
card in *BioStats Basics Online* [III. D-5] limits you to three groupings and three conditions simply for reasons of space.) Now the
plant data have three sets of variables: within-location site, soil
condition, and inoculation. The two-way ANOVA can test only
two of these variables against one another—say, the two inoculation conditions versus the three soil-condition conditions—so
we will ignore the other one; the within-location site isn't really
a consistent variable at all, since there is nothing that Site 1 at
the unplowed location has in common with Site 1 at the roadside or plowed locations. (These site data were gathered to judge
the role of small-scale variability in soil fertility, a factor we will
ignore.) Figure 7-7 rearranges the histogram of Figure 7-1 to
match the organization of the two-way ANOVA comparison.

The two-way ANOVA assumes that the data sets are parametric and have similar standard deviations; again, if there is
any doubt, the *F*-test is the critical filter for checking that the
SDs are similar. Moreover, the version of the test in *BioStats Basics Online* requires that the sample sizes in each set be the
same. (As we will discuss, there are ways of dealing with other
contingencies, though this equal-sample-size test is the most robust.) When, as is the case with the plant data, some samples
are slightly smaller (only 8 rather than 10 in the most extreme
case), the usual procedure is to discard data from the larger set
until all sets are equal. Of course, this has to be done in some
fair manner; you cannot simply discard the inconvenient outliers. Some researchers use a random-number table to decide
which data to discard; others prefer to drop the data in the middle of the sample range so as not to alter the mean, though this

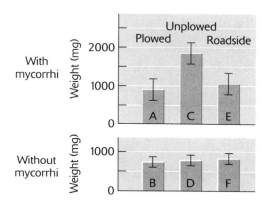

7-7 Histogram of the means of the six groups in Figure 7-1 rearranged to group treatments in the manner of two-way ANOVA. This version of ANOVA asks whether there is a difference between columns I and II, or a difference between rows 1, 2, and 3, or an interaction between the variables. To the eye, there is a consistent increase in weight associated with root fungi.

may reduce the accuracy of the estimate of variance. If there are major differences in sample size, you will need to use an ANOVA designed for this contingency.

The two-way ANOVA tests **three hypotheses**:

1. the "grouping" variable (column) has no effect;
2. the "treatment" variable (row) has no effect;
3. the "treatment" and "grouping" variables do not interact.

If an interaction is detected by the two-way ANOVA test, the results of the "grouping" and "treatment" tests must be ignored since the test cannot sort out the effects of single variables in the presence of interactions; if there is no significant interaction between the variables, then the results of the tests for the effects of the separate variables can be taken at face value.

When we apply this approach to our plant data, we find that there is a significant interaction of soil condition and fungal treatment that enhances growth. It also *seems* to suggest that

The **three hypotheses** tested by two-way ANOVA are that neither of the two variables considered separately has any effect on the distributions, and that the two variables acting together do not interact.

7 Multiple Parametric Distributions: ANOVA

Box 7.3

BioStats Basics Online   Two-way ANOVA test

The two-way ANOVA tests sample sets differing in two variables. One variable is arbitrarily called the "group" and the other the "treatment"; this is strictly a linguistic distinction of convenience. The two-way ANOVA test is found on the fifth ANOVA card (III. D-5). The group labels are A, B, and C, and correspond to the columns of the table on the test card; the treatment labels are 1, 2, and 3, and correspond to the rows. You may enter or import the data into the fields in the three-by-three table on the card. Remember that all sample sizes must be the same, or be made the same. When you click on ⟨**Compute**⟩, *BioStats Basics Online* will compute an *F*-statistic and df (degrees of freedom) value for three cases:

1. there is or is not a significant difference between the samples whose variable is labeled "treatment";

2. there is or is not a significant difference between the samples whose variable is labeled "groups";

3. there is or is not a significant difference arising from an interaction between a treatment and a grouping.

First, compare the *F*-statistic—"F (inter)"—for the third case against the scrolling field of critical *F*-values. You select the appropriate column by consulting the df value for the third comparison, "df(inter)"; you find the appropriate row by using the large "df(set)" at the bottom of the displayed list of calculated values. If your *F*-statistic is larger than the corresponding value in the table, there is a significant interaction ($P < 0.05$) between the two variables. In this case, the other values displayed have no formal relevance.

If the "F (inter)" value is not significant, then compare the "F (A,B,C)" *F*-statistic with the critical values in the table, using the "df(A,B,C)" and "df(set)" numbers to locate the appropriate entry. If the *F*-statistic is larger than the corresponding value in the table, then the "group" variable is having a significant effect on the distributions.

Next compare the "F (1,2,3)" *F*-statistic with the critical values in the table, using the "df(1,2,3)" and "df(set)" numbers to locate the appropriate entry. If the *F*-statistic is larger than the corresponding value in the table, then the "treatment" variable is having a significant effect on the distributions. Note that it is possible for both variables to have a significant effect simultaneously without the "F (inter)" statistic being significant if the variables do not interact.

there is a just-significant effect of having the root fungi present, and a not-quite-significant effect of unplowed soil condition; this would seem to imply that the fungi are more important. But remember that if you find an interaction effect in a two-way ANOVA, you cannot conclude *anything* about the roles of the individual variables. Thus you can't really say anything about the relative importance of fungi versus soil condition. But we *are* allowed to wonder, and we would probably want to repeat the experiment focusing on just this question. Indeed, this is one of the most powerful uses of ANOVAs: you can look at many different variables at once, ferret out which variables are likely to be important, and then design highly targeted tests (perhaps with smaller sample sizes) that use the powerful two-sample *t*-test to document putative cause-and-effect relationships. ("Putative," you will remember, because statistical tests can only indicate that there is an apparent effect; they cannot prove that you have correctly identified the actual cause.)

Need more than the basics? See **More Than the Basics:** *Two-Way ANOVA Computation.*

We are *not* saying that ANOVA is primarily a preliminary test, though many researchers use it in that way: ANOVAs are quite good at isolating subtle additive and synergistic interactions that *t*-tests would miss unless the investigator were looking specifically for them, and even then the *t*-test would require a separate experiment for each possible interaction.

Need more than the basics? See **More Than the Basics:** *Additivity Revisited.*

7.5 Hoary Extensions of the Two-Way ANOVA

We have dealt with the simplest and by far the most common kind of two-way ANOVA: several sample sets of equal sample size. If you were to design an experiment, this is the way you would probably set out to do it. If a datum gets lost (e.g., an animal dies), you toss out a datum from each of the other sets. However, it can happen that you do not have full data sets at all, but only their means; or you may have data sets with very different sample sizes, so that reducing the sample size in each set to that of the smallest set will require discarding most of your data. Let's worry both about how you could find yourself in such a fix, and then about what to do in the way of damage limitation.

7 Multiple Parametric Distributions: ANOVA

The unequal-sample-size contingency is the easiest to imagine actually happening to you. We once set out to compare the reaction times on a word-recognition test between right- versus left-handed males and females. We vastly underestimated the difficulty of recruiting left-handed individuals into the test, and thus wound up with very different sample sizes for our two-way ANOVA. Another time a brooder (a heated nursery with food and water for delicate baby birds, which serves as a substitute for incubating parent birds) containing all the chicks assigned to one part of an experiment failed to come back on after a power failure, and we lost a substantial part of our sample. Of course, the most common way to wind up with unequal sample sizes is not to run an experiment at all, but rather to gather data from published sources—the U.S. Census Bureau, for instance—where there may be no consistency in sample size.

Happily, there is a whole branch of ANOVA lore devoted to dealing with this kind of problem. We have not included the many special tests in *BioStats Basics Online* simply because no particular test is used by researchers with high frequency, but you will find them in most comprehensive statistical packages; you may find that the term "replications" is substituted for what we have called "sample size" in many tests. (We omit the math as well, presuming you will take that on faith.) The contingencies usually covered include cases in which:

The **statistical power** of ANOVA tests declines from the ideal state (all distributions have the same sample sizes), to cases in which the sample-size differences are slight (so that they can be made equal by dropping a few values), to cases in which sample sizes are consistently different in one variable only, through cases in which they are consistently different in both variables, to the worst situation in which the sample sizes are inconsistent.

1. there are consistent proportional differences in the sample sizes in groups *or* treatments, or

2. there are consistent proportional differences in the sample sizes of both groups *and* treatments, or

3. there is no consistent proportionality at all, the sample sizes being wildly different.

Need more than the basics? See **More Than the Basics: ANOVA III.**

The corresponding tests are less powerful as you go from condition (1) to condition (3), and the more usual two-way ANOVA we have already described—the one with equal sample sizes—is the most powerful of all.

If you are using someone else's two-variable data, and they come to you only as means of parametric distributions, you will be at a loss to plug them into a two-way ANOVA test: with only the means, you will not have the variances, and thus no obvious way to compute the all-important ratio of variances. In most cases, your pessimism is justified: there is no way to look for an interaction of variables, and if there is one you are on thin ice trying to make anything of the individual variables. But if an interaction is unlikely, a "two-factor ANOVA without replication" method is available in many statistical packages that uses the means to estimate the underlying variation. Fortunately, we have never needed to use this last-ditch approach to two-variable data.

Another, more interesting variation on the ANOVA approach involves paired data. You may have noted that there is no paired one-way ANOVA, and yet the paired *t*-test is far more powerful than its unpaired counterpart. But in some sense, two-way ANOVAs are vaguely analogous to paired tests, since each set (though not the individual samples themselves) is paired with another set with respect to a differing variable. Now in designing an experiment, you might be able to use the same individuals for testing under two or more sets of conditions. It stands to reason that such data ought to be more meaningful than the same number of samples collected from different individuals. Is there some way to exploit pairing in ANOVAs? If the order of testing of an individual has no effect (and you will usually be able to randomize test order for the variables involved), then there is a computationally intensive ***multivariate analysis of variance*** (or ***MANOVA***) is available in some of the more inclusive statistical programs that can make the fullest use of your paired design. *BioStats Basics Online* omits this test simply because it takes so long to run.

The **MANOVA** test applies the ANOVA logic to paired data.

In any sort of ANOVA situation, there is sometimes the temptation to pool data from different treatments or groups when there appears to be no difference between the groups. Now this will often seem reasonable when you are close to but not quite at significance and (if it is a two-way or higher-level ANOVA) you are prepared to use an ANOVA variant that deals with proportional differences in sample size. But all such pooling is post hoc, based on having already seen the data and *P*-values; as such, it is unwise, though few would go so far as to call pooling of apparently insignificant data actually dishonest. Our rule is to simply say no to post hoc pooling *unless* you are creating a sample distribution to be tested against a new sample that is yet to be collected. More learned practitioners of probability theory, however, can sometimes justify pooling, using arguments that are beyond the scope of this text. If it's really important, we suggest consulting a certified statistician.

7.6 Other Beyond-the-Scope ANOVAs

The most obvious higher-level ANOVA is the three-way ANOVA, in which the effects of three variables are compared. Although researchers once regularly performed one-way and two-way ANOVAs by hand (and thus needed the long and tedious documentation of the mathematical details that padded out older, and many current, statistics texts), the three-way ANOVA exceeds the limit of unaided pencil and paper, at least for most mortals. It is probably safe to assert that no one undertakes a three-way or higher-level ANOVA today without a computer to do the grunt work. For this reason, we can simply summarize here what you will find in a powerful statistical software suite if you ever need it. The same preconditions apply: the data sets are (fairly) parametric, and the variances are similar. The same titration of power also applies: paired data are better than unpaired; equal sample sizes are better than proportional ones in one or more variables; equal sample sizes in one or more variables and proportional ones in as few as possible of the other are better than proportional sample sizes in all variables; and nonproportional, unequal sample sizes are worse yet. You can also still do something with just the means, since there are higher-level ANOVAs that will estimate the variance from the distribution of sample means and wring what little statistical blood can be finally extracted from that particular data-based stone.

The other general style of ANOVA is quite different. These involve a nested (or hierarchical) analysis of variance. The two-way and higher-level ANOVAs we've already described are "crossed" analyses: you can construct a table with one variable on each axis (say, three soil conditions and two fungal inoculation states), generate an array with the corresponding number of "boxes" or "cells" for the sample sets—six in the plant example. But imagine cases in which some combination of variables is not possible. For example, in analyzing plant growth with respect to temperature and precipitation, you would be unable to find (or even in some cases to create) about half of the potential combinations (Figure 7-8). A similar situation arose when we attempted to test the responsiveness of female guppies to

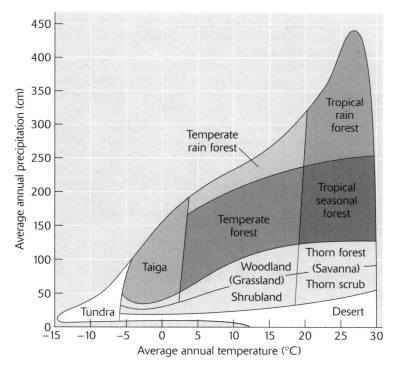

7-8 The combinations of average temperatures and precipitation found on our planet and the kind of vegetation characteristic of various combinations. Note that the upper left half of the graph (plus a small region just above the left half of the x axis) is blank: these combinations of precipitation and temperature do not occur over land. In large part this reflects the physical reality that the maximum moisture-holding potential of air declines with temperature. [Based on R. Whittaker, *Communities and Ecosystems* (New York: Macmillan, 1970).]

males with different tail sizes, colorfulness, and display rates: the range of display rates found in males was correlated with tail size, so that there were no low-rate males with large tails and no high-rate displayers with small tails. (In this case, which occurred in the far distant past when we were still ignorant of the ***nested ANOVA***, we created the missing categories by taking advantage of the cold-blooded physiology of fish: we warmed some small-tailed males to encourage higher display rates, and chilled some large-tailed males to generate lower rates.)

In ANOVAs with missing combinations of variables, the important variable can be analyzed by **nesting** the other variables and treating them as a source of random variation.

We assume you will take it on faith that the appropriate mathematical emendations have been developed to deal with cases in which there are missing combinations of variables, and thus a "nesting" approach is necessary to make the required comparison among variances. Most often, one of the variables is the main focus of interest; the other combinations of variables are nested within each adumbration of the primary variable and treated as a source of random variation. The same approach can be used when there are two important variables in experiments in which those variables are crossed, and the other variables are nested and, again, treated as "noise." Until you encounter this kind of challenge in your work, about all you need to know is that the appropriate tests are to be found in any high-end statistical package.

7.7 What to Do with Multiple Pairwise Comparisons

We have focused in this chapter on the common problem of having multiple comparisons within a data set. But what if you have performed 10 conventional pairwise tests, each from a separate independent data set, and found a pattern of significance? There are really two questions we need to consider. The first is: what is the chance of false positives with a given P-value threshold as the number of comparisons increases from 2? The second is: when is a corrected P-value appropriate, and how do we calculate it? Happily, both questions have fairly simple answers—ones which apply equally well to parametric, nonparametric, and categorical data.

Our intuition tells us that the odds of a false positive rise with an increasing number of tests, but not in the same way that the number of intraset comparisons mushrooms with added data categories in ANOVA. In ANOVA, the progression in the number of comparisons is 1 (a single pairwise comparison of two sets of data), 3 (three possible pairwise comparisons of three sets of data), 6 (six possible pairwise comparisons of four sets of data), 10, 15, 21, 28, 36, 45 (45 possible pairwise comparisons of 10 sets of data). Thus in ANOVA the number of comparisons increases exponentially, whereas for conventional independent tests—*multiple pairwise comparisons*—all that increases is the number of pairs being compared (Figure 7-9). Thus the equivalent

Experiments	ANOVA	Multiple pairwise comparisons (MPC)	Total comparisons ANOVA	MPC
3	Set 1 / Set 2 ↔ Set 3	Group 1a↔Group 1b Group 2a↔Group 2b Group 3a↔Group 3b	3	3
4	Set 1 ↔ Set 2 / Set 3 ↔ Set 4	Group 1a↔Group 1b Group 2a↔Group 2b Group 3a↔Group 3b Group 4a↔Group 4b	6	4
5	Set 1 / Set 2 ↔ Set 5 / Set 3 ↔ Set 4	Group 1a↔Group 1b Group 2a↔Group 2b Group 3a↔Group 3b Group 4a↔Group 4b Group 5a↔Group 5b	10	5

7-9 The number of data comparisons increases exponentially with the number of data sets in ANOVA tests; this is because in ANOVA situations each data set must be compared with each other set. For multiple pairwise comparisons, in which each group can be compared only to the other group in the pair, the number of comparisons increases linearly.

7 Multiple Parametric Distributions: ANOVA

progression is 1 (one pairwise comparison of two sets—one pair—of data), 2 (two pairwise comparisons, one for each of two pairs—four sets—of data), 3 (three pairwise comparisons, one for each of three pairs—six sets—of data), 4 (four pairwise comparisons, one for each of four pairs—eight sets—of data; see Figure 7-9), 5, 6, 7, 8, 9, 10 (10 pairwise comparisons, one for each of 10 pairs—20 sets—of data).

So back to our two questions: (1) what is the chance of false positives as the number of comparisons increases from 2; and (2) how do we compute a corrected *P*-value? At the outset of this chapter we listed (see Table 7-1) the chance of *at least one* false positive when multiple comparisons within a data set are being made. These same values apply to multiple independent comparisons. More relevant, useful, and intuitively satisfying, however, is the *net* probability of false positives—a value that corresponds to the average *number* of false positives in a set of comparisons. For $P < 0.05$ and other common thresholds, they are listed in Table 7-2. Clearly, you can generate this table by multiplying the *P*-value threshold by the number of pairwise data sets.

To adjust *P*-value thresholds to take into account the increased risk of false positives, you simply divide the desired overall threshold by the number of independent pairwise tests

Table 7-2 Probability of false positives

Samples	Number of paired comparisons	Level of significance 0.10	0.05	0.01	0.005	0.001
2	1	0.10	0.05	0.01	0.005	0.001
4	2	0.20	0.10	0.02	0.010	0.002
6	3	0.30	0.15	0.03	0.015	0.003
8	4	0.40	0.20	0.04	0.020	0.004
10	5	0.50	0.25	0.05	0.025	0.005
20	10	1.00	0.50	0.10	0.050	0.010
50	25	2.50	1.25	0.25	0.125	0.025

Table 7-3   *P*-value threshold in individual tests needed to give a desired overall level of significance

Number of paired comparisons	Desired level of significance				
	0.10	0.05	0.01	0.005	0.001
1	0.100	0.050	0.0100	0.0050	0.00100
2	0.050	0.025	0.0050	0.0025	0.00050
3	0.033	0.017	0.0033	0.0017	0.00033
4	0.025	0.013	0.0025	0.0013	0.00025
5	0.020	0.010	0.0020	0.0010	0.00020
10	0.010	0.005	0.0010	0.0005	0.00010
25	0.004	0.001	0.0004	0.0001	0.00004

7 Multiple Parametric Distributions: ANOVA

(Table 7-3). This operation is sometimes called the ***Bonini correction***. Another way to compute this is to decide on an overall *P*-value threshold, *P* (usually 0.05), call the number of tests, *c*, and solve for the individual-test threshold, α, using the following formula: $P = 1 - (1 - \alpha)^c$. Again, it should be clear how Table 7-3 was generated: the number of independent pairwise comparisons is divided into the desired overall *P*-value to produce the necessary per-test thresholds.

The **Bonini** correction for multiple pairwise comparisons requires dividing the *P*-value for a single set comparison by the number of pairs.

Now before you go off blindly dividing *P*-values, you should stop and ask yourself whether the *overall* chance of a false positive or the ***average number of false positives*** is what really matters. For instance, if you are looking for possible factors that might promote bladder cancer and test 25 different possible agents (e.g., coffee, lettuce, alcohol, raisins), then this correction is essential. Indeed, given that other people are looking at the same problem, and are unlikely to publish their results if they find nothing, you probably want to be doubly skeptical. On the other hand, if you are asking about a general pattern, this issue may be almost irrelevant.

When making multiple pairwise comparisons, it may be more important to consider the **average number of false positives** rather than the chance of a false positive.

For instance, we once asked whether mosquitofish females (live-bearing guppy-like fish in which the males control mating) had unexpressed preferences for males. The issue here is not whether a particular potential male feature is or is not attractive to females, but rather whether females have evolved preferences even though they are never allowed to express them. (This probably seems like a strange question, but tests on another species—short-finned mollies—had suggested that this bizarre state of affairs was possible.) When we ran 19 pairwise comparisons (asking females which of two model males, if either, they preferred), we found 15 instances in which females displayed a preference at $P < 0.05$. Here the relevant information is the pattern: females have preferences. We expect about one false positive (0.95 to be precise)—or false negative—in this data set. Since a false positive or two would not alter the pattern, dividing the P-values is both unnecessary and deeply misleading.

A note on generality here: as we said earlier, this analysis of false positives applies equally well to nonparametric comparisons. (Indeed, the mosquitofish example was drawn from a nonparametric data set.) Second, it should be obvious that you are always best off doing the minimum number of experiments. You may be tempted to do dozens of tests and report only the ones that appear to "work." This means your readers won't worry about corrected P-values. This well-worn tactic is both self-defeating and probably dishonest. It is one of the reasons that scientists generally perform a preliminary test before focusing on a few key comparisons that look promising. Your understanding of the logic of statistics should prevent you from being tempted into the well-populated pitfalls of mistaken "significance."

Points to remember

✔ Two-distribution comparisons such as the t-tests cannot be used directly to compare the means of more than two distributions with one another because the P-values for these tests presume only a single comparison.

✔ For comparisons of three or four distributions with one another, you can divide the t-test P-values by the number of comparisons (3 in the case of three distributions, 6 in the case of four distributions) to produce a simple but overly conservative correction.

✔ The one-way analysis of variance (ANOVA) has more statistical power than simple corrections; the difference is enormous with increasing numbers of distributions.

✔ ANOVA is less sensitive to variations from strictly parametric distribution of the data than are the *t*-tests.

✔ ANOVA compares the average variances of the distributions summed separately with (essentially) the variance of the data considered as a set. If all distributions are the same, the two values are about the same, and comparing the set variance to the average group variance produces an *F*-value near one; if not, one or more distributions has a significantly different mean.

✔ ANOVA deals with differing sample sizes by using weighted averages: each distribution mean contributes to the computation of the group mean in proportion to its sample size.

✔ ANOVA does not specify *which* distribution is apparently different. The Tukey-Kramer method is the most conservative way to identify the distribution that yielded a significant *F*-ratio.

✔ The two-way ANOVA test compares the effects of two different variables

in a suitably designed experiment. It can detect apparent effects of either variable independently or of the two variables interacting.

✔ ANOVA tests work best if all sample sizes are the same (or, by dropping data, can be made the same). They are less powerful when there are consistent differences in sample sizes associated with one variable. They are less powerful still when there are consistent differences associated with both variables. They are least powerful when the sample sizes are widely and inconsistently different.

✔ A variety of special-case ANOVA tests exist, including ones that handle paired data (MANOVA), three- or higher-level ANOVA comparisons, and instances in which some combinations of variables do not exist (nested ANOVA).

✔ The *P*-values for multiple *pairwise* comparisons (as opposed to the multiple *independent* comparisons that make the ANOVA approach necessary) are easily corrected by dividing the *P*-value by the number of pairs.

✔ When making multiple pairwise comparisons, it is often more important to determine the average number of false positives rather than the chance of at least one false positive.

Exercises

For each data set, check the data for normality. If the data are not normal, find the best transform to make them more nearly normal (the same transform must be applied to each data set). Select the two most different sets and compare them using the *F*-test. If the data pass both tests, analyze them with the one-way ANOVA, the Tukey-Kramer method (if there is a significant difference in variances as judged by the ANOVA test), and (if appropriate) the two-way ANOVA.

1. Table A lists the average number of sound bursts produced by dancing forager honey bees in each cycle of their dance to a food source 250 m from the hive. (The data are also found in file 7—Dance Dialects or 7-DIALECT.dat.) The number of sound bursts per cycle correlates with the distance to the food. Different subspecies of honey bees are said to have different distance "dialects." Is there any evidence in this set of data from three different European subspecies for such dialects?

Table A   Dance rates

Apis mellifera ligustic	A. m. carnica	A. m. caucasica
16.5	14.5	15.2
15.0	16.2	16.8
16.2	14.1	17.9
16.4	14.0	16.0
16.9	14.3	16.3
17.9	13.7	14.9
14.2	14.5	15.6
17.2	14.1	15.9
16.0	13.1	16.7
15.7	14.9	16.2

2. Table B lists 1990 starting salaries (in thousands of dollars per annum) for assistant professors in the natural sciences and in humanities at Freeman University. (The data are also found in file 7—Starting Salaries or 7-SALARY.dat.) The question at issue is whether there is any male-versus-female or science-versus-humanities difference in initial pay. In running the two-way ANOVA, justify your mechanism for equalizing sample sizes.

Table B Starting salaries

Natural sciences		Humanities	
Males	Females	Males	Females
38.5	39.5	34.5	32.0
35.0	34.0	31.0	33.5
41.5	36.5	32.0	34.0
37.0	39.0	32.0	35.0
36.6	42.0	33.0	33.0
39.5	41.5	34.0	34.5
37.5	38.0	33.5	36.5
38.0	37.5	35.5	34.0
36.5	38.5	33.5	35.5
36.0	37.0	34.0	34.0
37.0		32.5	
39.0		33.0	
36.5		34.0	
		33.0	
		34.5	

3. Most introductory biology courses serve a variety of constituencies, including premeds taking the course as a medical school prerequisite, other biology majors, other science majors, and students taking the course only to satisfy a university distribution requirement for a laboratory science. Keeping in mind that many highly qualified students place out of the course by scoring a 5 on the Advanced Placement exam, we wondered if there is any difference in grades earned by the various subgroups. For good measure, we recorded males and females separately. Partial data are shown in Table C (complete data are in file 7—Intro Biology Grades or 7-GRADES.dat); you may assume they are parametric since the course is graded on a curve. For this comparison we have omitted the premeds and included students only from the Monday and Tuesday labs. A = 4.0, B = 3.0, C = 2.0, D = 1.0, F = 0.0; a 3.3 is a B+ and a 3.7 is an A− on this scale.

Table C   Partial list of GPAs

Biology majors		Science majors		Distribution students	
Male	Female	Male	Female	Male	Female
3.0	2.7	3.3	4.0	2.3	3.7
3.7	4.0	3.7	3.0	2.7	3.3
2.7	4.3	2.3	2.7	4.0	2.7
3.3	1.7	1.3	3.3	3.0	3.7
2.3	2.3	2.7	3.3	3.3	3.0

More Than the Basics

The *F*-distribution

The *F*-distribution—the distribution described by the F table in Selected Statistical Tables—is common and useful in statistics. It describes the ratio of the variances of samples from a parametric parent distribution (as we saw in the simple *F*-test in Chapter 5), and is the basis for all analyses of variance.

Because variance is additive—the sum of two independent distributions has variance equal to the sum of their variances—ANOVA is able to subtract from the total variance the variance within the variances of the groups themselves ($V_{groups} = SS_{groups}/df_{groups}$) to get that part of the variance due to the differences between the groups ($V_{error} = V_{total} - V_{groups} = SS_{total} - SS_{groups}/df_{total} - df_{groups}$). The SS values are not subtracted directly due to the difference in degrees of freedom). Because the null hypothesis is that there is no variance between the groups, this variance suggests that the null hypothesis is in error. This deviation from the null hypothesis can be tested with the F-distribution. The ratio of the group and error variances will distribute according to the F-distribution.

This leads to the possibility of more complex ANOVA-based tests. In the two-way ANOVA, total variance will be made up by a much larger—but still additive—sum of variances.

One-way ANOVA computation

To perform the basic ANOVA, first calculate

$$C = \frac{\left(\sum_i \sum_j X_{i,j}\right)^2}{n_{total}}$$

Using this value (which is of no practical importance), calculate the "sum of squares" and "degrees of freedom":

$$SS_{total} = \sum_i \sum_j X_{i,j}^2 - C; \; df_{total} = n_{total} - 1$$

$$SS_{groups} = \sum_i \frac{\left(\sum_j X_{i,j}\right)^2}{n_i} - C;$$

$$df_{groups} = k - 1$$

$$SS_{error} = SS_{total} - SS_{groups};$$

$$df_{error} = df_{total} - df_{groups}$$

From these values, calculate the ANOVA statistic:

$$F = \frac{SS_{groups}}{df_{groups}} \frac{df_{error}}{SS_{error}}$$

where $X_{i,j}$ is the jth datum in the ith group,

n_i is the number of data in the ith group, and

k is the number of groups.

The degrees of freedom are

$$df_{numerator} = k - 1; \; df_{denominator} = n - k.$$

The Tukey-Kramer method

To perform the two-factor ANOVA with equal replication, use the formula

$$C = \frac{\left(\sum_i \sum_j \sum_l X_{i,j,l} \right)^2}{n_{total}}$$

$$SS_{total} = \sum_i \sum_j \sum_l X_{i,j,l}^2 - C;$$

$$df_{total} = n_{total} - 1$$

$$SS_{cells} = \sum_i \sum_j \frac{\left(\sum_l X_{i,j,l} \right)^2}{n_{i,j}} - C;$$

$$df_{cells} = ab - 1$$

$$SS_{error} = SS_{total} - SS_{cells};$$

$$df_{error} = df_{total} - df_{cells}$$

$$SS_{rows} = \sum_i \frac{\left(\sum_j \sum_l X_{i,j,l} \right)^2}{\sum_j n_{i,j}} - C;$$

$$df_{rows} = a - 1$$

$$SS_{cols} = \sum_j \frac{\left(\sum_i \sum_l X_{i,j,l} \right)^2}{\sum_i n_{i,j}} - C;$$

$$df_{rows} = b - 1$$

$$SS_{interaction} = SS_{cells} - SS_{rows} - SS_{cols};$$

$$df_{interaction} = df_{cells} - df_{rows} - df_{cols}$$

where n is the number of data in each cell of the matrix of treatments,

a is the number of "rows" (treatments manipulating variable A),

b is the number of "columns" (treatments manipulating variable B), and

$X_{i,j,l}$ is the lth datum from the ith row and jth column.

To test for difference by row:

$$F = \frac{SS_{rows}}{df_{rows}} \frac{df_{error}}{SS_{error}}$$

With degrees of freedom: $df_{numerator} = a - 1$; $df_{denominator} = ab(n - 1)$.
To test for difference by column:

$$F = \frac{SS_{rows}}{df_{rows}} \frac{df_{interaction}}{SS_{interaction}}$$

With degrees of freedom: $df_{numerator} = b - 1$; $df_{denominator} = ab(n - 1)$.
To test for interaction:

$$F = \frac{SS_{interaction}}{df_{interaction}} \frac{df_{error}}{SS_{error}}$$

With degrees of freedom: $df_{numerator} = (a - 1)(b - 1)$; $df_{denominator} = ab(n - 1)$.
    If the rows and columns were decided not by the experimenter but by chance, see More Than the Basics: ANOVA III for possible corrections.

Two-way ANOVA computation

To perform the two-factor ANOVA with proportional replication, use the formula

$$C = \frac{\left(\sum_i \sum_j \sum_l X_{i,j,l}\right)^2}{n_{total}}$$

$$SS_{total} = \sum_i \sum_j \sum_l X_{i,j,l}^2 - C;$$

$$df_{total} = n_{total} - 1$$

$$SS_{cells} = \sum_i \sum_j \frac{\left(\sum_l X_{i,j,l}\right)^2}{n_{i,j}} - C;$$

$$df_{cells} = ab - 1$$

$$SS_{error} = SS_{total} - SS_{cells};$$

$$df_{error} = df_{total} - df_{cells}$$

$$SS_{rows} = \sum_i \frac{\left(\sum_j \sum_l X_{i,j,l}\right)^2}{\sum_j n_{i,j}} - C;$$

$$df_{rows} = a - 1$$

$$SS_{cols} = \sum_j \frac{\left(\sum_i \sum_l X_{i,j,l}\right)^2}{\sum_i n_{i,j}} - C;$$

$$df_{rows} = b - 1$$

$$SS_{interaction} = SS_{cells} - SS_{rows} - SS_{cols};$$

$$df_{interaction} = df_{cells} - df_{rows} - df_{cols}$$

where n_{tot} is the total number of data,
$n_{i,j}$ is the number of data in the cell at the ith row and jth column,
a is the number of rows,
b is the number of columns, and
$X_{i,j,l}$ is the lth datum from the ith row and jth column.

To test for difference by row:

$$F = \frac{SS_{rows}}{df_{rows}} \frac{df_{error}}{SS_{error}}$$

With degrees of freedom: $df_{numerator} = a - 1$; $df_{denominator} = ab(n - 1)$.
To test for difference by column:

$$F = \frac{SS_{cols}}{df_{cols}} \frac{df_{error}}{SS_{error}}$$

With degrees of freedom: $df_{numerator} = b - 1$; $df_{denominator} = ab(n - 1)$.
To test for interaction:

$$F = \frac{SS_{interaction}}{df_{interaction}} \frac{df_{error}}{SS_{error}}$$

With degrees of freedom: $df_{numerator} = (a - 1)(b - 1)$; $df_{denominator} = ab(n - 1)$.
If the rows and columns were decided not by the experimenter but by chance, see More Than the Basics: ANOVA III for possible corrections.

Additivity revisited

The two-way ANOVA raises two intriguing questions: what does "interaction" mean, and why does interaction preclude testing for effects of the individual factors? The answers are even more interesting than is originally apparent.

Let us take the simplest case of the two-way ANOVA, two factors which each have two possible states, present and absent. This gives a two-by-two array of cells, one for each combination of factors. It would be reasonable to expect that each of these factors could have some effect on the mean of the measured variable, and that the two together would have a combined effect equal to the sum of their individual effects.

The two-way ANOVA, therefore, would provide three tests: one for each factor individually, and one for the two together minus the individual effects. This last—the interaction test—is really a test for additivity.

But wait! This is exciting news. This means that if we can isolate factors contributing to a variable, we can then test those factors for additivity. At long last, we can be specific about additivity and isolate nonadditive factors, a great theoretical victory which occasionally even has a practical use in the design of experiments to gather parametric data.

But the news is not unmixed good. Because additivity of factors is an assumption of every parametric test, and because if we have found significant interaction, we have proven the presence of a nonadditive factor, we must disregard all other parametric results on the data from experiments in which the nonadditive factor was not controlled. And this is why significant interaction prevents further parametric analysis of the whole data set. A researcher must then put the nonadditive effects under control and gather more data.

ANOVA III

The two-way ANOVA presented in More Than the Basics: The Tukey-Kramer Method is what is called a Model I ANOVA. It makes the assumption that the criterion for differentiating each row and each column was fixed by the experimenter; this is called the *fixed-effects*

model, or *Model I ANOVA*. Fixed-effects factors include gender, treatment, and other characteristics which by their nature cannot be chosen among randomly.

A random-effects factor is one in which the choice of categories is random. Generally, this happens when individuals are chosen at random so as to test for the presence of differences between individuals. For instance, if you were polling individuals at random to test the null hypothesis that people do not disagree on some question (the reality of evolution by natural selection, for instance), "person" would be a random-effects factor. The alternative to the fixed-effects model is the *random-effects model*. A two-way ANOVA with two random-effects factors is called a *Model II ANOVA*, and with one fixed- and one random-effects factor is called a *Model III ANOVA*.

For a random-effects factor, the *F*-statistic is computed differently.

To test for difference by column:

$$F = \frac{SS_{cols}}{df_{cols}} \frac{df_{interaction}}{SS_{interaction}}$$

With degrees of freedom: $df_{numerator} = b - 1$; $df_{denominator} = (a - 1)(b - 1)$.
To test for difference by row:

$$F = \frac{SS_{rows}}{df_{rows}} \frac{df_{interaction}}{SS_{interaction}}$$

With degrees of freedom: $DF_{numerator} = a - 1$; $DF_{denominator} = (a - 1)(b - 1)$.

Tests Covered in This Chapter

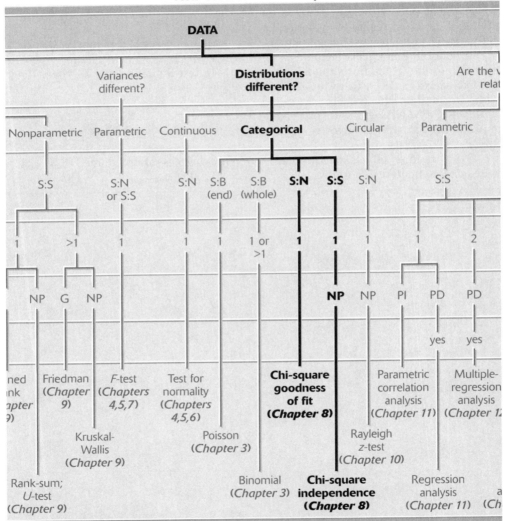

Categorical Data

8

The last four chapters have been devoted to dealing with continuously distributed parametric data. And we're not finished: we'll return to this convenient class of distributions when we show you how to ferret out correlations and putative cause-and-effect relationships between variables. For now, though, we want to continue describing how to compare distributions, considering cases that are either not continuous or not parametric.

8.1 Where Do Categorical Data Come From?

There are two ways for data to avoid being "normal":

1. the distribution can be continuous but so determinedly nonparametric that no mathematical transformation can rescue it, or

2. the data can be noncontinuous, or, as we have been calling them, *categorical.*

This chapter is devoted to analyzing categorical data; the next takes on terminally nonparametric distributions.

Categorical data fall into distinct, mutually exclusive categories that lack both quantitative and qualitative intermediates: response versus no response or male versus female, for instance, or mammals versus birds versus reptiles versus amphibians versus fish. Such data cannot usually even be said to have conventional distributions since there is often no logical way to plot

Categorical data fall into mutually exclusive categories that lack intermediates.

them except as arbitrarily ordered bar graphs of mean values; nevertheless we will call the ratio of data between each of the two or more categories a distribution. Clearly, this kind of data defies parametric analysis, and yet it is exceedingly common.

The breakthrough in dealing with categorical data came around 1900 when Karl Pearson developed the first chi-square comparison. Like Gosset and Pascal, he was driven to develop a test by a practical problem. For Pearson the pressing need was to analyze genetic data, for which traits are frequently categorical—wrinkled versus round peas, for example. Pearson's original insight led to other applications of the chi-square method, most notably (and almost inevitably) by R. A. Fisher. We will first describe the original chi-square test, which assumes you know the parent or null distribution. Then we will examine a powerful extension that will allow you to compare samples without knowing what the expected distributions ought to be.

8.2 Comparing a Sample to a Null Distribution

The **chi-square test for goodness of fit** compares categorical data to a known null or parent distribution.

The method for comparing a sample distribution of categorical data to a parent or null distribution is the ***chi-square test for goodness of fit***. Like the binomial test, Gauss test, and various *t*-tests, the goodness-of-fit test attempts to distinguish chance from causation by computing the probability that an observed (sample) distribution, or one more extreme, could be drawn from the parent (or null) distribution by chance; the null hypothesis is that the sample and parent distributions are the same. Perhaps the best way to picture the process is to imagine a box containing vast numbers of round and wrinkled peas, with the round ones outnumbering the wrinkled ones 3 to 1. The $3:1$ ratio is the null distribution predicted by the laws of inheritance discovered in the 1860s by Gregor Mendel. Mendel actually collected and categorized the pea type from each of 7324 plants and reported that 5474 produced round peas and 1850 had wrinkled peas. The goodness-of-fit test is designed to tell us the odds of drawing a $5474:1850$ ratio $(2.96:1)$, or one more different from the predicted $3:1$, at random from our box of mixed peas. (By "more different," we mean finding a ratio smaller than $2.96:1$ or larger than its mirror image, $3.04:1$.)

This one is easy: our intuition tells us that the chances of drawing a result outside the narrow range of 2.96–3.04 : 1 are very high; thus, the P-value will be high (it's actually >0.95), and there is no significant difference between what Mendel observed and what the null hypothesis predicted. (In fact, it seems unlikely that Mendel could have obtained such a good estimate of the 3 : 1 ratio in this and his other experiments just by chance; many modern researchers suspect he must have discarded some data that looked "wrong.") But if Mendel had found, say, 3500 round seeds and 3724 wrinkled seeds, you can readily guess that the odds of drawing that result from a null-hypothesis box filled with a 3 : 1 ratio of seed types would be vanishingly small (in fact, $P < 0.001$). Let's look at the actual mechanics of the test.

The goodness-of-fit test begins with a two-column table; one column is for the observed values and the other holds the expected values. Each row contains the values for one category of data. In the example here, we will use Mendel's results from crossing plants bearing round yellow seeds with plants producing wrinkled green seeds. His laws of inheritance predict that the second generation in this kind of cross will yield a 9 : 3 : 3 : 1 ratio of seed categories, the categories being round yellow, wrinkled yellow, round green, and wrinkled green, respectively. He collected peas from 556 plants in this experiment. Here is the table he would have created if Karl Pearson had invented the goodness-of-fit test 40 years earlier:

	Observed	Expected
round yellow (RY)	315	312.75
wrinkled yellow (WY)	101	104.25
round green (RG)	108	104.25
wrinkled green (WG)	32	34.75
Total:	556	556.00

A couple of points are probably immediately obvious to you. First, all the observed values are whole numbers, as they have to be because this is categorical data: all pea plants have one

The formula for computing chi-square for goodness of fit (χ^2) is

$$\chi^2 = \sum \frac{(O_i - E_i)^2}{E_i}$$

where O_i is the observed data count in the ith category, and E_i is the expected data count in the same category.

of four categories of seeds, and there are no partial plants. Second, the numbers in the expected column are not whole numbers; instead, they reflect the results of applying the $9:3:3:1$ ratio prediction to the sample size of 556. Third, the observed and expected columns sum to *exactly* the same value; if you make an error in your math, or round your expected values so that they do not come out exactly to the same whole number as the sum of the observed, computer-based versions of the test will usually balk.

The mechanics of the chi-square test are remarkably simple. You take the difference between each observed and expected value (d), square it (d^2), scale the difference by dividing by the expected value (d^2/exp), and then sum these scaled squared differences from each observed-expected comparison:

	Observed	Expected	Observed − Expected (d)	d^2	d^2/exp
RY	315	312.75	2.25	5.06	0.016
WY	101	104.25	−3.25	10.56	0.101
RG	108	104.25	3.75	14.06	0.134
WG	32	34.75	−2.75	7.56	0.216
			chi-square sum:		0.467

Need more than the basics? See **More Than the Basics:** *Chi-Square Goodness-of-Fit Computation.*

The next step is to compare this chi-square statistic to a table of critical values, using the typical degrees-of-freedom value (one fewer than the number of categories—three in this case since there were four categories of seed type). This chi-square statistic is far too small to be significant; even for a $P < 0.1$ the statistic would have had to have been larger than 6.25. Thus there is no significant difference between what Mendel observed and what is predicted by the laws of inheritance.

Here's another example with the opposite outcome. In this case researchers were asking whether the odor young salmon experienced in their home stream influenced their choice of streams to return to years later when they had grown to sexual maturity. They exposed substantial numbers of laboratory-hatched and -reared fry to one or the other of two obscure odors, morpholine and PEA. The fry were marked to reflect which chemical they had been "imprinted" on and then were released into Lake Michigan. As is usually the case with salmon, most perished before they were old enough to reproduce. When the time for reproduction came, the researchers monitored three very similar streams feeding into the lake. One was baited with morpholine, one with PEA, and one was unbaited. The data are categorical: individual fish entered one of three streams (categories), and there were no fractional fish.

A total of 953 salmon were recovered; 272 had been exposed to PEA, 681 to morpholine. The null prediction is that the imprinting will have had no effect, and the fish will distribute themselves randomly among the three streams. Thus we can set up two chi-square tables:

Stream bait	Morpholine-exposed fish		PEA-exposed fish	
	Observed	Expected	Observed	Expected
Morpholine	659	227	20	124
PEA	8	227	343	124
None	14	227	9	124
chi-square statistic:	1233.28		580.66	
critical value for $P < 0.005$:	10.60		10.60	

Clearly, these results are strongly at odds with the null hypothesis, and indicate that exposure to odors when young almost certainly affect stream-choice at maturity.

If you find that a chi-square statistic for a distribution with three or more categories is significantly different from the null distribution, it is perfectly valid to repeat the test with one category omitted (a strategy known as *column dropping*); this allows you to quantitatively isolate the category or categories that are responsible for the discordance between observed and expected. The largest set of categories that yield a chi-square statistic that is still consistent with the null hypothesis indicates that the omitted category or categories are the source of the

Box 8.1

BioStats Basics Online   **Chi-square goodness-of-fit test**

The chi-square test for goodness of fit is found on the second Chi-Square card in *BioStats Basics Online* (III. E-2). First enter or import your observed categorical data into the "observed" column; next compute the *precise* expected values for your total sample size from the ratios predicted by the null hypothesis. Enter these values into the corresponding fields in the "expected" column; be sure the total of your expected values *exactly* equals the sample size. Then click on the **Compute** button. *BioStats Basics Online* will compute the chi-square statistic and also display the number of degrees of freedom. Use this df value to choose the appropriate row of threshold values from the table on the card. Read across from left to right until you find a value greater than your calculated chi-square statistic. The *P*-value for your observation lies between the *P*-value at the top of that column and the one to its left.

The table of critical chi-square values is for a two-tailed test. If you can solidly justify an a priori one-tailed prediction, and the results support that expectation, then you can halve the listed *P*-values.

Both chi-square tests are limited to null (expected) data sets with no zero values and with no more than 20% of the expected values less than 5.

discrepancy. Thus in our example with a $9:3:3:1$ predicted ratio, a significant departure from the null hypothesis would lead us to try the $9:3:3$ subset of categories, or a $9:3:1$ subset, or the $3:3:1$ group.

The chi-square goodness-of-fit test has certain numerical limitations, which are discussed below in connection with the chi-square test for independence. Problems arise if any of the expected values equal zero (which will be obvious if you look at the math), or if more than 20% of the expected values are less than 5.

8.3 How Is the Goodness-of-Fit Test Different from the Binomial Test?

When we discussed the binomial distribution in Chapter 3, we pointed out that the distributions were derived from sets of two-alternative data states with known individual probabilities that summed to 1.0: heads versus tails, for example, where the probabilities are 0.5 and 0.5. Obviously, these are categorical data, and the individual probabilities are the bases for deriving null distributions. But this is just what we have been doing with the chi-square goodness-of-fit test. What's the difference?

The critical difference is that chi-square can deal with any number of categories, whereas the binomial can handle only two. But when there are only two categories—wrinkled versus round peas, for example—is it possible to use the binomial test? The answer is yes: the generalized binomial test—the one found on the Binomial Distribution card (II. E-1)in *BioStats Basics Online*, for example—can deal with two-category data for which the null probabilities are known. In fact, the binomial test has two advantages:

1. you don't have to compute the expected values, and
2. it's slightly more accurate, because the test assumes a known distribution of outcomes (the binomial curve) whereas chi square cannot.

This difference in accuracy is surprisingly small, however: for a $3:1$ null expectation from Mendel's laws and a sample size of 100, chi-square yields a $P < 0.01$ for a deviation from the expected $(75:25)$ of $87:13$, while the binomial deduces the same level of significance for a deviation of $86:14$. For larger sample sizes, the binomial test is markedly slower than chi-square, and though the difference between the power of the two in detecting deviations from the expected increases, it does so only modestly. (You do, however, get an exact P-value for your patience rather than a statistic that then must be referred to a table.) The more the expected ratios differ from $50:50$, the more likely it is that the binomial test will detect a significant difference (though, again, the degree of difference between the two tests is strikingly unstriking). In short, chi-square has a remarkably high ratio of statistical power to difficulty of use, which doubtless accounts for its continuing popularity even when, as you will see, it is clearly the wrong test to use.

8.4 Applications of Chi-Square When Category or Binomial Probabilities Are Estimated

If the null distribution is unknown, you will need to use the chi-square test for independence, which is described later in the chapter. In other cases, however, you may have some estimate of the null distribution. For instance, in Chapter 3 (page 65) we mentioned the possibility of having binomial data (say, number of males versus females) for which the proportion of individuals in each category could only be estimated from data. We pointed out that if these proportions are in the range of 0.1–0.9 and the sample size is large (>1000), you could go ahead and use the binomial test. Otherwise, the uncertainty in your estimate of these values is too great for the P-values generated by binomial tests to have much meaning. Similarly, you might have categorical data for which there is no theoretical expectation, but for which there is substantial data that might allow you to estimate the "expected." Clearly, your need to estimate this distribution is going to cut into the power of the test in some way, but the reduction in power is going to be small if your sample size is large. Exact treatment of this situation is quite complex, but excellent results can be obtained with a fairly simple modification of chi-square for goodness of fit.

If you are dealing with a binomial distribution, you must (for reasons that become obvious) subdivide your data to create a distribution of at least three outcome categories; then you use your estimates of the binomial probabilities to generate the corresponding expected categories. In doing this you must keep in mind that no expected values can equal zero, and no more than 20% of the expected values can be less than 5. This often dictates the outcome groupings necessary to generate your categories. The same rules apply to generating the expected distribution from conventional categorical data: a minimum of three categories, no expected value of zero, and fewer than 20% with an expected value less than 5. Let's assume you are looking at two-person pairings at lunch tables; the binomial states are male and female; the outcome categories are MM (i.e., two males sitting together), MF, and FF. We have an estimate of the male/female ratio (m/f, where m is the approximate proportion of males, and f is the estimated proportion of females) in the local population, and want to test the null hypothesis that the pairings are random with respect to sex. We can generate an approximate null distribution: MM $= m \times m \times T$; MF $= 2 \times m \times f \times T$; and FF $= f \times f \times T$, where T is the total number of pairs observed.

Next, perform the chi-square as usual. When you consult the table, however, use a degrees-of-freedom value one fewer than is automatically supplied. (Here is why you needed at least three categories: with two, the uncorrected df value is 1; after subtracting one, the df is zero, for which there is—and can be—no entry on the table.) The logic is similar to that in other degrees-of-freedom computations: the value is a measure of the number of independent values in play, which is usually one fewer than the sample size or category number because once you have that much of the data, you can infer the rest (which is thus not independent); in this case, you lose a degree of freedom because you have estimated the categorical probabilities from the data. If your sample size is large, the consequence of this reduction in degrees of freedom will be small for any particular trend.

Some statisticians prefer using the so-called G-test when dealing with estimated-category probabilities. There is considerable debate on this point, and no one is likely to be too hard on you if you stick with chi-square. Nevertheless the G-test (the log-likelihood ratio) is becoming more popular (especially in the binomial case). Here is how to compute it. You have several categorical groupings, each with an expected proportion of the total, estimated from data. The expected proportion for category i is f_i; the sum of all the f-values is 1.0. The observed proportion for category i is f_i, and these necessarily sum to 1.0. The G-statistic is

$$G = 2 \sum f_i \ln (f_i/f_i)$$

Compare this value to the table for the chi-square analysis using the corrected value for degrees of freedom.

The G-test is also preferred by some statisticians for all cases in which the absolute value of the difference between f_i and f_i is larger than f_i. We are not of that number.

8

Categorical Data

8.5 Chi-Square and the Quick-but-Dirty Approach

We pointed out in Chapter 2 that the lines between data types are more flexible than it is useful to admit to beginning students, for whom clear-cut distinctions are a conceptual blessing. Thus when we pointed out that researchers regularly lump continuous data into bins to facilitate plotting of graphs, we ran the (small) risk of implying that there is no essential and important difference between continuous and categorical data. What we can say with confidence is:

1. Some data are unambiguously categorical (e.g., heads versus tails) and some data are unambiguously continuous (height, for instance).

2. Continuous data are frequently collected or treated in categories (all the males taller than 1.70 m but no taller than 1.75 m, for example). When the categories are numerous and continuous, the data are analyzed as continuous data, with little loss of precision.

3. Discrete data (like number of children in a family, which is always a whole number—that is, it can never have an intermediate value such as 3.32—but nevertheless is ordered from low values to high ones) can be treated as categorical, or (if there are enough discrete categories) nonparametrically continuous, or (if there are enough discrete categories *and* the shape of the distribution is normal) parametrically continuous, as you will see.

4. Categorical data can almost never be treated as continuous or even discrete.

More surprising is the revelation that researchers often subdivide continuous data into a small number of wide categories and treat the data as categorical. Why would anyone want to do this? The answer is that categorizing continuous data allows you to make a quick chi-square check without having to type in detailed data or deal with means and SDs. For example, say you are wondering if male parents of biology majors are taller than their female spouses. Rather than entering the heights of all 70 parents of the students in a particular class, you scan down the list and count the number of males that are taller than their wives and the number of females that are taller than their husbands. You would notice that in 34 of 35 cases the fathers are taller than the mothers. Your null hypothesis is that there is no sex-specific difference; thus you can create the following chi-square table:

	Observed	Expected
Father taller	34	17.5
Mother taller	1	17.5
chi-square statistic:	31.11	
chi-square threshold for $P < 0.005$:	7.88	

Clearly, with this sample size you can see what's going on without resorting to the optimum but more tedious test (the paired t-test analysis).

In fact, given the lesser power of the chi-square test compared to continuous-parametric tests, even a quick-check result of $P = 0.10$ strongly suggests that you would *not* be wasting your time to perform the full-blown t-test. Note that we are not suggesting that you ever *substitute* chi-square for a continuous-data test when you have continuous data—that would make you look foolish in the eyes of your statistically well informed colleagues. But for a quick peek at the trends in your data, chi-square is a useful tool to have at hand.

8.6 Comparing Two or More Sample Distributions

The chi-square test for goodness of fit is the categorical version of the one-sample t-test: a sample distribution is compared to a null or parent distribution. But what about comparing two samples of categorical data (the analogue of the various two-sample t-tests), or multiple samples (as we did for continuously parametric data with the ANOVA)? Fortunately, there are equivalent methods for categorical data, and they even have the advantage that there is no need to distinguish between two-sample and multiple-sample cases. One method—the ***chi-square test for independence***—is fairly simple both to understand and to use, though it has some limitations with respect to sample sizes. The other, Fisher's Exact Test, is exceedingly tedious but always works. We will begin with the simpler of the two.

The **chi-square test for independence** compares two categorical sample distributions.

Just as the two-sample *t*-tests estimate SDs and means of the distributions from the data themselves, so too the chi-square test for independence must make the equivalent guesses. Consider this classic experiment by Niko Tinbergen, a pioneer in the study of animal behavior. Tinbergen wondered why gulls removed eggshells from their nests after the chicks hatched. After observing that the shell-removal behavior was missing in a species of gull that lives on sheer cliffs, where the risk of predation is negligible, he guessed that the shells in ordinary nests are removed because the white interiors of the eggs compromise the camouflage of the nest.

He tested this idea by creating 450 artificial nests, each with a set of chicken eggs painted a splotchy olive drab to look like gull eggs. In a third of the cases he set out a piece of broken eggshell, its white interior showing, 15 cm away; for another third the shell was placed 100 cm away; for another third it was 200 cm away, about the distance gulls typically carry off the remains of hatched eggs. He then determined which nests were discovered by predators by observing whether the painted eggs were taken or overlooked. His null hypothesis was that the distance of the shell from the nest would have no effect on predation frequency. Here are his results:

	Taken	Not taken
Shell 15 cm away	63	87
Shell 100 cm away	48	102
Shell 200 cm away	32	118

Analyzing these data with the goodness-of-fit test is impossible: there is no null distribution. This did not bother Tinbergen because he always trusted his data to the reader's judgment; alas, no journal today would publish such results without statistical analysis.

The test Tinbergen should have used is the chi-square test for independence, which works by creating a null distribution based on the sample data themselves. This is done by first summing each row (shell distance in this case) and then summing each column (nest fate); the data matrix can have any number of rows and columns. From these sums you (or your computer) will generate a set of "expected" values. For reasons that will be explained presently, if any of the numbers in your sets of observed data are small, you may need to look at the expected values before performing the test. Let's look at how you get those numbers.

Here are Tinbergen's data with the first steps of the test for independence performed:

The formula for the chi-square test for independence is

$$E_{i,j} = \frac{\left(\sum_B O_{i,B}\right)\left(\sum_A O_{A,i}\right)}{\sum_A \sum_B O_{A,B}}$$

$$\chi^2 = \sum_i \sum_j \frac{(O_{i,j} - E_{i,j})^2}{E_{i,j}}$$

where $O_{i,j}$ is the observed data count in the ith row, jth column.

Categories

Data sets:	Taken	Not taken	Totals
15 cm	63	87	→ 150
100 cm	48	102	→ 150
200 cm	32	118	→ 150
Total	143	307	→ 450

The null prediction is that if there is no effect of the treatment, the observed value should equal the column total times the row total divided by the sample size. The predicted distribution is shown in parentheses (see the table on page 202). If the rows had different sums, the ratio would have been applied separately

	Categories	
	Taken	**Not taken**
Data sets:		
15 cm observed	63	87
(expected)	(47.7)	(102.3)
100 cm observed	48	102
(expected)	(47.7)	(102.3)
200 cm observed	32	118
(expected)	(47.7)	(102.3)

to each row. The problem here is that the test for independence does not work well if

1. there are only two columns, and any *expected* value is lower than 5.0, or

2. there are more than two columns and any *expected* value is zero *or* more than 20% of the *expected* entries have values below 5.0.

If either of these caveats applies, you must do something else. Either the Fisher Exact Test (discussed presently) must be used, or (if the data are discrete and there are several possible values) a nonparametric test (discussed in Chapter 9) must be substituted. In this particular example, it was obvious from the outset there could be no entries with small numbers; in other instances you will need to check more carefully.

*Need more than the basics? See **More Than the Basics:** Chi-Square Independence Computation.*

Assuming your data pass the "5.0" barrier, the test then proceeds as a conventional chi-square using the synthetic expected values for each value in each row. The difference between the observed and expected is computed, squared, and divided (scaled) by its estimated expected value; all these scaled values are summed to generate a chi-square statistic. The degrees-of-freedom value is the product of one fewer than the number of rows times one fewer than the number of columns. In the case of Tinbergen's data, the chi-square statistic is 14.78; the

P-value threshold for 0.005 is 10.60. This strongly suggests that the distance between the shell and nest had a major impact on the survival of the eggs in the nest; no wonder gulls are programmed to recognize and remove broken eggs.

Though we have not gone into this point, the chi-square tables are based on a curve fitted to a histogram; at small sample sizes the histogram is increasingly erratic, and the curve is to some degree generously fitted to minimize the number of cases in which the *P*-values for various combinations of sample sizes and categories and treatments are underestimated. In these cases false positives are a risk. Look at Figure 8-1; this curve is a probability distribution just like the Gaussian curve, and is

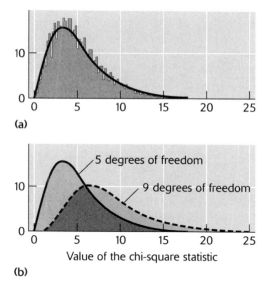

(a)

(b)

8-1 **(a)** The chi-square curve is only an approximation of the chi-square probability histogram. The curve lies above many histogram values, but below others. The chi-square curve is particularly inaccurate when some of the expected values are small. [Redrawn from D. Freedman et al., *Statistics*, 2nd ed. (New York: W. W. Norton, 1991), p. 480.] **(b)** The shape of the approximation depends on the number of degrees of freedom in the data.

Box 8.2

BioStats Basics Online   Chi-square independence test

The chi-square test for independence is found on the fourth Chi-Square card (III. E-4). It allows you to enter or import up to seven categories (columns) from up to 33 treatment groups (data sets). Unlike the chi-square test for goodness of fit, there is no need for the column values to have the same sum; but like the goodness-of-fit test, all observed values must be whole numbers. If any of your values are small, you will need to check to be sure the expected values synthesized by the test do not violate the rules for low values listed in the text.

When you click on the (Compute) button, *BioStats Basics Online* displays the number of categories and data sets (treatments), the number of degrees of freedom (df) this combination produces, and the chi-square statistic. To evaluate the chi-square value, click on the (Table) button, which will take you to the next card. There you will find the df value and chi-square statistic displayed, and a table of critical chi-square thresholds. As with the goodness-of-fit test, use the df value to choose the appropriate row of threshold values from the table on the card. Read across from left to right until you find a value greater than your calculated chi-square statistic. The *P*-value for your observation lies between the *P*-value at the top of that column and the one to its left.

The table of critical chi-square values is for a two-tailed test. If you can solidly justify an a priori one-tailed prediction, and the results support that expectation, then you can halve the listed *P*-values.

used in the same way: the area under the curve is integrated from the chi-square statistic to the end of the tail to obtain the appropriate *P*-value. But there are certain small-sample-size cases in which the value in the table is off by as much as a factor of 2 either high *or* low; these are the cases that are largely excluded by the no-less-than-5.0 rule above, and its "20%" equivalent for larger df values.

If you find yourself faced with categorical data that violate the rules allowing safe use of the test for independence, your first option is to see if you can reasonably combine some categories or treatments to get rid of the zero-value and less-than-5.0 limits. If so, you can proceed with the regular test. But if this is not possible, there is still hope. R. A. Fisher (who else?)

found the chi-square approximation annoying and unaesthetic, and so developed an excruciatingly precise test—Fisher's Exact Test—for categorical data. The exact test takes a brute-force approach to the problem, creating the same kind of contingency table we generated above for the gull-nest data, computing the probability of your observed result, and then, one by one, *computing every other possible result* more extreme relative to the synthetic set of "expected" values that you obtain by summing each row and column. For anything but the smallest total sample sizes, this test is extremely time consuming, even for a computer; a typical personal computer might need to crunch away all night in some cases. We have omitted Fisher's Exact Test from *BioStats Basics Online*, but you will find it in the more exhaustive statistical packages. (And remember, if the data are discrete, you may be able to use one of the nonparametric tests discussed in Chapter 9.)

Need more than the basics? See **More Than the Basics:** *Fisher's Exact Test.*

Also omitted from *BioStats Basics Online* are ways to analyze multivariate data, the categorical equivalent of the higher-level ANOVAs. For example, you could imagine adding six more test groups to Tinbergen's experiment to enable you to evaluate the effect of rotten egg odor added to the shells versus its absence on whether nests are taken. This would produce a three-dimensional array of results, the three axes being nest state, shell distance, and shell odor. Most workers analyze such data one variable at a time, but the higher-level categorical-contingency-table approaches can detect interactions that may not otherwise be apparent. These tests are so difficult to implement, however, that they are omitted from nearly all statistical packages.

Points to remember

✔ Categorical data fall into distinct, mutually exclusive categories that lack both quantitative and qualitative intermediates.

✔ When there is a null or parent (expected) distribution to compare the sample (observed) data to, the chi-square test for goodness of fit will test the hypothesis that there is no difference between the observed and expected distributions.

✔ When the null distribution is not precisely known, chi-square can still be used by reducing the number of degrees of freedom by one.

✔ When two or more sample distributions of categorical data are to be compared, the chi-square test for independence will test the null hypothesis that there is no difference between the distributions; this test produces an internally generated set of "expected" values.

✔ Chi-square tests do not work if any expected value is zero, or more than 20% of the expected values are less than 5.

✔ When there are three or more conditions or data sets, chi-square tells you only if there is a significant difference, not which condition of set is the source of the difference. You can use a column-dropping approach to establish where the likely source of the difference lies.

✔ Chi-square tests are so simple and relatively powerful that they are often used to make preliminary tests on continuous data by dividing the data into ranges and treating each range as a category.

✔ When a test yields a P-value just below threshold, it is worth trying Fisher's Exact Test, which avoids the overly conservative threshold estimates incorporated into chi-square tables.

Exercises

1. Table A lists the results of tests involving the lizard *Anolis carolinensis* (also called the American chameleon). This lizard is generally either brown or green, and the factors that cause an individual to change color are not well understood. The tests in Table A examined the effect of temperature and light on inducing a change in color. Thus "B >> G" represents a lizard that changed from brown to green after application of the stimulus, whereas "B >> B" indicates a lizard that did not respond. The column headed "Darkness, 23°C" represents lizards under the conditions that preceded the tests for a day; note that those that had elected to be green remained green, and those that opted for brown remained brown. In general, individual lizards that have adopted one of the two color morphs remain that color indefinitely unless some salient external cue changes. These data can be analyzed by the chi-square goodness-of-fit test. What are the expected values in each

Table A Color changes of *Anolis* lizards under various conditions; B = brown, G = green

Color before >> after	Treatment					
	Darkness	Darkness	Halogen	Halogen	Ultraviolet	Fluorescent
	23°C	37°C	23°C	37°C	23°C	23°C
G >> B	0	0	15	0	6	3
B >> B	8	1	18	12	8	10
B >> G	0	13	0	12	2	0
G >> G	4	11	0	8	3	6

case? Do any of the treatments yield significant results? What do the data suggest?

2. Early ethologists like Konrad Lorenz and Niko Tinbergen believed that chicks innately recognize the silhouettes of predators. Their tests were repeated using a variety of rearing, testing, and scoring techniques with contradictory results. In an attempt to sort out whether the various techniques were biasing the results, we repeated the tests using laboratory-hatched chicks tested indoors. Table B presents the results of varying the testing conditions (alone or in groups) on a particular response (crouching versus not crouching). Can any of these data can be tested using the chi-square goodness-of-fit test? What about the chi-square test for independence? Do any of the comparisons yield significant results? What do the data mean?

Table B Responses to silhouettes

	Tested alone			*Tested together*		
Silhouette	None	Goose	Hawk	None	Goose	Hawk
Response:						
Crouch	0	10	12	0	5	13
No crouch	28	18	16	28	23	15

Table C Student data

Student #	Gender	Cat or dog	Coke or Pepsi	Syllables in name
1	M	C	C	5
2	M	D	P	4
3	F	D	C	7
4	F	C	C	6
5	F	C	C	7
6	F	D	P	6
7	M	D	C	5
8	F	C	C	6
9	M	D	C	6
10	M	D	C	5
11	F	D	P	7
12	M	D	C	5
13	M	D	C	7
14	F	C	P	8
15	M	D	P	5
16	M	D	C	7
17	F	D	C	7
18	M	D	C	5
19	F	D	C	6
20	M	D	C	5
21	M	D	C	6
22	F	D	—	7
23	F	C	P	6
24	M	D	C	7
25	M	D	C	4
26	M	D	C	5
27	F	C	C	8
28	F	D	C	7
29	M	D	P	5
30	M	D	P	6
31	F	D	P	9
32	M	D	C	5
33	F	C	P	7
34	M	D	C	5
35	F	C	P	9

3. Table C (see opposite page) presents the categorical-answer portion of a questionnaire administered to biology majors. We know that on college campuses Pepsi outsells Coke at a 60:40 ratio. How about among biology majors? Which test did you use? Why? Is there any male/female difference in cola preferences? Which test did you use?

4. Is there any gender difference in pet preferences? How about syllable lengths of given names?

5. Table D shows the results of some imprinting tests. Ducklings were hatched in an incubator and approximately 18 hours after hatching were exposed to and allowed to follow the "imprinted object" for 30 minutes. The model had a particular color, A or B, and produced a particular sound, X or Y. One quarter were exposed to AX, one quarter to AY, one quarter to BX, and one quarter to BY. The ducklings were then tested with two objects in the same arena on day three; as indicated, the choice involved models with one or more of the imprinted characteristics altered. The table indicates the number of ducklings preferring each model, where "preference" meant spending more time following one model than the other. How can you use these data to determine whether sound or color is important in imprinting, and if both are, which is more important?

8 Categorical Data

Table D Imprinting data

	Choice of		Choice of	
	Imprinted object	Model with wrong color *and* sound	Imprinted object	Model with wrong sound
Followed	43	5	37	11
	Choice of		Choice of	
	Imprinted object	Model of wrong color	Model of wrong color	Model with wrong sound
Followed	32	16	30	18

6. After surgical removal of a breast tumor, patients are routinely subjected to radiation therapy of the site in the hope of killing any local cancer cells that might remain. If tumor cells are found in the lymph nodes in the shoulder on the side of the body where the tumor was found, a course of chemotherapy often follows. To justify such stressful treatments, a doctor should be sure that some substantial benefit to the patient is likely. In fact, since radiation is itself mutagenic, some patients wonder that this treatment is used at all. Here is a study comparing matched samples of women who had either extensive postoperative radiation treatments or none. The data in Table E represent the number of individuals diagnosed with a local recurrence of their cancer within six years. What is the proper test? One-tailed or two-tailed? Is there a significant difference?

Table E   Breast cancer data

	Women treated	Women not treated
Local recurrence	55	28
No local recurrence	441	111

7. Most breast cancer patients are treated with an estrogen antagonist called Tamoxifen. The drug is thought to slow tumor regrowth but may increase the incidence of uterine cancer. One study found evidence that Tamoxifen protected against heart disease. The data from the next study on the drug are given in Table F. Is there an effect of treatment on heart-related deaths? Does Tamoxifen decrease overall

Table F   Tamoxifen data

	Number of patients	Heart-related deaths	All deaths
Tamoxifen-treated patients	1435	19	280
Untreated patients	1450	22	338

deaths in breast cancer patients? Which test is appropriate? Two-tailed or one-tailed?

8. One of the classic tests of natural selection involved the release of two different forms of the peppered moth. One is mostly white, whereas the other "melanic" form is dark. During the course of the industrial revolution, the previously rare melanic forms came to dominate in industrial areas. Scientists assumed that the dark form was less visible to predators on the soot-darkened tree trunks. To test this idea, about 800 moths were marked and released in a wood near Birmingham, England; many of the moths were recaptured with traps. As a control, about 950 moths were released in an unpolluted forest in Dorset where the light-colored morph should be less conspicuous than the dark form. The results are shown in Table G. What test is appropriate here? One-tailed or two-tailed? Are the results significant?

Table G Mark-and-recapture studies of moths

	Birmingham results		Dorset results	
	Light-colored	Dark-colored	Light-colored	Dark-colored
Released	201	601	496	473
Recaptured	34	205	62	30

More Than the Basics

Chi-square goodness-of-fit computation

To perform the chi-square test for goodness of fit, use the formula

$$\chi^2 = \sum \frac{(O_i - E_i)^2}{E_i}$$

where O_i is the observed count of data in the ith category, and

E_i is the expected count of data in the ith category.

The degrees of freedom is one fewer than the number of categories.

Chi-square independence computation

To perform the chi-square test for independence, use the formula

$$E_{i,j} = \frac{\left(\sum_B O_{i,B}\right)\left(\sum_A O_{A,j}\right)}{\sum_A \sum_B O_{A,B}}$$

$$\chi^2 = \sum_i \sum_j \frac{(O_{i,j} - W_{i,j})^2}{W_{i,j}}$$

where $O_{i,j}$ is the observed count of data in the ith row, jth column.

The degrees of freedom is one fewer than the total number of categories.

Fisher's Exact Test

Data of the type analyzed by chi-square (independence) are called contingency tables—tables that show the observed occurrences of a variety of contingencies, or combinations of characteristics. It is possible to test the special case of 2×2 contingency tables for independence more precisely and flexibly with a computationally intensive test called Fisher's Exact Test.

This test is derived from the simple fact that for a contingency table such as the one below, the next most extreme table must have the same observed frequencies of each characteristic (the total in any given row or column) and different frequencies of each combination, and thus may be readily computed:

$T + 1 =$

A + 1	B − 1	A + B
C − 1	D + 1	C + D
A + C	B + D	A + B +
		C + D

$T − 1 =$

A − 1	B + 1	A + B
C + 1	D − 1	C + D
A + C	B + D	A + B +
		C + D

One of these tables, either $T + 1$ or $T − 1$, is more extreme than T_0; in this manner, all the tables more extreme may be computed (if $T − 1$ were more extreme, for instance, then the sum would be $T_0 + T_{-1} + T_{-2} + T_{-3}$, and so on, stopping as soon as the value in any cell reaches zero). The probabilities (obtained from a binomial calculation) of each table are then summed to give a (one-tailed) P-value:

$T_0 =$

A	B	A + B
C	D	C + D
A + C	B + D	A + B
		+
		C + D

$$P = \left(\frac{(A + B)!(C + D)!(A + C)!(B + D)!}{} \right)$$

$$\cdot \sum_{i=0}^{\min(A,B,C,D)} \frac{1}{(A - i)!(B - i)!(C - i)!(D - i)!}$$

Of course for any but the smallest values of A, B, C, and D, this calculation requires (and may even be well out of reach of) a computer.

Tests Covered in This Chapter

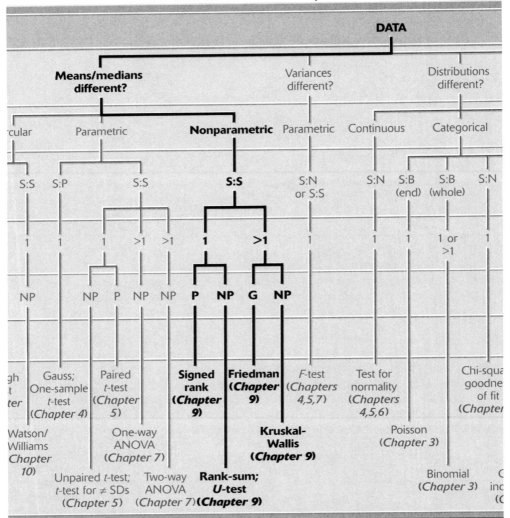

DATA

Means/medians
different?

Variances
different?

Distributions
different?

rcular   Parametric   **Nonparametric** Parametric   Continuous   Categorical

S:S   S:P   S:S   **S:S**   S:N   S:N   S:B   S:B   S:N
                                                        or S:S   (end) (whole)

1   1   1   >1   >1   **1**   **>1**   1   1   1   1 or   1
                                                                                   >1

NP   NP   P   NP   NP **P** **NP** **G** **NP**

gh   Gauss;   Paired   **Signed** **Friedman** *F*-test   Test for   Chi-squa
t   One-sample   *t*-test   **rank**   **(*Chapter*** (*Chapters* normality   goodne
ter   *t*-test   *(Chapter*   **(*Chapter*** **9)**   *4,5,7)*   *(Chapters*   of fit
        (Chapter 4) 5)**   **9)**   *4,5,6)*   *(Chapter*

Watson/   One-way   Poisson
Williams   ANOVA   **Kruskal-**   *(Chapter 3)*
Chapter   *(Chapter 7)*   **Wallis**
10)   **(*Chapter 9*)**

        Unpaired *t*-test; Two-way   **Rank-sum;**   Binomial
        t-test for ≠ SDs ANOVA   **U-test**   *(Chapter 3)*   inc
        (Chapter 5) *(Chapter 7)***(Chapter 9)**   (C

Nonparametric Continuous Data

9

The essential power of parametric tests lies in the consistent shape of the distributions: once you have subtracted the mean and divided by the SD, all parametric curves are exactly the same. (Oh, yes: if your data are in the form of actual numbers of observations rather than proportions of the total sample, you must also scale for n.) Thus any difference in mean or variance is readily detectable, and the mathematical apparatus for quantifying these differences is highly evolved and well oiled.

9.1 Why Nonparametric Tests Are Less Powerful

Consider now the problem posed by a set of nonparametric data. In this case, we have a test in which we are inducing death feigning (tonic immobility, or TI) in chicks by holding them in front of a test object (a stuffed hawk, for instance), then placing them on the ground and recording the number of seconds until the chick gets up and runs away. From working with lots of TI data, we know they are nonparametric (not the least of the reasons being the tendency of some chicks to fall asleep during the experiment); moreover, none of the common mathematical transformations described in the Chapter 6 can rescue the data (Figure 9-1). So what are we to do?

Let's look at one of the original methods for dealing with nonparametric data, the sign test. This test, developed in 1710 by the Royal Physician John Arbuthnott (in a paper entitled "An argument for Divine Providence taken from the constant regularity in the births of both sexes"), is no longer much used

9-1 Duration of immobility (shown here for the pressure-only treatment for one batch of chicks) is highly skewed and multimodal; moreover, the pattern varies from one batch of birds to the next. No standard transformation can convert this (or any other multimodal) distribution into a parametric curve.

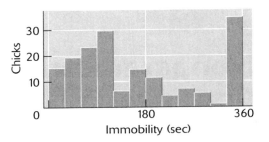

except as a quick-but-dirty check; a major improvement was developed by Frank Wilcoxon in 1945 (which we will describe presently). The value of the sign test for us is conceptual: it is a window into the logic and costs of nonparametric tests. It is designed to deal with paired data—typically the same individual under two test conditions. What it asks is which of the two conditions for each individual yields the larger effect; the null hypothesis is that there is no difference. So for this extract of TI data comparing the duration (in seconds) of immobility to two models (A is a stuffed hawk, B is the same bird with its "eyes" covered), we list the durations in the face of each model and score which was more effective:

Test chick #	Model A	Model B	Result A > B	Result A < B
1	89	22	+	
2	56	59		+
3	95	14	+	
4	35	9	+	
5	56	39	+	
6	59	60		+
7	77	11	+	
8	71	22	+	
9	48	32	+	
10	19	7	+	
$n = 10$			totals: **8**	**2**

Ties, if any, are dealt with by assigning half a point to each column.

The null hypothesis for the sign test is that there is no difference between the two results for the two treatments, and thus there should be $n/2$ cases of A > B and $n/2$ cases of A < B. The two-tailed probability for this result is $P = 0.0654$. How did we get this value? If you think about it, this is really a binomial test: there are two mutually exclusive outcomes (A > B versus A < B) whose individual probabilities (0.5) are known and sum to 1.0. Thus we just plug these numbers into any binomial test like the one in *BioStats Basics Online* and the rather disappointing *P*-value is quickly returned. What you see here is a nonparametric data set being made binomial, and then analyzed along conventional lines; other nonparametric tests often use more backhanded methods to fabricate something close to a parametric distribution.

The *P*-value is disappointing because your eyes tell you that the effect of the hawk-with-eyes model is really quite strong. If we had been able to treat these durations as parametric (by successfully transforming them), the paired *t*-test would have yielded a two-tailed value of $P < 0.01$, reflecting the very different sample means of 61.5 versus 28.2 seconds. What's missing from the sign test is any calibration for variance (the SD of parametric tests) and any weighting of degrees of difference (the 1-second "preference" of chick 6 for Model B receives just as much weight as the 81-second preference of Chick 3 for Model A).

9.2 Testing Paired Two-Sample Data

Wilcoxon's **signed-rank test** goes a long way toward weighting the relative differences in nonparametric paired data. This method (often also called the Wilcoxon test) begins by ordering the data according to the absolute difference between the two measures, then assigns a rank score to each data pair in order of difference. Thus in the case of our 10-sample TI test, the bird with the greatest duration difference (Chick 3) would get a 10 (since there are 10 data pairs and this is the most extreme) while the bird with smallest difference (Chick 6) would get a 1. The scores in each column are then summed (see the following table):

Paired nonparametric data involving a single comparison can be analyzed with the **signed-rank test**.

9

Nonparametric Continuous Data

Test chick (in order of difference)	Model A	Model B	Differ- ence	Rank score A > B	Rank score A < B
3	95	14	81	10	
1	89	22	67	9	
7	77	11	66	8	
8	71	22	49	7	
4	35	9	26	6	
5	56	39	17	5	
9	48	32	16	4	
2	56	59	−3		3
10	19	7	12	2	
6	59	60	−1		1
$n = 10$				totals: **52**	**4**

Ties are broken by assigning intermediate ranks, like 3.5 to each of two entries if the third- and fourth-highest differences were the same.

The null hypothesis for the signed-rank test is that the sum for each outcome (A > B versus A < B) will be the same, and thus there is no difference between the two distributions. For $n = 10$, the sum of $10 + 9 + 8 + \cdots + 3 + 2 + 1$ is 55, so we predict each column will sum to 55/2, or 27.5, if there is

Box 9.1

BioStats Basics Online   **The signed-rank test**

The test for evaluating paired two-sample nonparametric data is found on card III. F-2. To use it, simply enter or import your paired data and click on the (**Compute**) button. *BioStats Basics Online* will display the number of pairs, the smaller-sum test statistic, and graph the two distributions. Find the line in the table corresponding to the number of data pairs and read across from left to right until you find an entry smaller than your test statistic. The two-tailed *P*-value for your data lies between the *P*-value at the head of that column and the one heading the column to its left. If you can solidly justify a one-tailed prediction, and the data conform to this expectation, you can halve the listed *P*-value.

no difference resulting from the treatments. We then arbitrarily take the smaller sum (4 in our example) and compare it to a table to find the probability of obtaining a sum at least this different from the predicted null value (27.5) with our number of pairs (10) by chance. For the 10 pairs of death-feigning data, we find $P < 0.02$ (two-tailed). So the signed-rank test, at $P < 0.02$, clearly outperforms the sign test ($P = 0.0654$), but falls short of the parametric equivalent, the paired t-test ($P < 0.01$). (The signed-rank test is found on card III. F-2 in *BioStats Basics Online*.)

One final note about where the signed-rank P-values come from: they are derived from a conventional integrate-the-area-to-the-end-of-the-tail test using a probability histogram that reflects the chances of obtaining each value for a column sum. In this case, the probability curve looks fairly parametric (Figure 9-2). The curve is generated from the number of different ways

The formula for the signed-rank test is
$$dif_i = X_i - Y_i$$
$$S_{X>Y} = \sum_{i,dif>0} R_i$$
$$S_{X<Y} = \sum_{i,dif<0} R_i$$
$$T_S = min(S_{X>Y}, S_{X<Y})$$
where X_i is the ith result under the first condition, Y_i is the paired ith result under the second condition, and R_i is the rank of the magnitude of dif_i; ties are broken by assigning an intermediate value.

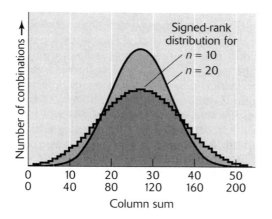

9-2 The probability histogram for the signed-rank test records the probabilities of every possible signed-rank outcome for two sample sizes ($n = 10$ and $n = 20$). The P-value is obtained by integrating from the observed smaller-sum value to the end of the distribution (doubling this value if, as is usually the case, the test is two-tailed), and then dividing by the total area under the curve to find the proportion of the area in the tail(s). For larger sample sizes, the probability distribution for the signed-rank test becomes increasingly parametric.

each sum can be obtained. In the TI example above, for instance, a sum of 1 only can result when there is only one "minority" sample and it receives a score of 1; but a smaller sum of 5 could result from one minority sample receiving a 5, or two receiving a 4 and a 1, or two receiving a 3 and a 2—five combinations in all. Thus there is nothing very mysterious about the source of the distribution used in the test, or the way this distribution is used to generate the *P*-value tables.

Need more than the basics? See **More Than the Basics: The Signed-Rank Computation.**

9.3 Evaluating Grouped Multiple-Sample Data

The signed-rank test is the nonparametric equivalent of the paired-sample *t*-test. As we showed with parametric comparisons, special problems arise when you have more than two sets of data: the number of possible sample comparisons skyrockets, and with it the chance of false positives. For parametric data we demonstrated in Chapter 7 that certain kinds of ANOVAs could deal with this worry. The (roughly) equivalent nonparametric method for grouped multisample data is the **Friedman test**. We would apply this test to our test chicks when we are comparing several different models to see what factors, if any, elicit death feigning; in the example we will look at here we compared three different eye colors (plus an eyeless control) to see if there is any "expectation" on the part of chicks about predator eyes. The null hypothesis is that all four distributions will be the same.

Multisample paired nonparametric data are analyzed by the **Friedman test**.

The formula for the Friedman test is $\chi_r^2 =$

$$\frac{12}{bk(k-1)} \sum_i \left(\sum_i R_{i,j}\right)^2 - 3b(k-1)$$

where b is the number of rows, k is the number of columns, and $R_{i,j}$ is the rank of the datum in the *i*th row, *j*th column among those in its row.

The Friedman test, like ANOVA, is less sensitive than a single two-sample test, and can be avoided when you are dealing with only three samples by making pairwise signed-rank comparisons and doubling the resulting *P*-values to account for the odds of false positives. The logic of the test resembles that of the signed-rank test: ranks are assigned to each datum, and the sum of ranks is used to compute a test statistic. But the procedural details are quite different:

1. The rankings are for each subject (chick in our case); thus for the four test situations, the longest TI duration for a given chick will receive a 4, the shortest a 1; ties are broken by giving intermediate scores; each other chick receives its own set of 1–4 scores.

2. The ranks are summed over all subjects for each treatment.

3. The null hypothesis is that the sum of the ranks will be the same or each column (treatment).

Here's a small subset of the eye-color TI data (duration of immobility in seconds) sorted for the Friedman test; note that these data do include a tie (Chick 8), which is dealt with as described for the previous nonparametric tests.

Test chick #	No eyes Time	(Rank)	Blue eyes Time	(Rank)	Yellow eyes Time	(Rank)	Green eyes Time	(Rank)
1	017	(2)	006	(1)	027	(4)	018	(3)
2	013	(1)	034	(2)	052	(4)	043	(3)
3	008	(2)	005	(1)	066	(4)	027	(3)
4	008	(2)	005	(1)	011	(3)	013	(4)
5	044	(3)	008	(1)	117	(4)	020	(2)
6	049	(3)	043	(2)	071	(4)	027	(1)
7	008	(2)	009	(3)	041	(4)	005	(1)
8	007	(1)	016	(3.5)	014	(2)	016	(3.5)
9	008	(2)	006	(1)	013	(4)	010	(3)
10	024	(2)	010	(1)	039	(4)	028	(3)
ranked totals:	20		16.5		37		26.5	
null expectation:	20		20		20		20	

A glance at the table and use of your native intuition tell you that yellow eyes are particularly effective in extending death-feigning behavior, but the other conditions are pretty similar. How does the Friedman test fare at quantifying this relationship? It computes the chance that all four columns are drawn from the same distribution at $P < 0.01$. (The Friedman test is found on card III. F-4 in *BioStats Basics Online.*)

Alas, like the ANOVA, what it doesn't do is tell you *which* treatment is creating the difference. Of course in this case it's obvious, but there is no objective test equivalent to the Tukey-Kramer method, which we used on one-way ANOVAs, for quantifying this

Box 9.2

BioStats Basics Online   The Friedman test

The test for evaluating grouped-sample nonparametric data is found on card III. F-4. To use it, simply enter or import your grouped data and click on the **Compute** button. *BioStats Basics Online* will display the number of subjects (the sample size, k), the number of test conditions (or treatments, b), the Friedman test statistic, and then graph the two distributions. The scrolling field is arranged by condition number, beginning with $b = 3$; for each condition number, there are several entries for various subject numbers (k). When the threshold value for the test statistic changes little between values of k, the entries are combined; indeed, for treatment variations greater than 4, the sample size doesn't matter given the way the test statistic is computed. Find the area in the table corresponding to the number of test conditions and the entry corresponding to your sample size. To be significant, the calculated test statistic must be larger than the entry in the table. The far-right column is for $P < 0.01$; the one to its left is for $P < 0.05$. These are two-tailed values; if you can solidly justify a one-tailed prediction, and the data conform to this expectation, you can halve the listed P-value.

difference. You must rely on the intuition of those reading your results. Sometimes the reader will be convinced by column dropping: eliminate the most unusual data set from the analysis and repeat the test (or use the paired test if dropping one set leaves you with only two); if this omission greatly alters the statistical significance, most readers will be convinced that the set you dropped was the one that was really different. (If the sorting and ranking of all these data seem tedious, rejoice that they are all now done automatically by statistical software.)

Need more than the basics? See **More Than the Basics:** *The Friedman Computation.*

Unpaired nonparametric data involving a single comparison can be analyzed by the **rank-sum test**.

9.4 Testing Unpaired Two-Sample Data

Just as the unpaired *t*-test was less sensitive than the paired *t*-test, so the nonparametric equivalent for unpaired data is less powerful than the signed-rank test. The approach for such data is the ***rank-sum test***, which was created in 1945 by Wilcoxon for the special case of equal sample sizes in the two groups being compared, and was generalized to all cases in 1947 by H. B. Mann and D. R. Whitney. As a result, the method is often called the Mann-Whitney test. Given that the later stages of the statistical analysis of larger sample sizes of unpaired nonparametric data often depart from the Mann-Whitney technique and make use instead of a powerful method (the U-test) developed in 1962 by G. S. Watson, any naming of this procedure other than the rank-sum test seems arbitrary and bound to offend someone.

The logic of the rank-sum test is easiest to understand when the sample sizes are the same. The data in the two groups are ordered and each datum is assigned a ranking number independent of group; thus if there are 10 individuals in each group, the rankings will run from 20 (the largest value) to 1 (the lowest). Here are some data on swimming speeds in 10 small- and 10 medium-sized female guppies; these are unpaired data because all 20 guppies are different individuals. The data have been ordered and ranked by the rank-sum method. (See the table on page 224.)

The null hypothesis for the rank-sum test is that the sums of the two columns should be the same if there is no effect of the difference between the two groups—body size, in this case—on swimming speed. The test statistic is the smaller sum; the P-value, as always, represents the odds of drawing data at least this different from a uniform null distribution. The result of this comparison is $P < 0.05$ by the rank-sum test—a surprisingly close call statistically considering the obvious difference in the data. But then this is the worst of all possible cases: unpaired nonparametric data. Ideally, no experiment should ever be performed unpaired or without heroic efforts to transform nonparametric data into something the t-tests can deal with. (The rank-sum test is found on card III. G-3 in *BioStats Basics Online*.)

When the sample sizes are different, the ratio of the smaller set size to the larger size—n_1/n_2, if n_1 is the smaller number—is used to "scale" the rankings of the larger group; thus the larger sample size is compensated for without discarding any of the data. You should be aware, however, that some software packages omit the rank-sum test and its multisample counterpart (discussed in the next section).

The formula for the rank-sum test is

$$T_X = \sum_X R_i; \quad T_Y = \sum_Y R_i$$

$$\begin{cases} n_X > n_Y \quad n_1 = n_Y; \\ \quad n_2 = n_X; \ T_1 = \sum R_i \\ n_X < n_Y \quad n_1 = n_X; \\ \quad n_2 = n_Y; \ T_1 = \sum_X R_i \end{cases}$$

$$T_2 = n_1(n_X + n_Y + 1) - T_1$$

$$T = \min(T_1, T_2)$$

where R_i is the ith rank, n_X is the number of data in the X column, and n_Y is the number of data in the Y column.

9 Nonparametric Continuous Data

Need more than the basics? See **More Than the Basics:** *The Rank-Sum Computation.*

Need more than the basics? See **More Than the Basics:** *The U-Test.*

	Swimming speed		Ranking	
Guppy #	Small	Medium	Small	Medium
61		0.538		20
68		0.467		19
62		0.449		18
63		0.436		17
69		0.425		16
64		0.389		15
70		0.379		14
41	0.364		13	
42	0.363		12	
47	0.354		11	
65		0.350		10
66		0.343		9
48	0.336		8	
67		0.325		7
46	0.322		6	
49	0.317		5	
43	0.310		4	
50	0.309		3	
44	0.303		2	
45	0.282		1	
			totals: 65	145
			null expectation: 105	105

Multisample unpaired nonparametric data is analyzed by the **Kruskal-Wallis test**. The formula is $H =$

$$\frac{12}{n_{total}(n_{total} - 1)} \sum_i \frac{\sum_i R_{i,j}}{n_j} - 3(n_{total} + 1)$$

where $R_{i,j}$ is the rank of the ith datum in the jth column, n_j is the number of data in the jth column, and k is the number of columns.

9.5 Evaluating Unpaired Multiple-Sample Data

The **Kruskal-Wallis test** is (roughly) the nonparametric equivalent of the ordinary one-way ANOVA. As with the rank-sum test, it assigns a rank to every datum. And then, like the Friedman test, the ranks for each treatment condition are summed. If the sample sizes are unequal, the ranks in that column are scaled. The null hypothesis is that the sum of ranks in each col-

Box 9.3

BioStats Basics Online   The rank-sum test

The test for evaluating unpaired two-sample data is on card III. G-3. Enter or import your data and click on the (Compute) button. The sample sizes of the two groups will be converted into two df values, one for each group; these will be displayed along with the test statistic–the smaller sum–and graphs of the distributions. Assuming your data are within the bounds of the Mann-Whitney method, you will need to use the two df values (labeled n_1 and n_2) to locate the threshold value in the table on the card. If your smaller sum equals or is less than this value, your data are significant at $P < 0.05$. In this event, click on the (P<.01) button; your dfs and smaller sum will be transferred to a new card with a table of threshold values for $P < 0.01$. If your smaller sum is equal to

or less than the value in the table, your data are significant at the 0.01 level. And, as always, if you can solidly justify a priori a one-tailed prediction, and your data correspond to that prediction, the P-value can be halved.

If your sample sizes fall outside the range of the Mann-Whitney tables, *BioStats Basics Online* will automatically use the U-test to generate a test statistic (z), and display a list of threshold z-values for your sample size. Reading from top to bottom, the P-value for your data lies between that for the first row whose threshold value is larger than your test statistic and the row above it. Again, these are two-tailed values; if you can solidly justify a priori a one-tailed prediction, and your data correspond to that prediction, the P-value can be halved.

umn will be the same; the P-value produced by the test gives the odds of drawing the observed distribution, or one more extreme, from a uniform distribution by chance.

The actual details of the mathematics are not very relevant—you've seen the logic before in other nonparametric tests. The main caveat is also familiar: even if the result is significant, the test cannot tell you which treatment is having an effect; eye-balling the graphs generated by *BioStats Basics Online* is your best guide. Because it must allow for false positives arising from multiple comparisons, the Kruskal-Wallis test is not very sensitive; the data must strike it a sharp, stinging blow before the test will concede that something nonrandom is going on. The

Box 9.4

BioStats Basics Online The Kruskal-Wallis test

The Kruskal-Wallis test evaluates unpaired multiple-sample data; it is found on card III. G-6. You should enter or import your data into columns for each experimental condition. When you click on the [Compute] button, *BioStats Basics Online* will display the number of treatments (k), the total sample size (n), a test statistic (H), and then graph the distributions. The table of threshold values will appear odd. For $k = 3$, it will list several combinations of sample sizes. Because of the way the test statistic is computed, if the sample size in each group is greater than 5, the threshold value is the same independent of increasing sample sizes. For values of k greater than 3, the sample sizes in the treatment do not matter at all, and the threshold table lists one value for each possible k. Threshold values are listed for two-tailed probabilities of 0.05 and 0.01; your test statistic must be larger than the value in the appropriate row in the table for your data to be significant. The usual a priori one-tailed logic and caveat apply to these values.

Need more than the basics? See **More Than the Basics:** *The Kruskal-Wallis Computation.* card in *BioStats Basics Online* describing the test offers some suggestions for reconfiguring multiple-sample nonparametric data so that other tests can have a go at it, perhaps with more chance of success. (The Kruskal-Wallis test is found on card III. G-6 in *BioStats Basics Online.*)

Points to remember

✔ Nonparametric tests are less powerful because the shape of the parent distribution is either not known or cannot be characterized by a simple, well-defined bell curve.

✔ All nonparametric tests depend on assigning rank orders to the individual data values and then evaluating the sums of the rankings.

✔ The most powerful test for nonparametric data, the signed-rank test, is used to compare a single set of paired data.

✔ Multiple sets of paired nonparametric data are compared with the Friedman test, which allows for the increased chance of false positives with the multiple comparisons. The

Friedman test does not identify which condition generates a significant difference, but column dropping can be used to ferret out the likely factor.

✔ A single set of unpaired nonparametric data can be compared with the rank-sum test.

✔ The Kruskal-Wallis test is used for multiple sets of unpaired nonparametric data. As with the Friedman test, it compensates for the increased risk of false positives, and if a significant result is obtained, column dropping may identify the treatment creating the effect. This is the weakest test covered in this text.

Exercises

1. Table A is a partial list of display times elicited from male Siamese fighting fish (*Betta splendens*) when presented with a model of one or other of two colors. (The complete data are found in file 9—Betta Blue vs Green or 9-BETTA.dat.) Is there a difference in the degree to which the two colors stimulate the aggressive spirit of male fighting fish? Which test did you use, and why?

Table A Partial list of fighting-fish display times

Male #	Display to blue model	Display to green model
1	26	10
2	18	30
3	107	97
4	50	82
5	65	67
6	19	16
7	67	49
8	86	77
9	22	52
10	35	29

9 Nonparametric Continuous Data

2. Table B presents partial data on the stimuli that elicit prey-capture behavior in toads. (The full data set is in file 9—Toad Strikes or 9-TOAD.dat.) Is there any significant difference between the responses? Which test did you use, and why?

What factors about the models seem to be important? (You may want to do some selective column dropping here; in *BioStats Basics Online* this requires shifting data columns to the left to fill any empty columns.)

Table B   Toad responses

Responses in strikes/10 sec directed at computer-generated, animated models; each model was presented twice and responses were summed. Possible model characteristics were: v = oriented vertically; md = medium; sm = small; bn = brown; gr = green; wl = with legs; wm = with moving legs; xm = with static legs; wh = with highlight; xh = without highlight; wo = with outline; xo = without outline. The seven models were:

$A = md\text{-}bn\text{-}wl\text{-}wm\text{-}wh\text{-}wo$ $E = v\text{-}md\text{-}bn\text{-}wl\text{-}wm\text{-}wh\text{-}wo$

$B = md\text{-}bn\text{-}wl\text{-}wm\text{-}xh\text{-}xo$ $F = md\text{-}gr\text{-}wl\text{-}wm\text{-}wh\text{-}wo$

$C = md\text{-}bn\text{-}wl\text{-}xm\text{-}wh\text{-}wo$ $G = md\text{-}bn\text{-}xl\text{-}xm\text{-}wh\text{-}wo$

$D = sm\text{-}bn\text{-}wl\text{-}wm\text{-}wh\text{-}wo$

				Model			
Toad	A	B	C	D	E	F	G
George	12	0	14	3	0	11	9
Chelsea	11	0	9	2	0	10	9
Ross	14	1	9	4	0	13	9
Sox	15	0	12	5	0	7	7
Barbara	13	0	8	3	0	14	8

3. Table C presents partial data on the A (academic) and P (personal) ranking awarded by an admissions committee to subsequent biology majors, as well as cumulative GPAs. (The complete data are found in files 9—A Ranks, 9—P Ranks, and 9—GPAS, or 9-ARANK.dat, 9-PRANK.dat, and 9-GPA.dat.) Does either sex score significantly higher than the other on any of these three measures? Which test did you use, and why? Because the A-rank and P-rank data are both categorical and ordered-discrete, you might have been able to use the chi-square test for independence on them. Remembering the restrictions on sample numbers, could you use either as chi-square data? If either A-rank or P-rank is significant by the nonparametric test you used, and is suitable for chi-square, compare the sensitivity of the two tests by running the data through the chi-square alternative.

Table C Partial list of GPAs and rankings

	Males			Females	
GPA	A-rank	P-rank	GPA	A-rank	P-rank
3.7	1	3	3.3	3	3
3.2	3	4	3.0	2	4
2.9	2	3	3.6	2	3
3.7	2	3	3.4	1	3
3.3	2	3	3.2	3	2
3.7	2	3	3.2	3	4
3.2	2	3	3.1	2	4
2.5	4	3	3.6	1	4
2.7	4	1	3.6	2	4
1.7	3	4	3.4	2	3

4. We saw in the last chapter that the gender comparison of syllable number in given names was impossible using the chi-square test of independence. We could have used Fisher's Exact Test, but since the data are discrete and ordered, ranging from 4 to 9, try a

9 Nonparametric Continuous Data

nonparametric comparison. The data are displayed in Table D (and found in file 9–Syllable Number or 9-SYLL.dat.) Is there a difference? What test did you use?

Table D Syllable numbers

Male given names, syllable numbers				Female given names, syllable numbers			
5	4	5	6	7	6	7	6
5	5	7	5	6	7	8	7
7	5	5	6	6	7	6	8
7	4	5	5	7	9	7	9
6	5	5					

5. Platys are tropical fish closely related to swordtails. Female platys were offered a choice between a model of a normal male of their species and one with a sword. In one case the sword was mounted as a ventral extension of the tail, just as it is in swordtails; in another test the sword extended from the dorsal margin of the tail. The number of seconds the female spent actively approaching each model is given in Table E (and in files 9–Dorsal Sword Platy and 9–Ventral Sword Platy, or 9-DORSAL.dat and 9-VENTRL.dat). Did the females show any preference? What test did you use?

Table E Partial list of approach times (sec)

Test 1, approach times (sec)			
Normal male	Dorsal sword	Normal male	Ventral sword
420	561	2	558
392	3	252	124
69	0	121	391
132	246	380	345
0	46	1	207
230	29	41	224
204	1	67	74

6. Table F gives a partial list of fuel-economy data for various classes of vehicles. (The complete file is named 9–Fuel Economy or 9-FUEL.dat.)

The question at issue is whether vehicle size has an effect on gasoline use. Does it? Which test did you use?

Table F Partial list of fuel economy data (mi/gal)

Subcompacts	Compacts	Mid-size	Large	Sport-utility
31	25	23	28	17
24	23	18	27	23
30	23	31	25	25
28	20	29	25	23
18	24	29	26	22
20	23	28	25	22
28	26	25	26	25
25	26	29	25	23
34	24	26	28	26
28	23	27	26	23
28	31	26	26	22

7. Correlation studies have shown a lower incidence of certain estrogen-related cancers (e.g., those of the breast, uterus, and prostate) in individuals who consume larger than average quantities of cruciform vegetables, such as broccoli and cabbage. The hypothetical mechanism favored by many researchers is that the putative active chemical in cruciforms, indole-3-carbinol (I3C), breaks down estrogen in the blood, thus "detoxifying" a hormone that encourages the growth of certain tumors. However, the observed correlation could easily be a consequence of some other factor; for instance, high cruciferous-vegetable eaters are also less likely to be overweight, are more likely to exercise, and tend to be from higher socioeconomic groups. One experiment sought to test the effect of I3C in individuals who did not normally consume cruciferous vegetables, monitoring the level of digested estrogen fragments in the urine. Measurements were taken

9 Nonparametric Continuous Data

before and after a measured dose of 13C, and are shown in Table G. (The creatine mentioned in the table is an energy intermediary in muscle cells; its level in the urine is used to standardize the concentrations of other molecules, compensating for differences in water intake among other factors. The data are also in file 9–13C Effects or 9-13C.dat.) Are these data parametric? If so, do they pass the F-test? If not, what is the correct test? Are the distributions significantly different?

Table G   Estrogen metabolite excretion (nmol/mmol creatine)

Subject #	Before 13C	After 13C
1	1.15	1.40
2	0.80	0.25
3	0.65	0.45
4	0.55	0.35
5	0.55	0.25
6	0.50	0.45
7	0.40	0.20
8	0.35	0.30
9	0.20	0.15

More Than the Basics

The signed-rank computation

To perform the signed-rank test, use the formula

$$\text{dif}_i = X_i - Y_i$$

$$S_{X>Y} = \sum_{i,\text{dif}>0} R_i$$

$$S_{X,Y} = \sum_{i,\text{dif}<0} R_i$$

$$T_s = \min(S_{X>Y}, S_{X<Y})$$

where X_i is the ith result under the first condition,

Y_i is the ith result under the second condition, and

R_i is the rank of the magnitude of dif_i, with ties broken by assigning an intermediate value.

The Friedman computation

To perform the Friedman test, use the formula

$$\chi_r^2 = \frac{12}{bk(k-1)} \sum_j \left(\sum_i R_{i,j} \right)^2 - 3b(k-1)$$

where b is the number of rows,

k is the number of columns,

and $R_{i,j}$ is the rank of the datum in the ith row, jth column among those in its row.

The degrees of freedom is df $= k - 1$.

The rank-sum computation

To perform the rank-sum test, use the formula

$$T_X = \sum_X R_i; \ T_Y = \sum_Y R_i$$

$$\begin{cases} n_X > n_Y & n_1 = n_Y; \ n_2 = n_X; \\ & T_1 = \sum_Y R_i \\ n_X < n_Y & n_1 = n_X; \ n_2 = n_Y; \\ & T_1 = \sum_X R_i \end{cases}$$

$$T_2 = n_1(n_X + n_Y + 1) - T_1$$

$$T = \min(T_1, T_2)$$

where R_i is the ith rank,

n_X is the number of data in the X column, and

n_Y is the number of data in the Y column.

If you have more than 30 data, use the U-test instead.

The U-test

To perform the U-test, use the formula

$$U_X = n_X n_Y + \frac{n_X(n_X + 1)}{2} + \sum_X R_i$$

$$U_E = \frac{n_X n_Y}{2}$$

$$\sigma_U = \sqrt{\frac{n_X n_Y(n_X + n_Y + 1)}{}}$$

$$z = \left| \frac{U_X - U_E}{\sigma} \right|$$

where R_i is the rank of the ith datum,

n_X is the number of data in the X column,

and n_Y is the number of data in the Y column.

If you have fewer than 30 data, use the rank-sum test instead.

The Kruskal-Wallis computation

To perform the Kruskal-Wallis test, use the formula

$$H = \frac{12}{n_{\text{total}}(n_{\text{total}} - 1)} \sum_j \frac{\sum_i R_{i,j}}{n_j} - 3(n_{\text{total}} + 1)$$

where $R_{i,j}$ is the rank of the ith datum in the jth column,

n_j is the number of data in the jth column, and

k is the number of columns.

The degrees of freedom is df $= k - 1$.

9 Nonparametric Continuous Data

Tests Covered in This Chapter

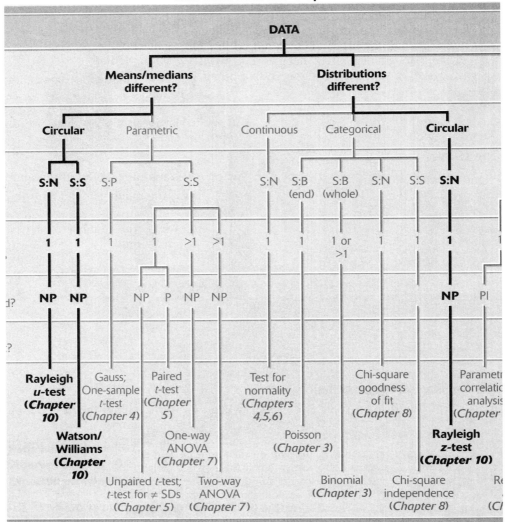

DATA

Means/medians different?   **Distributions different?**

Circular   Parametric   Continuous   Categorical   **Circular**

S:N S:S   S:P   S:S   S:N S:B   S:B   S:N   S:S   **S:N**
                                    (end) (whole)

1   1   1   1   >1 >1   1   1   1 or   1   1   **1**
                                          >1

NP   NP   NP   P NP NP   **NP**   Pl

Rayleigh Gauss;   Paired   Test for   Chi-square   Parametr
u-test   One-sample t-test   normality   goodness   correlatic
(Chapter t-test   **(Chapter**   **(Chapters**   of fit   analysis
10)   (Chapter 4) 5)   4,5,6)   (Chapter 8)   (Chapter

Watson/   One-way   Poisson   **Rayleigh**
Williams   ANOVA   (Chapter 3)   **z-test**
(Chapter   (Chapter 7)   **(Chapter 10)**
10)
              Unpaired t-test; Two-way   Binomial   Chi-square   R
              t-test for ≠ SDs ANOVA   (Chapter 3)   independence
              (Chapter 5)   (Chapter 7)   (Chapter 8)   (C

Circular Data

10

In all the continuous distributions we have looked at so far, whether parametric or nonparametric, the value of one datum could at least be said to be larger or smaller than another: a height of 157 cm is greater than one of 143 cm. **Circular distributions** are fundamentally different: the parameter being measured is periodic, and thus there is no logical way to establish whether one point on such a distribution is "larger" than another.

Circular distributions arise when data are collected from a periodic scale.

10.1 Where Do Circular Distributions Come From?

To take a simple example, suppose you are measuring departure bearings of homing pigeons, or estimates of north by blindfolded undergraduates. We typically define north as 0°, with angles measured clockwise from north; thus 25° is technically smaller than 45°. But if you chose to measure by going counterclockwise from north (and why not?), 25° would instead be 335°, and 45° would become 315°; and so the first angle would now be "larger." Worse yet, what distinguishes 90° and 450°? Both point the same direction, but one involves taking more than a full turn. This also makes the point that we will not be talking about "tails" in this chapter: since distributions extend infinitely in both directions from the mean, there is no way to tell if a given datum in a circular distribution that actually is, say, just 10° to the right of the mean, might not instead be a distant outlier to the far far left of the mean (350° to the left), or for that matter even farther (370°) to the right.

The same ambiguities arise with any periodic data. Days are repeating periods for which midnight is arbitrarily defined as the starting point on land; but in many navies noon is the beginning of the day—or worse yet, in many submarines and a number of other military units, noon in Greenwich, England, starts the day regardless where the ship or installation happens to be on the globe. January 1 is the first day of the Western New Year, but other cultures begin on quite different dates. Other periodic measures include weeks, tidal cycles, and lunar months.

Circular data need to be analyzed in some way analogous to linear data—at the minimum, you need to be able to calculate means and some measure of dispersion (the equivalent of variance or SD in linear data), and you need to be able to compare a sample distribution to a null or parent distribution, and two samples to each other. In this chapter we'll see how that is done.

10.2 Determining the Mean Bearing and Degree of Dispersion

The first step is to decide how to plot and summarize circular data. There are three methods in use. The "rose-diagram" approach was invented by Florence Nightingale in 1858 to analyze the annual distribution of deaths from various causes (Figure 10-1a). She took the precaution of using a nonlinear (square-root) radial scale so as not to overrepresent larger values with disproportionately larger areas; later workers (particularly in ad agencies) have not always been so careful. An alternative is to plot a circular histogram (Figure 10-1b). Neither of these methods captures the mean or dispersion in any quantitative way, but they can display large amounts of data efficiently.

The more popular method is to plot the data points themselves at the periphery of a "unit circle" and then compute a mean and some measure of dispersion. (If there are large numbers of data points, the plot itself can become difficult to create, though the analysis remains the same.) This approach was

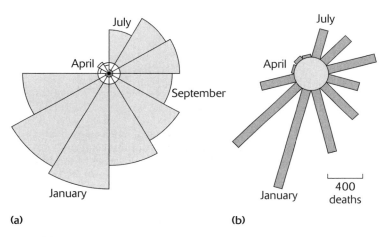

10-1 **(a)** The "rose" display pioneered by Nightingale represents the data with proportional areas; thus the radius of each sector is proportional to the square root of the sample. The data show the monthly mortality of soldiers stationed abroad in connection with the Crimean War; few of these deaths are related to combat. The small dotted circle in the middle is the death rate in Manchester during the same period. **(b)** An alternative way to present the same data is the circular histogram. [Redrawn from F. Nightingale, *Notes on Matters Affecting the Health, Efficiency, and Hospital Administration of the British Army* (London, 1858).]

developed first by Lord Rayleigh, who was beginning around 1870 to analyze the scattering of light; here, however, we will consider primarily cases of animal orientation. Figure 10-2 shows a set of departure bearings for pigeons released on sunny days; each pigeon's "vanishing bearing" (its direction when last seen by the observer) is plotted round the circle, and a vector has been drawn inside the circle. The direction of the vector is the ***mean bearing*** of the pigeons, and its length is proportional to the ***clustering*** of the bearings—that is, it would just touch the unit circle if all the bearings were identical, but would have a length from the center of zero if the bearings were evenly distributed around the circle. Let's look at how this ***"mean vector"*** is created, first intuitively and then mathematically.

A circular distribution has a **mean** and a **degree of clustering**, which are summarized by the direction and length of its **mean vector**.

10 Circular Data

10-2 Departure bearing of a group of pigeons released individually and tracked until out of sight. The arrow from the center is the mean vector, which represents both the mean bearing of the birds (162°) and the degree of clustering (0.75) on a scale of 0–1.0. [Based on data from W. T. Keeton.]

The formulas for the mean vector are

$$\bar{X} = \sum \sin(X_i)$$

$$\bar{Y} = \sum \cos(X_i)$$

$$r = \sqrt{\bar{X}^2 + \bar{Y}^2}$$

$$\Theta = \arcsine(\bar{X})$$

where X_i is the x-axis value of the ith point, Y_i is the y-axis value of the ith point, r is the length of the mean vector, and Θ is the angle of the mean vector (measured clockwise from straight up).

Imagine a massless disc oriented horizontally. On the periphery of this disc you place 1-g weights in the direction of each departure. Once all the data are in place, you move a pin under the disc until you find the point at which it just balances—the center of gravity. Draw a line from the center of the disc to this center of gravity and you have the mean vector: the direction reflects the mean of the angles of placement of the weights, and the position from the center corresponds to the dispersion of the weights (that is, the center of gravity would be at the edge if all the weights were piled up at a single bearing, but would be at the center if the weights were evenly distributed around the compass).

The mathematical analogue of this process is to take the sine and cosine of each angle; this generates an x- and y-value for each point. If, as is usual, up is 0°, the sine is the x-axis value and the cosine is the y-axis value (Figure 10-3). (Sometimes the data are plotted with 0° to the right; in this event, the cosine is the x-axis value and the sine is the y-axis value.) Now you sum all the sines and then all the cosines and get the mean value of each ($\bar{X}$ and $\bar{Y}$) by dividing by the sample size. The result is the x-y position of the tip of the mean vector. Taking the arcsine of $\bar{X}$ (or the arcsine of $\bar{Y}$) will give you the angle of the mean vector; solving the Pythagorean formula will give you its length: (vector length)$^2 = \bar{X}^2 + \bar{Y}^2$. *BioStats Basics Online* does this automatically (card IV. E-2), but you should bear

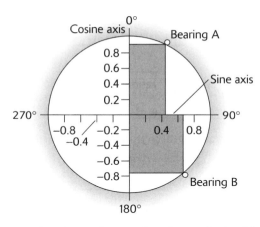

10-3 The mean bearing and vector length are obtained by first taking the sine and cosine of each sample bearing (here shown as the projection of the bearing on the x axis for the sine and the y axis for the cosine). For Bearing A the values are 0.42 for the sine and 0.91 for the cosine; for Bearing B the values are 0.68 for the sine and -0.74 for the cosine. All the sines are summed and divided by the sample size to yield a mean sine; similarly, all the cosines are summed and divided by the sample size. The resulting point—the mean sine on the x axis and the mean cosine on the y axis—is the tip of the mean vector.

in mind that not all statistical software can deal with circular data.

If your data are not actual angles, you will have scaled them at the outset (e.g., [hour of the day $\times$ 360]/24) into degrees—unless, of course, you are a physicist, in which case you will have done the whole thing in radians (probably even if the original data were in degrees to begin with); if you are in ROTC, you may even have used "grads."

The length of the mean vector is commonly called r. It is a measure of clustering; thus $(1 - r)$ is the equivalent measure of dispersion. For what it's worth, angular variance is $s^2 = 2(1 - r)$; angular deviation is $s = (180/\pi)\mathbf{v}\ 2(1 - r)$. You

10 Circular Data

Need more than the basics? See **More Than the Basics:** *Circular Distributions.*

rarely need these values since they are generally only used within analyses performed by computer (and hardly ever even there).

10.3 Testing for Clustering

Circular data analysis is somewhat unusual in that very often the only test necessary is one to judge whether the data are nonrandomly clustered; the null hypothesis is that they are not. There is no equivalent *t*-test: that is, ordinary (*non*circular) parametric data always have a mean and SD, and these can be tested against some other specific mean. But if you are determining, say, which time of day has the highest humidity, measured over many days, of course, you may have no null-hypothesis time; it will be quite enough to begin with to see if there is any consistent pattern at all, and, if so, how strong it is. The null hypothesis is that there is no regular high-humidity time.

The **Rayleigh *z*-test** is used to determine if circular data are significantly clustered. The formula is $z = nr^2$, where n is the sample size and r is the length of the mean vector.

You can look for clustering with the ***Rayleigh z-test***, which computes the odds of obtaining a distribution with the degree of clustering (the value of *r*—the length of the mean vector) ob-

---

Box 10.1

BioStats Basics Online   The Rayleigh *z*-test

The test for clustering is found on card IV. E-2. The box at the top center labeled "Predicted mean bearing" should be unchecked. Enter or import your data (in 0°–360° format) and click on (**Continue**). *BioStats Basics Online* will plot your data, draw a mean vector, display the mean vector's direction and length, and list *n* and the *z*-statistic. Compare this computed value with the table of threshold values on the left side of the card by locating the row corresponding to your sample size (*n*);

the computed statistic must be larger than the values shown for your data to be significant at the level indicated at the top of the column.

Because of the nature of circular data, there is no choice of one- versus two-tailed tests. Note that in plotting your data, *BioStats Basics Online* lumps angles and will plot two points with the same bearing on top of each other; this graphical limitation does not compromise the accuracy of the calculations.

served in the sample, or a higher value, by drawing data at random from a uniform circular distribution. To compute this you are clearly going to need the original data (to calculate the value of r), and your sample size will be critical. The test will also tell you the mean bearing, and if you take the trouble to compute s, you can calculate some confidence limits (though this is rarely done). The actual test statistic, z, is simply nr^2; the value obtained is then compared against a table of threshold values. (The z-test is found on card IV. E-2 in *BioStats Basics Online*.)

In the case of homing pigeons released on cloudy days (Figure 10-4), the clustering is significant at $P < 0.01$. But for birds released wearing magnets under the same overcast conditions, $P > 0.05$. Apparently pigeons have more difficulty getting their bearings, at least early on, when wearing magnets under overcast skies.

There are also tests that allow you to combine the mean bearings of several different sets of measurements (with differing sample sizes) to generate a second-order mean bearing. This is sometimes necessary when the phenomenon under study is

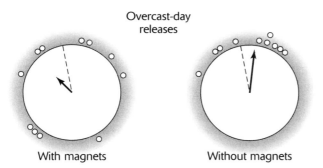

Overcast-day
releases

With magnets   Without magnets

10-4 Vanishing bearing of homing pigeons under overcast skies wearing either magnets or equivalent brass weights. The expected bearing (the direction of the home loft) is shown by the dotted line. There is no significant orientation displayed by the birds wearing magnets. [Based on data from W. T. Keeton.]

10 Circular Data

Need more than the basics? See **More Than the Basics:** *The Rayleigh z-Computation.*

too weak to yield a significant difference from one set of data alone. For example, in the study of magnetic-field orientation of caged birds, each bird generates a set of bearings over the course of a night (either in the form of landing rates on a set of radially arrayed perches, or footprints left as the bird attempts to escape from a funnel-shaped well with an ink pad at the bottom). Second-order analysis could combine the data from all these birds. In fact, conventional first-order mean bearings yield no clear pattern in this highly artificial situation, but second-order analysis produces significant, reproducible effects. Since you are unlikely ever to need second-order statistical analysis, however, the test is omitted from *BioStats Basics Online*.

10.4 Testing a Specific Hypothesis

Out of necessity, Rayleigh extended his analysis to include cases in which there is an expected bearing—a specific hypothesis, rather than the usual generic no-clustering null hypothesis for circular distributions. This is roughly like the one-sample *t*-test, in which you are expected to supply an expected mean but not the distribution's SD. The underlying assumption of this test (and the next, for two-sample comparisons) is that the data are parametrically distributed; this is almost always true for circular data. Here are two examples. In the first, students have been taken blindfolded in a bus to a site some distance from the campus and then asked to point first to the university and next toward north (Figure 10-5a). As you would expect with so many individual tests, when we look just for "significant" clustering we sometimes see it, though not necessarily in the homeward direction; this is just what you would expect by chance using a $P < 0.05$ threshold. But when we use the ***Rayleigh u-test*** with the home bearing as the predicted bearing, thus excluding false correlations in other directions, the values are random (Figure 10-5b). (The Rayleigh *u*-test is found on card IV. E-2 in *BioStats Basics Online*.)

The **Rayleigh *u*-test** is used to determine if circular data are significantly clustered and not significantly different from an a priori prediction. The formula is

$$u = n \cdot r \cdot \cos(\Delta\Theta)\sqrt{2/n}$$

where r is the length of the mean vector and $\Delta\Theta$ is the angular difference between the sample and null means.

On the other hand, when we kidnap honey bees leaving the hive on their way to a particular food source, carry them in darkness to a release site 150 m from the hive, and then

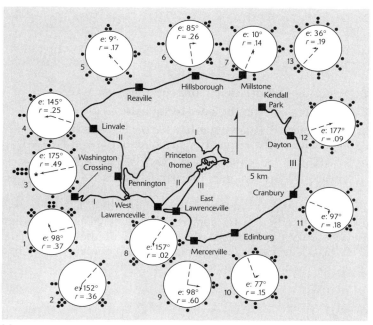

(a)

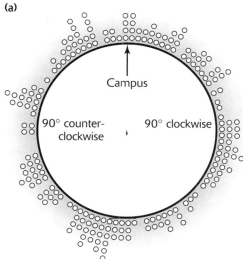

(b)

10-5 (a) Routes and homeward-pointing data from several human homing tests. (b) When all the data are combined with vertical as the homeward direction, the mean vector length is 0.05 and the mean bearing is 179°.

release them one at a time, we see a significant clustering of departure bearings (Figure 10-6). There are four initial predictions:

1. confusion and disorientation, and thus no clustering;
2. confusion, causing the bees to fly back to the hive;
3. insensibility that anything has happened, and thus departure in the original flight direction; and
4. an ability to accommodate the transportation and fly toward the original goal (unseen behind a ridge of trees).

Need more than the basics? See **More Than the Basics:** *The Rayleigh u-Test.*

The experiment was designed to separate these last three predictions by 60° each. By applying the Rayleigh *u*-test, we find the bees were oriented in the direction of the unseen goal ($P < 0.01$), suggesting that they had worked out where they had been taken and then planned a novel route.

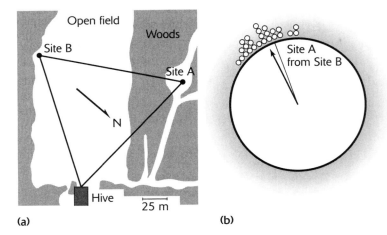

(a)   (b)

10-6 **(a)** Forager honey bees leaving the hive for Site A were captured and transported in a dark container to Site B, where they were released. **(b)** Departure bearings of released foragers suggest that these bees were able to set a course for Site A even though it is not visible from Site B. (Mean vector angle was 324°, versus a true bearing of 330°; mean vector length was 0.96.)

Box 10.2

BioStats Basics Online The Rayleigh *u*-test

The test for orientation in a predicted direction is found on card IV. E-2, the same card containing the Rayleigh *z*-test. The box at the top center labeled "Predicted mean bearing" should be checked and the predicted mean bearing typed in. Enter or import your data (in 0°–360° format) and click on **Compute**. *BioStats Basics Online* will plot your data, draw a mean vector, display the mean vector's direction and length, and list *n*, the *z*-statistic, and the *u*-statistic. Compare the computed value for the *u*-statistic with the table of threshold values on the right side of the card by locating the row corresponding to your sample size (*n*); the computed statistic must be larger than the values shown for your data to be significant at the level indicated at the top of the column.

Because of the nature of circular data, there is no choice between a one-versus a two-tailed test. Note that in plotting your data, *BioStats Basics Online* lumps angles and will plot two points with the same bearing on top of each other; this graphical limitation does not compromise the accuracy of the calculations. Note that the *u*-test automatically supplies the *z*-statistic, so you can look for nonrandom clustering without running the test again.

In each case the test statistic, u, is $n \cdot r \cdot \cos(\Delta\Theta) \sqrt{2/n}$, where $\Delta\Theta$ is the angular difference between the sample and null means. The result is compared to a table of threshold values.

10.5 Comparing Two Samples

Sometimes you may need to compare two circular distributions to see if they have different means; this is roughly like the un-paired two-sample *t*-test, though there is no presupposition that the dispersions are similar. (Unfortunately, there is no equiva-lent of the powerful paired two-sample *t*-test for circular distri-butions; this is a surprising omission, and must mean that no probability-literate researcher has ever desperately needed this comparison. There *is* a technique for comparing *nonparametric* circular data; in the unlikely event you ever find yourself with such data, consult a more exhaustive statistics text for the

10 Circular Data

Watson U^2-test.) For example, suppose you are comparing the departure bearings of pigeons, one group of which has been "clock-shifted" (kept in an isolated room with the lights coming on and going off out of phase with normal dawn and dusk). The idea here is that when the sun is available, the birds might use it to determine direction. Since the sun moves from east to west over the course of the day, it follows that to use the sun as a compass, they must know the time of day. Thus if their internal clock says "noon," then the sun must be in the south, and the pigeons can use that fact to select the correct orientation for departure (Figure 10-7).

If, on the other hand, the birds have been exposed to an artificial day advanced 6 hours, then local noon will occur at about 6 P.M. artificial-day time; such birds should look at the sun, deduce from their internal clocks that it lies in the west, and thus depart in a different direction from the unshifted birds.

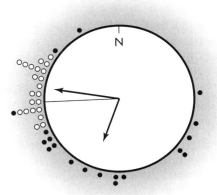

10-7 Homing pigeons maintained in rooms with an artificial day/ night cycle in phase with the actual times of dawn and dusk are well oriented when released (open circles). Birds kept in rooms with the day/night cycle advanced six hours (filled circles) depart about 90° to the left of normal birds. [Based on data from W. T. Keeton.]

Box 10.3

BioStats Basics Online   The Watson-Williams test

The test for comparing the mean bearings of two distributions is found on card IV. E-3. Enter or import your two distributions (in 0°–360° format) and click on [**Compute**]. *BioStats Basics Online* will plot your data, draw two mean vectors, and then list the sample sizes, df value, and the test statistic, *F*. The list of threshold values for the *F*-statistic is found in a scrolling field. Scroll down to the appropriate df line and compare your statistic with the values in that line; for significance at the level indicated at the top of the column, your *F*-statistic must exceed the value in the table.

As with the *u*-test and *z*-test, the nature of circular data means that there is no choice of one- versus two-tailed test. Note, again, that in plotting your data, *BioStats Basics Online* lumps angles and will plot two points with the same bearing on top of each other; this graphical limitation does not compromise the accuracy of the calculations. The two sets of data are plotted with slightly different symbols: short right- versus left-slanting lines. Note also that the length and angle of the mean bearings are not listed on this test card; if you need these numbers, transfer your data, one set at a time, to the preceding card for analysis.

The discrepancy should be about 90° counterclockwise, but the exact value depends on the sun's actual path through the sky, which in turn varies by date and latitude. The ***Watson-Williams test*** allows you to test for a difference in means between two distributions without needing to know how different they ought to be; the null hypothesis, naturally, is that the two distributions have the same mean direction. The details of the calculation are immaterial since they are programmed into *BioStats Basics Online*. (The Watson-Williams test is found on card IV. E-3 in *BioStats Basics Online*.) In the case of the pigeons, the distributions are different ($P < 0.01$).

The **Watson-Williams test** is used to see if two distributions of circular data are clustered and have significantly different means.

There are multisample versions of the Watson-Williams test that permit testing several mean bearings against one another; these are roughly equivalent to the one-way ANOVA for linear data. Thus they incorporate corrections for the effects of testing multiple means.

Need more than the basics? See **More Than the Basics:** *The Watson-Williams Computation.*

10 Circular Data

Points to remember

✔ Circular statistics must be used to evaluate data taken from a periodic scale.

✔ Circular data are almost always parametric, but computing a precise SD is often impossible because the tails of the distribution may overlap. Circular statistics depend instead on a measure of clustering (the length of the mean vector).

✔ The Rayleigh z-test is used to see if circular data are clustered.

✔ The Rayleigh u-test is used to see if circular data are clustered around a predicted value; the data must be both significantly clustered and have a mean not significantly different from the predicted value.

✔ The Watson-Williams test is used to see if two circular distributions are significantly different; both distributions must be clustered and have means significantly different from one another.

Exercises

1. In 1990, an article in the *Journal of the National Cancer Institute* reported that Hodgkin's disease is most likely to be diagnosed in the spring, when immune cell levels fall. The data are listed in Table A. (They are also in file 10–Hodgkin's Diagnoses or 10-HODG.dat, with the months converted into angles, such that January is 0°, April 90°, and so on; because there is essentially no change in significance thresholds beyond $n = 50$, the data are in units of 5 cases, yielding an n of about 200.) The report failed to apply circular statistics to see if the spring increase is significant. Is it? Which test did you use? Why?

Table A Diagnoses in males per month

January:	95	May:	70	September:	80
February:	90	June:	85	October:	90
March:	90	July:	65	November:	85
April:	110	August:	55	December:	70

2. Animals, like bees and pigeons, that use the sun's direction as a compass must allow for its movement from east to west during the day. The rate of direction change varies with the latitude, time of year, and time of day. Table B presents the results of two experiments in which bees trained to a feeder in a particular direction were trapped in the hive for two hours. (The data are also found in files 10–Sun Compensation A and 10–Sun Compensation A, or 10-SUNA.dat and 10-SUNB.dat; the tests differ in exactly when the period of entrapment began.) When released, the foragers set off along bearings that represented their estimate of the food direction relative to the sun, allowing for the sun's movement during the intervening two hours. Three hypotheses had been put forward to explain sun compensation. One posited that animals use an annual average rate of 15°/hour, and live with the

Table B   Test A Predictions: sun's arc, 36°; 15°/hr, 90°; extrapolation, 63°

	Observations		
Angle	Foragers	Angle	Foragers
26°	0	66°	7
34°	0	74°	6
42°	0	82°	3
50°	1	90°	1
58°	2	98°	0

Test B Predictions: sun's arc, 52°; 15°/hr, 90°; extrapolation, 21°

	Observations		
Angle	Foragers	Angle	Foragers
10°	3	60°	0
20°	7	70°	1
30°	13	80°	0
40°	5	90°	0
50°	2	100°	0

discrepancies. Another supposed that animals learn the precise time-to-direction relationship between sun and true direction. Yet a third imagined that bees extrapolate the most recently measured rate of sun movement. The predictions for each hypothesis are listed in the table. Which is the best match? Which test did you use? Why?

3. The original data suggesting that humans could tell direction are shown in Table C (and are found in files 10–Manchester Brass A, 10–Manchester Brass B, 10–Manchester Magnet A, and 10–Manchester Magnet B, or 10-MANBSA.dat, 10-MANBSB.dat, 10-MANMGA.dat, and 10-MANMGB.dat). These tests were performed by Robin Baker at the

University of Manchester in the late 1970s. Analyze these data to see if the four groups are oriented in the predicted direction, and if the two groups at each site (magnet versus brass) are significantly different.

4. The idea that the direction correlation in the honey bee dance was used by recruit bees to find advertised food sources was severely attacked in the late 1960s. New experiments seemed to indicate that bees used odor cues and that careless techniques had given the impression of language use where none was involved. A later test used a trick that reoriented the forager dances so that, if recruits used the direction information, they would be sent off in arbitrarily chosen

Table C Manchester Tests

Site A home: 60°

Brass	315°	315°	0°	0°	0°	45°	45°	45°	45°	45°	45°	90°
	90°	90°	225°									
Magnet	270°	270°	270°	165°	315°	315°	315°	315°	0°	0°	0°	45°
	45°	90°	90°									

Site B home: 305°

Brass	0°	0°	90°	135°	180°	245°	270°	270°	270°	270°	270°
	270°	285°	315°	315°	315°	315°					
Magnet	45°	90°	90°	135°	180°	225°	225°	270°	270°	270°	270°
	315°	315°	315°	315°							

directions. Table D shows the results of four tests with differing "aiming" direction. (The data are also in files 10–Recruit Capture 35A, 10–Recruit Capture 60, 10–Recruit Capture 35B, and 10–Recruit Capture 72, or 10-BEE35A.dat,

10-BEE60.dat, 10-BEE35B.dat, and 10-BEE72.dat.) Are the bees nonrandomly oriented? If so, is it in the direction predicted by the dance? Is there anything odd about these numbers?

Table D   Dance orientation vs recruit behavior

Dance direction	Recruits captured at					
	22°	35°	48°	60°	72°	85°
35°	7	31	7	0	3	12
60°	0	11	6	28	3	1
35°	3	10	4	0	1	0
72°	3	0	0	1	10	3

5. Table E presents data from beetles orienting to a windborne odor. (The data are also found in file 10–Beetle Orientation or 10-BEETL.dat.) In one case, the beetles are intact; in the other, the right antenna is missing. (The antennae of insects are their major olfactory organs.) Is each group well oriented? Is there a significant difference between their mean bearings?

Table E   Beetle orientation

	Angle											
	240°	290°	295°	300°	305°	310°	320°	330°	335°	340°	345°	360°
With right antenna	0	0	0	0	0	0	0	1	5	8	1	0
Without right antenna	1	2	2	8	1	16	4	0	0	0	0	1

10 Circular Data

More Than the Basics

Circular distributions

The mathematics of circular distributions are not as well studied as those of ordinary distributions. For instance, at least four different formulas for circular variance exist, no two in agreement. This is primarily because the mathematical formula for the circular normal distribution is prohibitively complex:

$$y = \frac{1}{\sigma\sqrt{2\pi}} \, e^{\frac{\theta^2}{2\sigma^2}} \sum_{k=-\infty}^{\infty} (e^{\theta + 2\pi k})^{-\frac{\pi k}{\sigma^2}}$$

This is a formula that one would not wish on one's worst enemies, even if all the terms were well defined (which, of course, they aren't).

In fact, there is an entire set of nonparametric circular tests. *BioStats Basics Online* does not use these because in most cases circular data may be analyzed as if parametric, and because the nonparametric circular tests require truly huge amounts of data.

The Rayleigh z-computation

To perform the Rayleigh z-test, use the formula

$$\overline{X} = \sum \sin(X_i)$$

$$\overline{Y} = \sum \cos(X_i)$$

$$r = \sqrt{\overline{X}^2 + \overline{Y}^2}$$

$$z = nr^2$$

where X_i is the ith datum, and n is the number of data.

The Rayleigh u-test

To perform the Rayleigh u-test, use the formula

$$V = n \cdot r \cdot \cos\left(\arctan\left(\frac{\overline{X}}{\overline{Y}}\right) - \mu_0\right)$$

$$u = V\sqrt{\frac{2}{n}}$$

where n is the number of data,

r is the magnitude of the mean vector,

$\overline{X}$ is the x-component of the mean vector,

$\overline{Y}$ is the y-component of the mean vector, and

μ_0 is the expected mean bearing.

The Watson-Williams computation

To perform the Watson-Williams test, use the formula

$$r_w = \frac{n_1 r_1 + n_2 r_2}{}$$

$$F = K \frac{(n_{total} - 2)(n_1 r_1 + n_2 r_2 - n_{total} r_{total})}{}$$

where n_1 is the number of data values in the first sample,

n_2 is the number of data values in the second sample,

n_{total} is the total number of data values,

r_1 is the magnitude of the mean vector of the first sample,

r_2 is the magnitude of the mean vector of the second sample, and

r_{total} is the magnitude of the mean vector of all the data.

K is a correction factor based on r_w.

Tests Covered in This Chapter

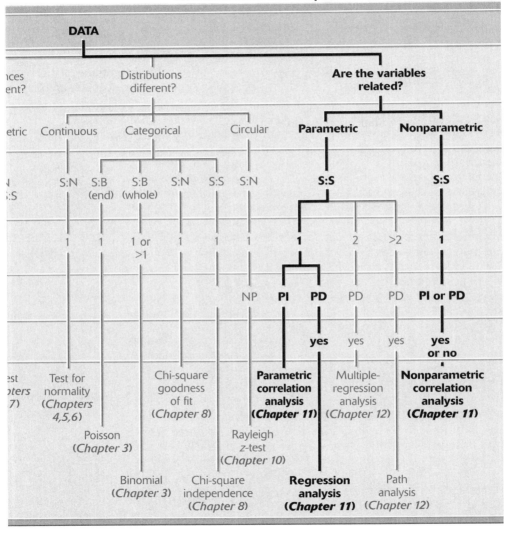

Variables I: Correlation and Regression

11

Up to now, we have concentrated on how to compare two or more distributions with each other; sometimes we have compared means, other times variances, and other times the distributions as a whole. In making these comparisons, we have often varied one parameter between the various test conditions (e.g., eye color on a predator model) and recorded a second variable (duration of death feigning, for instance). So in some sense, we have already asked how one variable affects another, but not in any very systematic way—we've merely determined whether or not a change in an *experimental* variable leads to a significant difference in a *measured* variable.

Our focus now changes to dealing more explicitly with how two (or more) variables are related to one another. Instead of eye color, which is a categorical variable, we might have asked how the degree of eye separation on predator models is related to the duration of death feigning. Both are continuous variables, and it could be that, as we systematically increase eye separation, the length of immobility will gradually increase, decrease, or remain unchanged.

11.1 Correlation versus Cause and Effect: When to Draw the Line

There are two general methods for relating two or more variables: *correlation analysis* and *regression analysis*. Both techniques were first developed around 1880 by Francis Galton (a

Secular trends are systematic changes in mean values of a variable. In most populations, height, weight, and IQ are slowly rising, creating secular trends that make comparisons across age groups more difficult.

Regression to the mean is the tendency of the offspring of an unusually tall or short, heavy or light, smart or dull individual to be less unusual in that trait— usually because (1) the trait is only partly genetic, and (2) the other parent is likely to be closer to the mean.

cousin of Charles Darwin) to help him understand human inheritance. Figure 11-1 shows one of the most famous comparisons from a study by his student Karl Pearson, inventor of the chi-square test: each point represents the height of a father versus that of his son. You will see at a glance that taller fathers tend to have taller sons, but there is a great deal of scatter. A closer analysis reveals that sons are, on average, slightly taller than their fathers; this is known as a *secular trend*, and continues even today as each generation is slightly taller than the last. A careful look also reveals that the sons of especially short and particularly tall fathers are less unusually short or tall than their male parent; this is the phenomenon of *regression to the mean*; it results in part from the genetic contribution of the other parent, the mother, who is more likely to be of average height than especially tall or short. Obviously there is a wealth of information in this graph, but how can a researcher best summarize and quantify it?

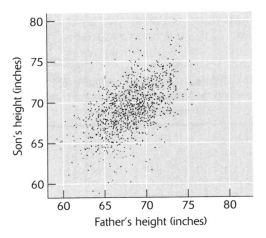

11-1 Pearson's graph of father's versus son's heights. Each point represents one father/son combination. If you let your eye draw a mental line through this cluster of points, and then examine the line, you will probably find that the line generally lies above the diagonal connection 58/58 to 80/80—the line of equal father/son heights.

Correlation analysis is nonjudgmental: it asks only how changes in one variable are related to the value of another; it does not touch on the issue of cause-and-effect relationships. At its most basic level, it asks how often an increase in the x-axis variable is associated with an increase in the value of the y-axis variable. (To enhance statistical power, the analyses also consider the relative degree of increase.) If there were a perfect correlation (defined as $+1.0$), every increase in X would inevitably be accompanied by an *increase* in Y; if there were a perfect negative correlation (-1.0), every increase in X would be accompanied by a *decrease* in Y. If there were no correlation between X and Y (0.0, the null hypothesis), no pattern would be evident. Weaker, more probabilistic associations would lead to intermediate correlation values (e.g., $+0.5$) and a scattering of points (Figure 11-2). The correlation statistic is used as the basis for calculating the odds of drawing a set of data of your sample size with that degree of correlation (or one more different from zero) from a parent distribution of uncorrelated data. As we will show, correlation analysis can be performed on either parametric or nonparametric data, though two different mathematical approaches are involved.

Regression analysis is much more aggressive, but it is limited to a narrower range of variable pairings: (1) there must be a plausible cause-and-effect relationship; (2) the value of the putative cause variable must be well known—that is, there must be negligible uncertainty in any cause-variable values; (3) the putative effect variable must be parametrically distributed for each value of the cause variable. If these three conditions are met, then you can also plot a best-fit regression line to your data. Figure 11-3 (see page 259) does this for scores on the verbal section of the Scholastic Aptitude Test taken by students who later became biology majors against their first-year college GPAs; the plausible cause $\rightarrow$ effect relationship is cognitive ability (as the Educational Testing Service would have us believe is measured by the vSAT in high school) $\rightarrow$ grades in college. You could also compute a correlation and its *P*-value, and derive an equation relating values of Y (GPA) to the corresponding value of X (vSAT). The null hypothesis is that there is no systematic relationship between putative cause and effect. (Again, we say

Correlation analysis determines how changes in the value of one parameter are related to changes in the value of another measure; there is no attempt to determine cause and effect.

Regression analysis determines the way in which changes in the well-known value of a cause variable alters the magnitude of the parametrically distributed effect variable.

11 Relationships between Variables I

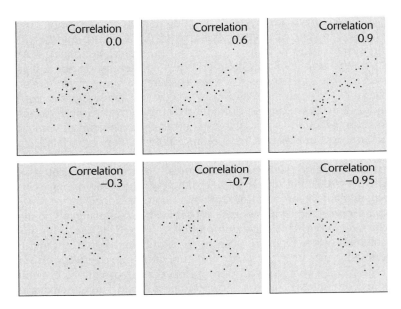

11-2 Various degrees of correlation. Positive correlations slope up to the right, while negative correlations slope down to the right. Higher absolute values of the correlation coefficient are the result of lower scattering in the data points. If all the points fell on a single line that sloped to the upper right, the correlation would be 1.0. When nonparametric data are analyzed, a perfect correlation requires that every change in the value of the x-axis variable be associated with a corresponding change in the y-axis variable; as we will see, when this condition is not met, the number and size of the discrepancies are considered. For parametric data, a perfect correlation also requires that the degree of change in the x-axis value correspond to the degree of associated y-axis change. [Based on D. Freedman et al., *Statistics*, 2nd ed. (New York: W. W. Norton), 1991, p. 121.]

"putative" because you cannot be certain that the variable whose value seems to control another is actually the true cause; we will return to this issue presently.)

The equation derived by regression analysis quantifies much of what is going on to yield the observed pattern. But so strin-

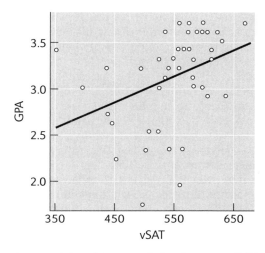

11-3 There is a partial positive correlation between vSAT scores in high school (*x* axis) and first-year college GPA (*y* axis) among biology majors; the value of the correlation is 0.36. Because vSAT is thought to measure a plausible cause variable (intellectual ability) that ought to have an impact on the effect variable (first-year GPA), and because first-year GPAs are parametrically distributed, we can fit a regression line to the data. We can also derive an equation that relates the two variables using statistical data computed in any regression analysis; in this case the relationship is GPA = 1.66 + (0.00227) × vSAT; this equation is valid only for the region covered by the data, and thus we cannot extrapolate to a vSAT of 200 (if we did, we might infer that the freshmen with the minimum-possible SAT score would have an average GPA of 2.1).

gent are the preconditions listed above, and so tempting is it to draw a line through data, that the academic literature (not to mention the popular press) is filled with lines fitted to data that are wholly ineligible for regression analysis. In other cases, regression analysis is applied to marginally appropriate data for which the independent variable is not precisely known; the researcher presumes that the values are "close enough," but who is to say? Actually, you *can* roughly quantify the likely error and thus set some reasonable limits on the preconditions. The SAT example is a case in point: the results of any single administration of the vSAT to an individual are generally within 15 points of the average of many administrations to that same person. With a range of about 300 points in the data set, and assuming there is no systematic bias involved, this 5% uncertainty (15/300) translates (trust us) into a regression correlation that is about 1–2% too low. This illustrates two critical points: variability tends to lower correlation values, thus providing a conservative bias (that is, more false negatives), and 5% variability is hardly different in practice from "precisely known" (zero variability).

Even height, which is typically rounded to the nearest inch, has an uncertainty of about the same magnitude relative to the range within a sex; this ambiguity arises from errors in measurement, variations in posture while being measured, and rounding. Thus a 5% uncertainty (relative to the range of x-axis values) is quite tolerable in practice; a 10% uncertainty is marginal, and likely to substantially lower the correlation; larger uncertainties are unacceptable. Although methods do exist to fit lines to data with greater independent-variable uncertainty, and to data not meeting the other conditions, they are not nearly as statistically meaningful, and we have nothing further to say about them.

11.2 Correlating Nonparametric Data

So let's say you have two variables: they can either be related in a plausible cause-and-effect manner, or not; they can both be parametrically distributed, or one can be, or neither may be. Here is how these alternatives titrate out for parametric correlation analysis (PCA) versus nonparametric correlation analysis (NPCA) versus linear regression analysis (LRA; we'll discuss nonlinear cases later):

	Cause-and-effect relationship?	
	Yes	No
Data types:		
Both parametric	LRA	PCA
One parametric		
Cause parametric	NPCA	NPCA
Effect parametric	LRA	NPCA
Neither parametric	NPCA	NPCA

Clearly, the use of parametric correlation analysis looks pretty rare, but it is worth knowing how to do because it is potentially much more powerful than its nonparametric counterpart. Moreover, it is an essential step to a wonderfully informative technique—multiple correlation—that we cover in Chapter 12. Note also that the opportunities for regression analysis are only slightly more common. Clearly, the nonparametric correlation analysis is the most widely applicable test for relating variables, and we will take it up first.

We said that the basic fact of correlation analysis is that a correlation statistic measures the degree to which an increase in the value of X is associated with an increase in the value of Y. (The choice of which variable to assign to the x versus y axis is arbitrary in correlation analysis.) But as the data are rarely perfectly correlated, how do you handle the scatter? You could simply calculate the proportion of points for which an increase in X leads to an increase in Y, ignoring the degree of change. This would be the correlational equivalent of the sign test for nonparametric paired data (described in Chapter 9)—not very powerful since quite a bit of information is ignored.

The actual nonparametric correlation statistic now most commonly computed uses an improved method developed by C. E. Spearman around 1905. It is calculated by ranking all the x-axis values of the points and then all the y-axis values. For a given datum, D_i (the ith data point in the set), its x-value gets a rank and its y-value gets a rank. If there are n data points, then the highest X value will be n and the lowest 1; the same

The **Spearman rank correlation** is

$$d_i = R_{X,i} - R_{Y,i}$$

$$r_s = 1 - \frac{6\sum d_i^2}{n^3 - n}$$

where $R_{X,i}$ is the rank of the ith X-value among the set of X-values, $R_{Y,i}$ is the rank of the ith Y-value among the set of Y-values, and n is the number of pairs of values.

range applies to the Y values. For datum D_i, let's call the pair of ranks r_{Xi} and r_{Yi}. For a correlation of 1.0, of course, r_{Xi} and r_{Yi} will be equal. But if they are not, you need somehow to weight their degree of difference and then combine that with the equivalent values from all other data points. The procedure begins by simply taking the difference for each ranking pair, which we will call d_i for this pair: $d_i = r_{Xi} - r_{Yi}$. For perfect correlations, the d_i values should sum to zero: $\Sigma d_i = 0$. For the null hypothesis of no correlation, on the other hand, $\Sigma d_i = n/2$. Next it squares the various individual differences (d_i^2 in this case) and adds them all together (Σd^2). Finally, it computes the correlation, which is known as r_s for nonparametric data (the subscript refers to Spearman); the formula is: $r_s = 1 - [(6 \Sigma d^2)/(n^3 - n)]$.

Need more than the basics? See **More Than the Basics:** *Calculating Nonparametric Correlation Values.*

Let's apply this analysis to data on cigarette consumption and subsequent death rates from lung cancer. You might suppose that we are using correlation analysis because the tobacco industry has convinced us that there is no plausible cause-and-effect relationship between smoking and lung cancer. The real reason is that the death rates do not appear to be parametrically distributed:

Country	1930 consumption		1950 deaths		d	d^2
	Rate	Rank	Rate	Rank		
Australia	480	7	180	6	1	1
Canada	500	5	150	8	−3	9
Denmark	380	8	170	7	1	1
Finland	1100	2.5	350	2	0.5	0.25
Great Britain	1100	2.5	460	1	−1.5	2.25
Iceland	230	11	60	11	0	0
Netherlands	490	6	240	4	2	4
Norway	250	10	90	10	0	0
Sweden	300	9	110	9	0	0
Switzerland	510	4	250	3	1	1
United States	1300	1	200	5	−4	16
		$n = 11$				d^2: 34.5

Putting this value into the formula, we get:

$$r_s = 1 - [(6 \sum d^2)/(n^3 - n)]$$

$$= 1 - [(6 \times 34.5)/(11^3 - 11)]$$

$$= 1 - [(207)/(1320)]$$

$$= 1 - 0.157 = 0.843.$$

The data are plotted in Figure 11-4. A correlation of 0.843 is quite high, and would be significant at $P < 0.01$ even for a sample size of 8. Thus there is a clear relationship between these two variables. We suppose that tobacco companies might argue that lung cancer causes smoking.

The good news is that *BioStats Basics Online* automatically computes the correlation and lets you compare the r_s for your sample size with a table of threshold correlation values. And this is not the end of potential analysis: if n is larger than 9 and your correlation is within the range from -0.9 to $+0.9$, then you can also compute a standard error of the coefficient and

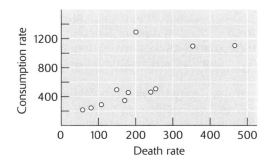

11-4 A correlation plot of cigarette consumption rates against death rates from lung cancer in Western countries. Your eye tells you that there is a correlation and also probably some background rate of lung cancer deaths in the absence of smoking.

Box 11.1

BioStats Basics Online **Nonparametric correlation analysis**

The test for analyzing nonparametric data for possible correlations is found on card IV. A-2. Enter or import your paired data and click on [**Compute**]. *BioStats Basics Online* will display the sample size (*n*) and the correlation coefficient (r_s); then it will plot the data. If the data cross either the *x* or *y* axis, you will see a vertical and/or horizontal line indicating their position. Compare the correlation coefficient with the line for your sample size on the scrolling field in the upper right of the card. The calculated correlation must be larger than the corresponding value in the table for the correlation to be significant at the *P*-value at the head of that column. *BioStats Basics Online* does not calculate the SE; the formula for this simple calculation is given in the text.

compare coefficients—which you might do to see if the correlation between cigarette smoking and lung cancer is significantly stronger than the one between coffee drinking and lung cancer, for instance (which, of course, it is). It seems odd that you can compute an SE having used nonparametric data, but the fact is that you are not basing that on a mean (which would not work for nonparametric data) but on a correlation; this is equivalent to the repeated-sampling techniques we showed you how to use on nonparametric distributions in Chapter 5. Like repeated-sampling means, correlations drawn from nonparametric data are themselves parametrically distributed. The standard error is $\sqrt{(1 - r^2)/(n - 2)}$.

You can also quantify the degree of difference between two parametric correlation coefficients (beyond merely checking to see if their SEs overlap), or even compare several different correlations; these methods, however, are beyond the scope of this text. Be aware, however, that there are small differences between the scaling factors used when the original data were parametric versus nonparametric; the text you consult will probably deal prominently with the parametric case, and so you may need to read carefully, check another section, or scan the footnotes to find the nonparametric correction.

11.3 Parametric Correlation Analysis

Parametric correlation analysis is potentially far more powerful than nonparametric analysis. The reason is that you know that the distributions are each parametric, with means, SDs, and well-understood quantitative relationships that add mathematical muscle and certainty. As we shall see, it also opens the door to multiple regression.

Parametric regression begins with a transformation of all the data into standard parametric units: for the x-axis values, a mean is found, and that value subtracted from all x-axis entries (yielding a mean of zero); the x-axis SD is computed and then all x-axis entries are divided by the SD (to yield data scaled to SD units); the same transformation is then applied to the y-axis data. Once all the data are in standard units, the parametric analysis can begin. The parametric correlation, known as $r_{1,2}$, is usually obtained through the ***product-moment method***; this improvement on Galton's original development was the work of Karl Pearson around 1895. All the n data pairs are multiplied (i.e., for the ith datum, $x_i \cdot y_i$) and summed ($\sum xy$). The sum of the products is then scaled by dividing it by the square root of the sum of the squares of the data: $\sqrt{\sum x^2 \sum y^2}$.

Figure 11-5 shows the results of this analysis applied to heights and shoe sizes for some biology majors. Both height and shoe size are parametrically distributed, but there is no plausible cause-and-effect relationship between the two. The correlation is obvious to the eye, and the product-moment calculation yields a value for $r_{1,2}$ of 0.735, which is highly significant (indeed, even at a sample size of 11, $P < 0.01$).

As with nonparametric correlations discussed in the previous section, you can compute standard errors for parametric correlations. The standard error is, again, $\sqrt{(1 - r^2)/(n - 2)}$. And as with nonparametric correlations, you can also quantify the degree of difference between two parametric correlation coefficients (beyond merely checking to see if their SEs overlap), or even compare several different correlations; for parametric correlations, there are no limits on the sample size or range of correlation values. The details of these rarely needed methods are beyond the scope of this text.

The product-moment correlation is

$$x_i = \frac{X_i - \mu_X}{\sigma_X}$$

$$y_i = \frac{Y_i - \mu_Y}{\sigma_Y}$$

$$r_{1,2} = \frac{\sum(x_i y_i)}{\sqrt{\sum x_i \sum y_i}}$$

where X_i is the X-value of the ith datum, Y_i is the Y-value of the ith datum, μ_X is the mean of the distribution of X, μ_Y is the mean of the distribution of Y, σ_X is the SD of the distribution of X, and σ_Y is the SD of the distribution of Y.

Need more than the basics? See **More Than the Basics:** *Additivity Once More.*

Need more than the basics? See **More Than the Basics:** *Calculating Parametric Correlation Values.*

11 Relationships between Variables I

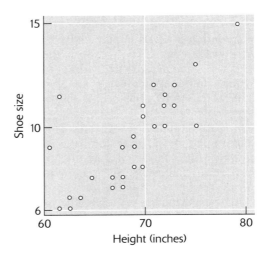

11-5 A parametric correlation analysis of height versus shoe size of biology majors (standardized to female sizes) yields a relatively high correlation of 0.735. Parametric correlation analysis is appropriate because there is no cause-and-effect relationship, and both variables are parametrically distributed.

Box 11.2

BioStats Basics Online   Parametric correlation analysis

The test for analyzing parametric data for possible correlations is found on card IV. A-3. Enter or import your paired data and click on (**Compute**). *BioStats Basics Online* will display the sample size (n), the means and SDs of the data, and the correlation coefficient ($r_{1,2}$); then it will plot the data. If the data cross either the *x* or *y* axis, you will see a vertical and/or horizontal line indicating their position.

Compare the correlation coefficient with the line for your sample size on the scrolling field in the upper right of the card. The calculated correlation must be larger than the corresponding value in the table for the correlation to be significant at the *P*-value at the head of that column. *BioStats Basics Online* does not calculate the SE; the formula for this simple calculation is given in the text.

11.4 Linear Regression Analysis

As we said earlier, if your variables have a plausible cause-and-effect relationship, and the ***effect*** variable is parametrically distributed, you can perform regression analysis. The "***cause***" variable, which is conventionally known as the ***independent variable***, is plotted on the x axis; the effect variable, or rather the ***dependent variable***, is plotted on the y axis. (This variable is called "dependent" because, to the extent there is a cause-and-effect connection, its value *depends* on the magnitude of the independent variable. "Cause," we hasten to say yet again, is in quotes as a reminder that you cannot be sure that you have identified the true cause as opposed to a secondary or a correlated one.) The values of the independent variable must have negligible uncertainty.

The **cause**, or **independent**, **variable** is plotted on the x axis, and is presumed to partially control the value of the **effect**, or **dependent variable**, which is displayed on the y axis.

Linear regression provides not only a correlation coefficient but also fits a line to the data and derives a formula to describe how the value of X affects the value of Y. Let's look at Pearson's father/son height data again, this time with a regression line fitted to it (Figure 11-6). The dashed line on the graph connects all points of equal height—the line expected if sons were exactly as tall as their fathers. You can see at a glance that the regression line has a gentler slope than the equal-height line; this is regression to the mean, a phenomenon mentioned above. You can also see that more points fall above the equal-height line than below it; this is the secular variation alluded to earlier. You should also note the scatter. The correlation coefficient, r, is 0.51; n is 1078.

The formula for this regression line is $Y = 35.0 + 0.51X$; 35.0 is the value of Y when X is zero—the point at which the regression line crosses the y axis—while 0.51 is the slope of the regression line. This means you can compute any point on the line (within the range of data used to create the line) given either the independent variable (X, the father's height) or the dependent variable (Y, the son's height). Thus if you take a father 60 inches tall, on average his son should be $35.0 + (0.51 \times 60) = 35.0 + 30.1 = 65.1$ inches tall. The son is taller in consequence of a secular increase of about an inch and a regression to the mean of 50.1%. To put it another way, if there were

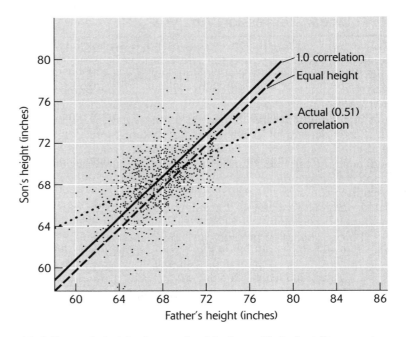

11-6 Pearson's height data again, this time with its best-fit regression line (solid) and the line expected if each father/son pair had the same height (dashed). Most points fall above the equal-heights line, reflecting a secular trend; the regression line (dotted) has a gentler slope than the equal-heights line, a consequence of regression to the mean.

no secular variation and no regression to the mean—a perfect correlation between the generations—then the son of a 60-inch male would also be 60 inches tall; if only secular variation were a factor, the son will be 61 inches tall; because the average height of fathers was about 68 inches, and the correlation was only 0.51, about half of the 8-inch difference between the father's height and the mean was "regained" by the son (about 4 inches), and thus the average son ought to be roughly $1 + 4 = 5$ inches taller than a 60-inch father.

So how do you draw this "best-fit" line? One way to visualize the process is shown in Figure 11-7a. Each x-axis (independent-variable) value has a parametric distribution of y-axis (dependent-variable) values associated with it. Ideally, the regression line connects the mean of each distribution (assuming, of course, that a straight line provides the best fit). But since each of these is a sample distribution, the sample mean is only an approximation of the true mean for each value of X. So in fact we are actually trying to fit a line through a set of means bounded by standard errors (Figure 11-7b). On average, the regression line should pass between 68.26% of the pairs of SE brackets (and 95% of the 2-SE brackets).

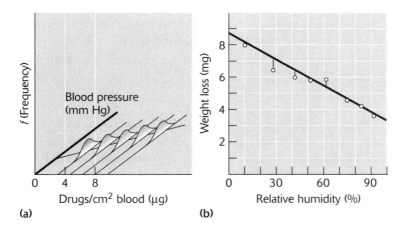

(a)   (b)

11-7 (a) Each independent-variable value is associated with a parametric distribution of dependent-variable values; in this three-dimensional representation, the distributions and their means are represented as curves rising out of the conventional two-dimensional independent-versus-dependent variable plane. The regression line passes through these distributions in a way that best approaches the means. **(b)** Another set of data plotted conventionally with the mean and SE of each independent-variable value shown and the best-fit regression line drawn in. [**(a)** and **(b)** Redrawn from R. R. Sokal and F. J. Rohlf, *Biometry*, 2nd ed. (New York: W. H. Freeman, 1981), pp. 459 and 463, respectively.]

The **least-squares method** fits a line to a set of points in a way that minimizes the sum of the squares of the deviation of each point from the line.

The method used to fit the regression line to the data is known as the ***least-squares method***, which was independently developed by Karl F. Gauss and the French mathematician A. M. Legendre around 1800. The idea is that the best-fit line should be the one that minimizes the sum of the squares of the y-axis differences between the Y-value of each datum and the Y-value of the regression line. The use of summed squares should be familiar to you by now, having seen it in test after test (e.g., in computing the SDs of parametric distributions and the chi-square statistic for categorical data). The process is pretty tedious without a computer if there are many data, but is faster (and can work about as well) if you simply use the Y-value means for each value of X.

We say "can work" because there is a potential problem if:

1. there are few means contributing to the line,

2. the sample sizes contributing to some means are small, and

3. the small-sample-size means are at the extreme limits of the independent variable.

For instance, the data in Figure 11-6 have only sparse representation at the short and tall ends of the father's-height axis; any means there would be subject to considerable uncertainty. The consequence is that the means in the center of the x-axis can act as a pivot for the regression line, and the unreliable means at the end can create an anomalous slope as a result of ordinary sampling error. This "end-effect" is a risk in correlation analysis too when means rather than raw data are used; the only solution is to be alert to the problem and conservative in your interpretations when forced to use means.

Assuming that you've not been careless, and your regression line is therefore not distorted by a means-based end effect, you can compute a P-value. The one calculated by *BioStats Basics Online* is the F-statistic, with the null hypothesis being that b (the slope of the line) $= 0$. There is also a t-test method some researchers use, but it gives the same result. The standard error of the regression correlation is given by $(\sqrt{1 - r^2})SD_y$, where SD_y is the standard deviation of the dependent variable.

The 95% confidence interval, in case you ever need it, is $b \pm SE \times t$, where t is the value for $P = 0.05$ with $n - 2$ degrees of freedom shown in the table in *BioStats Basics Online* for the *t*-test.

You can readily see from Figure 11-7 why the *y*-axis values (the dependent variable) must be parametrically distributed to draw a best-fit line through the means. What may not be so obvious is why you cannot fit a regression line to a parametric correlation. Remember that the only difference is the absence of a putative cause-and-effect relationship. The problem is that the best-fit line depends on which variable is assigned to the *x* axis—that is, which is considered to be the "cause." Figure 11-8 shows the two very different regression lines that are computed if you assume that height causes weight versus the supposition that weight causes height. Clearly, there cannot be two best-fit lines; there must be an unambiguous way of deciding which variable is the probable cause.

Having warned you against drawing a regression line in the absence of a cause-and-effect relationship and thus making yourself look foolish to your colleagues, we must qualify the prohibition slightly: you *can* draw a regression line if you are only trying to use one variable to predict the value of the other—

Need more than the basics? See **More Than the Basics:** *Calculating Linear Regression Values.*

11 Relationships between Variables I

Box 11.3

BioStats Basics Online Linear regression analysis

The test for dealing with linear regression is found on card IV. B-2. Enter or import your paired data, taking care that the independent variable (the possible cause) is assigned to the *x*-axis column and the dependent variable (the putative effect) is consigned to the *y*-axis field. When you click on **Compute**, *BioStats Basics Online* will compute and display the degrees of freedom, the correlation coefficient, and a test statistic (*F*). It will

also display the values for the slope of the best-fit line and its *y*-axis crossing. Finally, it will plot the points and draw the least-squares regression line. To see if the correlation is significant, compare the calculated *F*-statistic to the df values on the line in the scrolling field corresponding to ($n - 2$); the *F*-statistic must be larger than the value in the table for the correlation to be significant at the *P*-value listed at the top of the column.

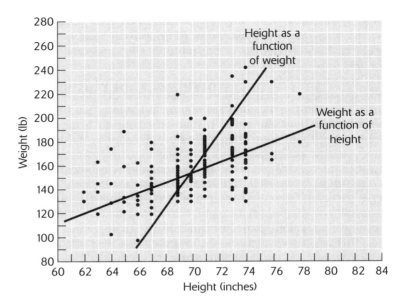

11-8 If there is no clear dependent and independent variable, then it is not clear which variable should be regressed on the other. Here height and weight are regressed, first with height as the independent variable, and then with weight as the independent variable. [Based on R. J. Herrnstein and C. Murray, *The Bell Curve* (New York: Free Press, 1994), p. 565.]

using a person's weight, for instance, to predict that person's height. In this case, the variable you are using for prediction is treated as though it were the independent variable; the one you are trying to predict is then treated as dependent. As pointed out earlier, you must be careful to get the pseudo-cause and pseudo-effect variables straight, because regression yields different lines depending on which is the dependent and which the independent variable.

Now for a general remark about correlation and regression values: each test, as we have seen, generates a test statistic r, which is the correlation coefficient. This statistic ranges between -1.0 and $+1.0$. It quantifies an intuitive perspective on the data: how much the two variables track one another. It is computed in essentially the same way we calculate standard deviations for conventional sample means, except here the sample mean is a line. As we will explain in Chapter 12, r^2 is the equivalent measure for variance in correlations (just as SD^2 is the variance in a parametric distribution). Clearly, r^2 ranges from 0 to 1.0, and is inevitably smaller than r. The value of r^2 is the fraction of the variance in the data that that is shared between the two variables; the rest of the variance arises from other sources. This number is important in partitioning the blame for an effect between putative alternative independent causes.

11.5 Which Hypothesis to Test?

We said in Chapter 3 that we would return to the space shuttle O-ring question in this chapter. We saw that Poisson analysis confirms NASA's assumption of independence in O-ring failure. NASA also attempted to find some sort of correlation between failures and a number of variables, including temperature before launch (which is now understood to have been the problem—the O-rings become brittle at low temperatures). Could NASA have known? The first step to answering that question is to ask what sort of an analysis is proper. If temperature is the independent variable, then failures/launch would have to be parametrically distributed to allow proper regression analysis. They are not, since zero was by far the most common value. Thus not even parametric analysis is possible without some transform that would normalize these data, and no such effort was made.

When NASA calculated the nonparametric correlation for the failure data, it *should* have obtained a value of $-.18$ for 24 launches; the threshold correlation for this small sample size is $-.41$ (Figure 11-9a). (The negative correlation reflects the tendency for failures to increase with decreasing temperature.) In fact, however, NASA improperly performed this calculation by inexplicably considering only the launches for which at least one failure had been observed; this yielded a *positive* correlation of 0.09, suggesting that launches at cold temperatures might actually be slightly safer (Figure 11-9b). Nevertheless, even if

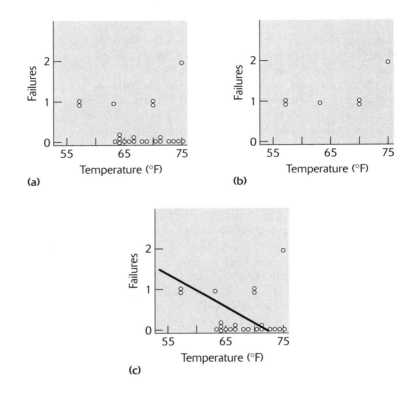

11-9 **(a)** The nonparametric correlation between launch temperature and O-ring failures/damage; the correlation is -0.18 for the 24 launches that preceded the *Challenger* disaster. **(b)** The nonparametric correlation is $+0.09$ if only the seven launches in which an O-ring failure/damage occurred are considered; there is, however, no plausible justification for omitting the zero-failure launches. **(c)** Had the failure/damage data been normalized so that regression analysis could have been justified, a correlation of -0.57 would have emerged. The regression line assumes a linear relationship between the variables. Later information suggests that the effect should increase more steeply at lower temperatures, worsening the risk.

NASA had done its calculations correctly, it could not have rejected the null hypothesis that temperature has no effect.

But should a safety-conscious agency like NASA be in the business of seeking to avoid false positives, or should it worry more about false negatives? As we have said before, our usual null hypothesis as dispassionate scientists is that nothing is going on, and we require strong evidence to convince us otherwise. A safety analysis should usually adopt a guilty-until-proven-innocent approach, considering a null hypothesis that every factor analyzed is a hazard, or at least employing a very different P-value threshold. And did NASA make any effort to normalize the data, without worrying that it had only 10 O-ring failures spread across seven launches to work with? Had it normalized the failures and temperatures, the equivalent calculation should have yielded a parametric correlation of about $-.6$, which is significant at $P < 0.01$; here is a dramatic illustration of the importance of normalization techniques. Again, had the failure data been normalized and subjected to regression analysis, the regression line would have predicted something like 3.2 failures at the prelaunch temperature of 1°C, an uncomfortably high number when failure of two rings in the same joint could be lethal (Figure 11-9c).

11.6 What If the Data Are Nonlinear?

We have been assuming that the best-fit regression line will be linear. The post hoc data now available on O-rings suggests that the failure rate does not rise linearly with decreasing temperature, but more nearly exponentially; in fact, it may be surprising that only one joint in *Challenger* failed at 1°C. This observation underscores the point that the relationship between two variables may not be linear. True, statisticians would prefer linearity because it is easier to deal with mathematically. But if your eyes tell you that there is a relationship between two variables that meet the criteria for regression analysis in an interaction that is obviously not linear, what are you to do? In Figure 11-10a, for instance, the number of amphibian and reptile species on islands in the West Indies seems to fit a log transform rather than a straight line.

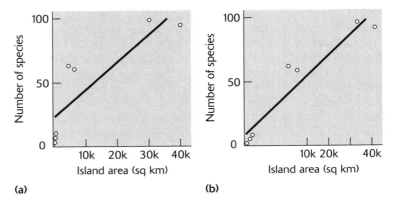

(a)   (b)

11-10 Ecological theory predicts that the number of species should increase with island size. **(a)** When the number of species of reptiles and amphibians in the West Indies are regressed as the dependent variable against island area, the points fit a regression line only poorly. **(b)** When the independent variable is transformed by taking the square root of island area, the fit is much better. Subsequent analyses using a variety of species and island sizes show that this transformation inevitably yields a linear fit. [Based on data in R. H. MacArthur and E. O. Wilson, *The Theory of Island Biogeography* (Princeton, N.J.: Princeton University Press, 1967).]

First of all, it should be obvious that you cannot do anything to the dependent-variable data: its parametric nature is a precondition for analysis. Thus you are left with somehow adjusting the independent-variable values. The usual approach is to try one of the data-transformation techniques discussed in Chapter 6, with special attention to finding an e^x value that yields a straight line. Now if there is an obvious a priori reason for assuming that there should be a particular e^x scaling involved—a typical characteristic of growth, for instance—this transform can be accomplished with a minimum of worry. In the case illustrated by Figure 11-10, the underlying theory of island biogeography predicted that species numbers should increase with the square of the area, and area is the independent variable.

When theorists transformed the area axis by taking its square root, the regression-line fit was much improved (Figure 11-10b).

Without an a priori justification, you must worry that you are just fitting data, a post hoc process. Thus the *P*-value we ultimately obtain is just a measure of fit, not a true measure of the significance of the correlation against a null hypothesis of no correlation. For a true measure of the correlational probabilities, you must perform your computation *before* you transform the *x*-axis values. Consider the island-area case: the first regression (Figure 11-10a) yielded a correlation coefficient of 0.85, significant at $P < 0.05$, and therefore the researchers already knew that there is a relationship between the variables. Had they been curve-fitting at random, the eventual best fit (which yields a correlation of 0.95, $P < 0.01$) could serve only as a hypothesis to guide further experimentation; but since the a priori theory predicted a square-root relationship, you can take the later statistics at face value.

11.7 Cause and Effect?

If you find a significant correlation when you run a regression analysis, does this mean you have demonstrated a causal relationship between the independent and dependent variables? As you have probably guessed from the repeated use of the word "putative" and the placing of "cause" in quotation marks, the answer is not necessarily (and perhaps never) yes. There are two general problems regression analysis is heir to.

The first involves cases of **common causes**. Suppose you had data on species numbers for reptiles and carnivorous birds on islands. You might hypothesize that the factor controlling predator numbers is prey numbers, and thus regress them against each other with reptile species number as the independent variable and carnivorous bird species number as the dependent variable. You would see a high correlation and an excellent least-squares fit, but there is in fact no significant cause-and-effect relationship between the two variables; instead, they have a common cause (island area, as we just saw).

When the value of an effect variable appears to depend on the magnitude of a putative cause variable, it can be because their values both depend on another factor, which is the **common cause** for each.

When the value of an effect variable depends on the magnitude of a putative cause variable, it can be because the apparent independent variable itself is an effect that depends on another higher-level factor; in that event, the "cause" variable is actually an **intermediate cause**.

The second class of potential problems involves ***intermediate causes***. Our favorite example of intermediate causes was uncovered in the mate-choice behavior of female guppies. We run these tests by placing a female in the center section of a three-compartment tank, with a male in each end section; we score the amount of time the female spends at each end, analyzing the results as paired nonparametric data. We measured female preference for tail size and found a significant correlation between choice strength and tail area. Cause and effect? Later, we found a much larger preference for display rate, and then discovered that the display rate is highly correlated to tail area. We might have been tempted by the first results to conclude that there is a cause-and-effect relationship between tail area and female preference. Being wise in the philosophical ambiguities inherent in the concept of "cause," however, we would never have said this in print, but we might have been tempted to think in those terms.

The wisdom of our encultured reluctance to accept the common-sense interpretation of these results was powerfully demonstrated by the display-rate data, and again by the correlation between area and rate. (In this particular case, we were able a few weeks later to show that tail area and display rate are independently evaluated by females, but it could just as easily have turned out that only display rate matters, and that tail area determines display rate.) Thus it is possible to measure a significant correlation between two variables without being sure that the "cause" variable is the immediate cause, the ultimate cause, or something intermediate along a chain of causal events.

When the value of an effect variable depends on the magnitude of a putative cause variable, it can be because the two variables are not independent, but instead **covary** with each other.

A third kind of problem involves ***covariance***. Consider Pearson's height data: he measured independent (father's height) and dependent (son's height) variables, regressed the two, and deduced with some further calculations the magnitude of the apparent cause-and-effect relationship (the heritability, in this case). But Pearson's estimate is incorrect because he did not know that humans (mostly unconsciously) consider height (among many other things) when they choose spouses. Thus tall males marry, on average, females that are taller than the mean for the female sex—a phenomenon known as *assortative mating*. This unrecognized source of variance served to inflate Pearson's estimate of heritability for height to well above its true value. If he had been working with rats, he would have been able simply to choose a thousand males, mated them randomly to females (ideally, all identical twin sisters), and then have had "clean" data on heritability of rodent characteristics.

One lesson here is that finding a significant regression correlation means that the variables are probably related, but does not prove that the independent variable is the sole or immediate cause of the measured effect. You should always be cautious and conservative in your interpretation of data. Lead your readers to the conceptual cliff, but let them make the jump on their own. The same advice, of course, applies to all statistical procedures: a significant P-value not only doesn't prove your hypothesis (it only gives you the odds that one distribution is not drawn from another), it cannot prove that your experimental (independent) variable is the actual cause of any difference in the distributions.

The other lesson is that you must be ever alert to the chance that the world is more complex that your hypotheses allow. In particular, you must look for evidence that there are multiple causes at work, and that the causes themselves may interact. The example of Pearson's height data, where two sources of genetic input themselves interact, is an especially sobering example. But there are statistical ways of dealing with this sort of complexity and reducing it to a set of quantifiable effects; we turn to these techniques in the next chapter.

Points to remember

✔ Two measures for each of several organisms or events may appear to vary together in some systematic way. Depending on the nature of the data, the degree of correlation can be quantified.

✔ If the two measures of the data have no obvious cause-and-effect relationship, correlation analysis is appropriate.

✔ If both measures are parametrically distributed and there is no cause-and-effect relationship, parametric correlation analysis will quantify the degree of correlation.

✔ If one or both variables are nonparametrically distributed and there is no cause-and-effect relationship, the less powerful nonparametric correlation analysis will quantify the degree of correlation.

✔ If there is a cause-and-effect relationship, the values of the cause (independent) variable are well known, and the values of the effect (dependent) variable are parametrically distributed, the powerful regression analysis can quantify the degree of correlation and compute a line of best fit.

✔ Even if there is a cause-and-effect relationship, if the values of the cause variable are not well known,

or if the values of the effect variable are not parametrically distributed, only nonparametric correlation analysis can be applied to the data.

✔ A significant correlation between a pair of cause-and-effect variables cannot prove a cause-and-effect relationship. The two variables could depend on an unknown common cause, and thus both be effect variables; the apparent cause variable could actually be an intermediate cause, dependent on an unknown variable farther "upstream" in the chain of cause and effect; or the two variables may not actually be independent, and instead covary in some way.

Exercises

1. Table A presents a partial list of the win : loss records for Major League Baseball teams in 1992 and 1993. (The full list of data is found in file 11−1992 vs 1993 Baseball Wins or 11-WINS.dat. The data are somewhat inconsistent because two expansion teams were added in 1993; the win : loss records of these teams do not appear in the data because they cannot be compared between 1992 and 1993. The win : loss records of the other 1993 teams include games against the expansion teams, which means the

Table A   Partial list of win : loss records

Team	1993	1992
Atlanta	104 : 58	98 : 64
Baltimore	85 : 77	89 : 73
Boston	80 : 82	73 : 89
California	71 : 91	72 : 90
Chicago AL	94 : 68	86 : 76
Chicago NL	84 : 78	78 : 84
Cincinnati	73 : 89	90 : 72
Cleveland	76 : 86	76 : 86
Detroit	85 : 77	75 : 87
Houston	85 : 77	81 : 81

average number of games won per team considered in these data is not a whole number.) Assuming team records are not just a matter of luck, we might suppose that a team that is good one year ought to tend to be good the next. Is there a significant correlation between years? To find out, you will need first to determine whether the data are parametric to see if you can use the more sensitive parametric analysis. Why is regression analysis inappropriate?

2. Species characteristics often change in a systematic way with latitude or altitude. The presumption is that average temperature (which varies with latitude and altitude) selects for different traits. Table B lists the

average height of yarrow plants at various altitudes in the western United States. (The data are also in file 11–Yarrow Heights or 11-YAROW.dat.) You may assume that the heights listed represent the means of parametric distributions. Is there a significant correlation? Which analysis is correct here?

3. The ease of inducing death-feigning behavior in chicks seems to increase with age and mobility. The mean number of induction attempts one each day are listed in Table C. (The data are also in file 11–Induction Attempts or 11-INDUC.dat.) As you can probably guess, the distribution of induction attempts is not parametric. Is there a significant correlation? What test is appropriate?

Table B Plant heights at various altitudes

Location	Altitude (m)	Height (cm)
Mather	1250	75
Aspen Valley	1750	50
Yosemite Creek	2000	49
Tenaya Lake	2300	32
Tuolomne Meadows	2500	21
Big Horn Lake	3100	15
Timberline	2800	20
Conway Summit	2400	25
Leevining	2000	43

Table C Susceptibility to tonic immobility

Day #	Attempts
1	6.7
2	5.3
3	4.6
4	2.9
5	1.8
6	1.4
7	1.2
8	1.3

4. We saw in Figure 11-3 that the correlation between vSAT score and first-year grades of biology majors was 0.36. Table D has a partial list of mSAT scores and GPAs for the same students. (The complete data are in file 11—mSAT vs GPA or 11-MSAT.dat.) Try regressing these numbers against each other. Which is the independent variable? Is there a significant correlation? Is it larger or smaller than the vSAT correlation?

Table D   Partial list of scores and GPAs

Student #	mSAT	GPA
1	640	3.7
2	600	3.2
3	620	2.9
4	560	3.7
5	680	3.3
6	500	3.7
7	670	3.2
8	630	2.5
9	530	2.7
10	530	1.7

5. Some researchers believe that homing pigeons know where they are because they can measure slight variations in the earth's magnetic field, variations that provide good clues to latitude and perhaps to longitude as well. One line of evidence offered is that in pigeon races, the birds return to the home loft more slowly on days of "magnetic storms" (caused by excess ions in the earth's ionosphere, which in turn are produced by solar flares). A more controlled test is to release homing pigeons at magnetic anomalies, sites with iron deposits that increase the earth's field strength slightly (not enough to deflect a magnetic compass) and irregularly. Table E presents mean vector lengths for pigeons released at sites of different strengths; the field values listed are strengths in excess of the background level of 50,000 gamma at the latitude of the tests. (The data are also in file 11—Magnetic Anomalies or 11-ANOM.dat.) The mean bearing lengths can be taken as parametric.

Table E   Effect of anomalies on orientation

Magnetic field variability (gamma)	Length of mean vector
30	0.65
100	0.70
150	0.35
200	0.40
300	0.37
700	0.22
2800	0.17

Try a regression analysis. If you find a correlation but are unhappy with the best-fit line, try transforming the independent variable.

6. See if there is a significant correlation between shoe size and the number of syllables in the given names of biology majors. There is a partial list in Table F. (The complete data are in file 11–Shoe Size vs Syllables or 11-SHOE.dat.) What is the appropriate test? (All values are in terms of female sizes.)

Table F Partial list of student data

Student #	Shoe size	Syllables
1	12	5
2	9	4
3	6.5	7
4	7	6
5	7.5	7
6	6	6
7	11.5	5
8	9	6
9	9.5	6
10	10	5

7. Computer prices vary greatly, but then so do features and effective speed. Perhaps speed is the major factor in pricing. Table G (page 284) shows a sample of prices and effective speeds for a variety of systems using Motorola RISC processors at the end of 1997. Speeds are relative to Intel Pentium I

processors; the Pentium II has an effective speed about 50% faster, with another 50% increase for Pentium III. Processor speeds have thus undergone a huge increase in speed since these data were collected. (The full set of data is found in files 11–Desktop Data or 11-DESK.dat, and 11–Laptop Data or 11-LAP.dat.) All prices have been adjusted to a standard configuration (32MB RAM, 2MB VRAM, 1GB hard drive, 12× CD-ROM, 15-inch monitor) by adding or subtracting the appropriate costs. The costs for any given effective speed are parametrically distributed. How should these data be analyzed? Is there a significant correlation for desktop systems? For laptop systems? If you were able to regress the data, does a speed of 0 MHz cost $0? How do the per-MHz costs differ between desktops and laptops? Analysts say that the newest (i.e., fastest) systems are always overpriced to take temporary advantage of a marginal speed edge; is this piece of conventional wisdom borne out by the data?

8. The Nurse's Health Study tracked more than 50,000 nursing personnel over a 20-year period looking for health correlations. Other possible correlations in the study are also of interest. One variable measured was related to hours of weekly exercise. The nurses were divided into five exercise levels with

Table G   Effective speed vs adjusted prices for computers

Desktop systems

Effective speed (MHz)	Adjusted cost ($1000s)	Effective speed (MHz)	Adjusted cost ($1000s)
240	1.90	300	2.10
300	1.85	240	1.45
200	1.45	250	2.05
275	2.15	530	2.45
465	2.25	530	2.95
270	2.05	450	3.45
300	2.15	522	4.15
375	2.40	200	1.65
450	3.05	240	1.80
450	2.85	240	1.90
200	1.85	280	2.35
240	2.05	275	2.05
240	1.90	300	2.15
225	1.80	315	2.05
250	1.85	335	1.35
270	2.35	375	1.55

Laptop systems

Effective speed (MHz)	Adjusted cost ($1000s)	Effective speed (MHz)	Adjusted cost ($1000s)
133	2.20	240	4.40
166	2.50	200	3.75
160	3.15	180	3.20
500	5.50		

thresholds that divided the sample roughly evenly between groups. Table H presents the average personal characteristics for each exercise group. (The data are also found in file 11–Nurse Data or 11-NURSE.dat.) Body mass is not parametrically distributed; fat intake and fiber intake, on the other hand, are. What can you do with these

Table H Weekly exercise and other personal characteristics of nurses

Exercise group*	Body mass (kg/m²)	Smokers (%)	Vitamin use (%)	Fat intake (g/day)	Fiber intake (g/day)	Colon cancer rate[†]
I	26.0	26.4	38.4	60.3	16.0	0.74
II	25.3	22.5	39.6	59.3	16.8	0.51
III	25.0	19.6	41.3	58.7	17.3	0.54
IV	24.5	16.4	44.2	57.6	18.0	0.48
V	24.0	16.8	46.1	56.3	18.7	0.39

*Group I: <2 hours/week; Group II: 2–4 hours/week; Group III: 5–10 hours/week; Group IV: 11–21 hours/week; Group V: >21 hours/week.

[†]Per thousand person-years.

data? What test or tests are appropriate? What can you say about the well-known hypotheses that fat ingestion promotes colon cancer, that fiber ingestion helps prevent this cancer, that smoking promotes the disease, that vitamins C and E help prevent cancers, and that exercise reduces the risk of colon cancer?

9. Conventional wisdom holds that inbreeding is bad, largely because it raises the odds of expression of formerly hidden (recessive) lethal or deleterious genes. Data are in general scarce. One study of great tits (small songbirds), however, did analyze egg failure as a function of the relatedness of parent birds. The results are shown in Table I. (The

Table I Inbreeding vs egg viability

Parental relatedness	Number of eggs that fail to hatch
0.250	2.1
0.125	1.9
0.063	0.6
0.031	0.5
0.015	0.6
0.008	0.3

data are also found in file 11– Inbreeding Data or 11-EGGS.DAT.) The failures are parametrically distributed. What is the correct test? Are the results significant?

More Than the Basics

Calculating nonparametric correlation values

To perform Spearman's rank correlation analysis, use the formula

$$d_i = R_{X,i} - R_{Y,i}$$

$$r_s = 1 - \frac{6\sum d_i^2}{n^3 - n}$$

where $R_{X,i}$ is the rank of the ith X-value among the set of X-values,

$R_{Y,i}$ is the rank of the ith Y-value among the set of Y-values, and

n is the number of pairs of values.

Additivity once more

Correlation answers the recurring question, what do we do with nonindependent variables? As we know, the variance of a sum of independent variables is equal to the sum of the variances—if $Z = X + Y$, then $V_Z = V_X + V_Y$. But if the variables are not independent, we must revise this sum.

If $Z = X_1 + X_2$, then $V_Z = V_1 + V_2 + 2r_{1,2}\sqrt{V_1 V_2}$, where $r_{1,2}$ is the correlation between the two variables. This is the basis of multiple regression analysis.

In fact, $r_{1,2}\sqrt{V_1 V_2}$ is a quantity called the *covariance*. The covariance is analogous to the variance, except that instead of calculating a sum of squares, we calculate a sum of products:

$$V_{1,2} = \sum \frac{(X_{1,i}) - (\mu_1)(X_{2,i} - \mu_2)}{}$$

Calculating parametric correlation values

To calculate the product-moment correlation coefficient, use the formula

$$x_i = \frac{X_i - \mu_X}{}$$

$$y_i = \frac{Y_i - \mu_Y}{}$$

$$r_{1,2} = \frac{\sum (x_i y_i)}{\sqrt{\sum x_i \sum y_i}}$$

where X_i is the X-value of the ith datum,

Y_i is the Y-value of the ith datum,

μ_X is the mean of the distribution of X,

μ_Y is the mean of the distribution of Y,

σ_X is the SD of the distribution of X, and

σ_Y is the SD of the distribution of Y.

Calculating linear regression values

To perform parametric regression analysis, use the formula

$$SS_{total} = \sum Y_i^2 - \frac{\left(\sum Y_i\right)^2}{n}; \ df_{total} = n - 1$$

$$SS_{regression} = \frac{\left(\sum (X_i Y_i) - \frac{\sum X_i \sum Y_i}{n}\right)^2}{\sum X_i^2 - \frac{\left(\sum X_i\right)^2}{n}}; \ df_{regression} = 1$$

$$SS_{residual} = SS_{total} - SS_{regression}; \ df_{residual} = df_{total} - df_{regression}$$

$$F = \frac{SS_{regression}}{df_{regression}} \frac{df_{residual}}{SS_{residual}}$$

where X_i is the X-value of the ith datum, Y_i is the Y-value of the ith datum, and

n is the total number of data,

with degrees of freedom: $df_{numerator} = 1$; $df_{denominator} = n - 2$.

The slope of the best-fit line is given by

$$M = \frac{\left(\sum (X_i Y_i) - \frac{\sum X_i Y_i}{n}\right)^2}{\sum X_i^2 - \frac{\left(\sum X_i\right)^2}{n}}$$

The y-axis intercept of the best-fit line is given by

$$a = \sum \frac{Y_i - mX_i}{ }$$

Techniques Covered in This Chapter

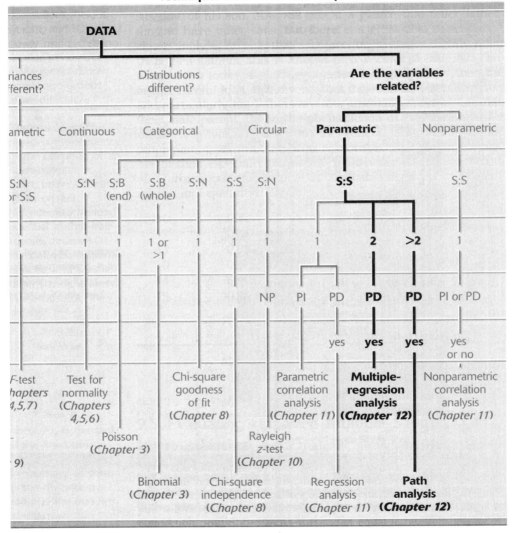

between Variables II: Multiple Regression

12

We said in Chapter 11 that Pearson's analysis of the heritability of height in humans was flawed because of interacting variables. (Heritability studies attempt to partition the sources of variance in a trait like height between various genetic and environmental variables; they provide a good model for multiple-variable analysis in general.) Let's construct a diagram representing the way independent variables contribute to the dependent variable (son's height) as Galton and Pearson envisioned it (Figure 12-1a) and as researchers now understand it (Figure 12-1b). We are going to use the usual verbal shorthand of heritability analyses: we will talk in terms of independent traits and other factors combining to generate dependent traits; however, the actual analysis deals strictly with *variance* (the familiar SD^2, which you may recall can be added), and thus we are in fact looking at how variance in independent traits and variance in other (generally environmental) factors add to generate the observed variance in dependent traits.

12.1 How Multiple Variables Interact: Path Analysis

You might imagine that three truly independent variables contribute to the dependent variable: father's height (or rather, the father's height genes), mother's height (genes), and environmental influences. These three variables, together, account for all the observed variance. Thus if we encode these variables as

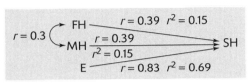

(a)

Father's height: FH ⸺ $r = 0.51$ $r^2 = 0.25$
Mother's height: MH $\frac{r = 0.51}{r^2 = 0.25}$ ⟶ SH: Son's height
"Environment": E ⸺ $r = 0.71$ $r^2 = 0.50$

FH ⸺ $r = 0.39$ $r^2 = 0.15$
$r = 0.3$ ⤻ MH $\frac{r = 0.39}{r^2 = 0.15}$ ⟶ SH
E ⸺ $r = 0.83$ $r^2 = 0.69$

(b)

12-1 (a) A path diagram showing the correlations (r) and proportions of the total variance (r^2) linking the dependent variable, son's height (SH), to the three putative independent variables, father's height (FH), mother's height (MH), and "environment" (E, which includes chance and all other effects). The variances in such diagrams must sum to 1.0. This path diagram assumes no correlation between any of the independent variables. **(b)** A more realistic path diagram that allows for the modest degree of positive assortative mating ($+0.3$) with respect to height among parents. Taking this factor into account alters the values for the true correlations and variances. It is important to remember that path diagrams assume additive interactions and deal only with variances—the variance in MH, FH, E, and SH—rather than with the traits themselves.

FH (father's height), MH (mother's height), E (environment), and SH (son's height), you could compute correlations of each independent variable with the dependent variable: $r_{FH,SH}$, $r_{MH,SH}$, and $r_{E,SH}$. We will simply assert (see More Than the Basics: Variance, Additivity, and Standard Error in Chapter 5 and Accounting for Variance in this chapter for the logic) that correlation values must be squared before they are combined; the reason is that correlations are in terms of standard deviations (SD), whereas only variances (SD^2) can be added. Since together the three factors we've mentioned account for everything, it follows that $r_{FH,SH}^2 + r_{MH,SH}^2 + r_{E,SH}^2 = 1.0$, where "1.0" means

"100% of the observed variance." (Or at least that is what nearly everyone assumes; it's always possible, however, that the interactions are not additive but combine in some other way—perhaps as a synergistic effect. The standard assumption is based on two somewhat dubious facts: (1) in the few cases that this axiom can be quantitatively tested, it's proved to be roughly true, and (2) if the interactions are not additive, the math would be nearly impossible. We are, of necessity, assuming additivity.)

Pearson found that $r_{FH,SH} = 0.51$; assuming that $r_{MH,SH}$ is of about the same magnitude (a reasonable assumption in most cases of inheritance), you can compute the values in this pathway. (For mathematical simplicity, we will assume from this point on that the apparent correlations are $r_{FH,SH} = r_{MH,SH} = 0.5$.) Since the total variance is 1.0 and is composed of $r_{FH,SH}^2 + r_{MH,SH}^2 + r_{E,SH}^2$, and since you know that $r_{FH,SH}^2 = r_{MH,SH}^2 = (.5)^2 = .25$, then you have $1.0 = .25 + .25 + r_{E,SH}^2$, which is to say $r_{E,SH}^2 = 0.5$. (And, therefore, $r_{E,SH} = \sqrt{0.5} = 0.71$; using the actual measured correlation of 0.51 instead of the rounded value gives a total heritability of 0.52 rather than 0.5.) Thus half the variance in height is genetic in origin; the other half is environmental (or a result of chance effects, which are traditionally lumped with environment since the object of the inventors of heritability analysis was to quantify only the genetic component).

But as we mentioned in Chapter 11, Pearson overestimated the heritibility of height because he did not know that humans mate assortatively with respect to height—that is, a tall male on average is slightly but significantly more likely to marry a tall female. The correct form of the pathways contributing to height (Figure 12-1b) allows for this correlation between parental heights ($r_{FH,MH} = 0.3$). You will notice that there are new values for the correlations between each independent variable and the dependent variable. Where did these come from?

Path analysis constructs diagrams like this and evaluates the flow from "independent" variables (in quotation marks for the moment to remind us that they are not in this case independent of each other, but rather covary because of assortative mating) to dependent variables. The flow from father to son is via two paths: the direct FH → SH route and the indirect FH → MH → SH route. You know the apparent FH:SH and MH:SH

Path analysis is used to quantify the correlations in a web of interacting variables; it infers unmeasurable correlations from observed correlations, and can yield true values of cause-and-effect interactions that are otherwise obscured by hidden interactions.

correlations and the actual FH:MH correlations. We say "apparent correlations" because the $r_{FS,SH}$ 0.5 that Galton measured (and the 0.5 we infer for $r_{MH,SH}$) are not the true correlations, because each is compounded of the direct *plus* the indirect route. The true values are the ones shown in Figure 12-1b.

So how do you get these true correlation values? Let's call the true correlation p to distinguish it from the measured correlation, r. The measured correlation of 0.5 for FH $\rightarrow$ SH is the sum of p (the true value of the direct path) plus 0.3 p (the indirect path from the mother, which is the product of the FH:MH and FH:SH correlations). Thus $0.5 = p + 0.3\ p; p = 0.38$. Thus the true values are $p_{FH,SH} = 0.38$ and $p_{MH,SH} = 0.38$, and thus each parent's direct genetic contribution is $(0.38)^2$, or a bit more than 0.15; together these sum to a genetic contribution of 0.31. You must subtract 0.31 from 1.0 to get the environmental component, which is 0.69. This corrected genetic component of 0.31 is substantially less than the value of 0.52 that Pearson initially calculated; clearly, it is important to look for correlations between so-called independent variables and correct apparent interactions to take into account these hidden ones.

Need more than the basics? See **More Than the Basics:** *Accounting for Variance.*

Path analysis can be far more elaborate than this relatively simple example. Even just having different correlations between the two independent variables and the dependent variable ($r_{FH,SH} \neq r_{MH,SH}$) makes the equations much more difficult to

12-2 A path diagram for a situation involving four independent variables (X_1, X_2, X_3, X_4), three of which are correlated with one another. Computing the relative variances that contribute to the variation in the dependent variable (Y) is far more complex than in the height example.

solve. And adding additional independent variables is even worse. For instance, there could be four "independent" variables, three of which interact with each other (Figure 12-2), generating six correlations to sort out; if the three independent-to-dependent variable correlations have different values, the equation occupies about half a page. Really complex problems, like the one illustrated in Figure 12-3 (analyzing the factors that account for the variance in the job status of two brothers) generate extraordinary webs of interactions; nevertheless, the approach is the same. What *is* missing here is any sort of way of

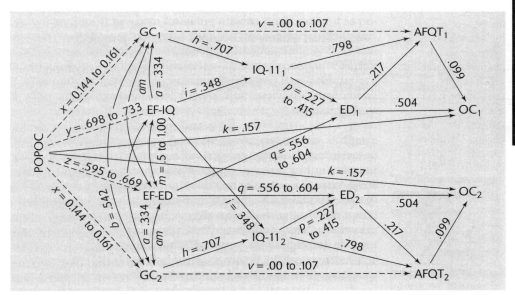

12-3 An elaborate path diagram accounting for the correlation between occupational status (OC) of brothers. The factors that are important include IQs, amount of education (ED), genetic contributions (GC), family environment (EF), parent occupation (POC), and various combinations of these variables. In several cases, the correlations are listed as inferred ranges, which highlights the difficulty of deriving precise values when so many variables and correlations are in play and few apparent correlations can be measured directly. [Redrawn from C. Jencks, *Inequality* (New York: Basic Books, 1972), p. 343.]

evaluating standard errors of the correlation values when they are derived indirectly (as many inevitably are when correlations between independent variables require recomputing the measured independent-dependent correlations). Moreover, you cannot readily derive a regression-like formula relating the variables.

12.2 The Goal of Two-Variable Multiple Regression

Multiple regression is used to quantify the regression values between various independent variables and a single dependent variable even when the independent variables covary. Though all variables must be parametrically distributed, the technique yields precise regression values, probabilities, standard errors, and other statistical values that path analysis cannot usually provide.

Need more than the basics? See **More Than the Basics:** *Calculating Multiple Regression.*

When one special condition is met, the interactions and effects of two independent variables on a dependent variable can be quantified with all the precision we hope you've come to expect of statistical analysis. The condition is this: all the variables, dependent and independent alike, must be parametric. The reason is that the interaction between the *independent* variables must be computed as a parametric correlation, which naturally requires that these variables be parametric. For normal regression analysis, you may recall, the independent variable could be nonparametric.

In the last chapter, we showed that a linear regression line follows a least-squares path through the data points. In the case of two-variable **multiple regression**, try to imagine a three-dimensional array of points, each the intersection of two independent-variable values and one dependent-variable value (Figure 12-4). Through this set of points passes a best-fit plane— a plane drawn by the same least-squares method. If we call the two independent variables X_1 and X_2, and the dependent variable Y, then the equation describing this plane is $Y = a + b_1 X_1 + b_2 X_2$, where a is the value of Y when the best-fit plane passes through the Y-variable axis (that is, where X_1 and X_2 are both zero); b_1 is the slope of the plane when X_1 is the only independent variable in play (that is, when X_2 is zero); b_2 is the slope of the plane when X_2 is the only independent variable at work (that is, when X_1 is zero). The problem for multiple-regression analysis is to find this best-fit plane and then to quantify its degree of fit—that is, the proportion of the variance in the data the chosen independent variables account for.

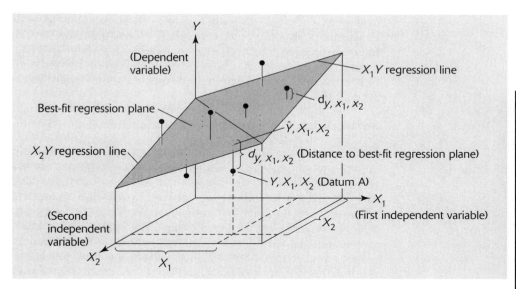

12-4 Multiple regression finds a best-fit plane that minimizes the sum of the squares of the deviations between each data point and the regression plane. In this three-dimensional representation, the dependent variable is plotted on the vertical (Y) axis; the two independent variables, X_1 and X_2, are plotted at the bottom. A particular value, datum A, is shown located at coordinates Y_A, X_{1A}, X_{2A}; its deviation from the best-fit plane is shown as a line projecting onto the plane. The interaction between each independent variable and the dependent variable is shown as the line created by the intersection of the regression plane and the plane of the $X_1 Y$ axis and the $X_2 Y$ axis. In this example, the $X_1 Y$ correlation is positive (yielding an upward-sloping line), while the $X_2 Y$ correlation is negative (producing a downward-sloping line). [Redrawn from R. R. Sokal and R. J. Rohlf, *Biometry*, 2nd ed. (New York: W. H. Freeman, 1981), p. 621.]

12.3 Admit or Reject? The Problem

Let's take an example from the kind of problem that faces college admissions committees. Say you want to predict the degree of probable success of each applicant. You know there are many variables involved, many of which (e.g., motivation) cannot be directly measured. You are presented with several imperfect (perhaps even irrelevant) sets of data, some of which cannot be readily quantified or calibrated: verbal SAT scores, quantitative SAT scores, scores on advanced-placement tests, high school GPAs, degree of challenge in the student's choice of high school courses, letters from teachers, information about family background, and so on.

One approach to this problem would be to attempt to quantify each kind of information (independent variable) for each student previously admitted, then to regress each against college GPA (a plausible dependent variable, and one which college administrators optimistically equate with college success). If the independent variables are *actually* independent, the strength of each correlation (or rather, the square of each correlation) will tell us the relative weighting to place on each variable in the admissions process. The degree to which the sum of the squared correlations fail to add to 1.0 will tell us how much uncertainty remains in accounting for college success (as measured by GPA).

But this doesn't work. In fact, if you try it, the total correlation seems high (indeed with enough factors, it far exceeds 1.0), and yet many seemingly highly qualified students insist on doing badly, and a surprising number of "marginal admits" shine. The error, of course, lies in the intrinsic correlation of these independent variables with each other. For instance, to the extent that high school grades depend on intellectual aptitude and that Scholastic Aptitude Test scores may capture that quality, the two should be to some degree correlated. Thus, just as with Galton's height data, there will be multiple pathways contributing to each measured correlation, with the result that they overestimate the actual correlations.

So let's try to work out a simple problem for our admissions committee: how much weight, if any, should be put on vSAT and/or mSAT scores in predicting the GPA of biology majors? If scholastic aptitude matters in biology students, and the SAT to some extent measures that aptitude, you should expect to see some positive correlation of each score with GPA. And if some basic sort of "general intelligence" underlies both quantitative and verbal skills (as many general-intelligence psychologists believe), the vSAT and mSAT scores of individual students should also show some degree of correlation. On the other hand, if they represent different kinds of talent (as many modular-mind psychologists maintain), they might be uncorrelated, or even negatively correlated. Thus you should try to sort this out, and estimate the degree of remaining uncertainty by applying two-variable multiple-regression analysis to the data. In this analysis we are using a sample size of all 47 biology majors at Freeman University, consisting of 24 males and 23 females. Later we will want to analyze the sexes separately to see if the correlations are different.

Let's try first a naïve attempt to correlate the two SAT tests separately with GPA. Surprisingly perhaps for a science, we find a correlation of 0.37 for vSAT → GPA but only 0.17 for mSAT → GPA (Figure 12-5a). This roughly 2 : 1 ratio in apparent correlation is interesting: the National Merit Scholarship qualifying score is compounded of the mSAT score on the PSAT

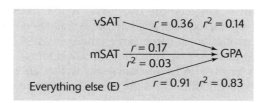

(a)

(b)

12-5 **(a)** Path analysis for predicting GPA from SAT scores ignoring any correlation between vSAT and mSAT. **(b)** The same path analysis allowing for the vSAT : mSAT correlation. The predictive value of the SAT scores is lower as a result of the correlation between the independent variables.

(an SAT-like test scored on a 20–80 scale rather than the 200–800 scale of the SAT) plus *twice* the vSAT score. But these are only correlations; we must square their values to get the proportions of the variance that each factor seems to account for: vSAT $= (0.37)^2 = 0.14$; mSAT $= (0.17)^2 = 0.03$. Perhaps the verbal part of the PSAT test should have a fivefold rather than a twofold weighting.

For GPAs, the two SAT scores together seem to account for a mere 17% (that is, $0.14 + 0.03$) of the variation in college GPAs among biology majors at Freeman. But this straightforward interpretation is correct *only* if vSAT and mSAT are not correlated with each other. Perhaps they are *negatively* correlated, and thus this exam, which supports a huge test-preparation industry and commands the angst of millions of parents and students annually, actually *does* have some serious predictive value for science and humanities GPAs when considered separately.

You can find out if this is true by running a parametric correlation analysis between the two SATs. When you do, you will find a *positive* correlation of 0.28 between vSAT and mSAT. Since this correlation is much larger than the mSAT : GPA correlation, there is just a chance that the indirect path from vSAT $\rightarrow$ mSAT $\rightarrow$ GPA is responsible for the entire apparent mSAT : GPA correlation. If so, the admissions office ought to ignore mSAT scores altogether (at least for potential biologists). Or perhaps the true correlation between mSAT and GPA is *negative*, and the vSAT : mSAT correlation is what is responsible for an apparent mSAT $\rightarrow$ GPA effect. Let's see.

12.4 Sorting Out the Influence of Multiple Variables

Multiple-regression analysis allows you to quantify this set of test and performance interactions with as much precision as is possible given the sample size. The method was pioneered by Karl Pearson at the end of the nineteenth century, and much refined by his colleague G. U. Yule in the early 1900s. The analysis requires knowing the parametric correlation between the independent variables, as well as the parametric correlation between each independent variable and the dependent variable.

The test performs a parametric form of path analysis, generates a true set of correlations, an equation for the best-fit plane, and a pair of pure regression lines. Figure 12-5b presents the correct path diagram; Figure 12-6 shows these two true regression lines for our example. The true regression lines describe the effect of one independent variable when the other is held constant. The correct correlations are: vSAT:GPA = 0.35 (versus an apparent correlation of 0.37); mSAT:GPA = 0.07 (versus an apparent correlation of 0.17). The corresponding variances are 0.12 for the vSAT and 0.005 for the mSAT.

Thus mSAT scores do have *some* predictive value, but it is only one twenty-fourth that of vSAT scores; together, SAT scores account for only 13% of the variation in GPAs. In short, SAT scores are responsible for only an eighth of the variance in college GPAs of biology majors at Freeman University, and the mSAT is essentially useless as a predictor. (Interestingly, mSAT scores have more predictive value for social science majors at Freeman. And, to be fair, the range of SAT scores at Freeman University is very narrow, which necessarily minimizes these correlations.) Multiple-regression analysis also determines the *P*-values for the two independent variables, both together and separately.

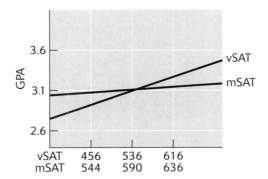

12-6 Multiple regression yields these two corrected regression lines, one for mSAT versus GPA when vSAT is held constant, and the other for vSAT versus GPA when mSAT is fixed. The axes are in terms of standard deviations (+1.0 SD, the mean, and −1.0 SD are marked); thus the scale for vSAT is different from the one for mSAT.

Box 12.1

BioStats Basics Online   Multiple regression

The multiple-regression test is easily the most complex in *BioStats Basics Online*. There are two ways to perform the test: you can either begin with the raw data (the method we describe here) on card IV. D-3, or you can enter directly on the card the following summary information: the means of X_1, X_2, and Y, the SDs of X_1, X_2, and Y, and the parametric correlations $r_{X1,Y}$, $r_{X2,Y}$, and $r_{X1,X2}$. (We provide this direct-entry option because you may have access only to data summaries, or may already have obtained these data from the Parametric Correlation card (IV. A-3), or may not have all three variables in one data set, but rather as three separate sets of paired data.)

Enter or import your data into the Multiple Regression: Step One card (IV. D-3) and click on **Compute**. *BioStats Basics Online* will compute and display the parametric correlations between each independent variable and the dependent variable, and between the two independent variables; it will also

calculate and display the means and standard deviations of all three variables, as well as the sample size. There is a table on this card if you wish to check the significance level of any of the correlation values.

Next, click on the **Transfer Data** button; this will take you to the Multiple Regression: Step Two card (IV. D-4) and automatically enter all the computed values from the previous card. When you click on the **Compute** button on this card, *BioStats Basics Online* will calculate a host of values and plot two true regression lines. The most important values are the slopes of the two true regression lines, which are each presented in two forms: the left-most value is in the units of the variables, as are the *b*-values that go in the regression-plane formula; the right-most values are in SD-scaled units, and are the true unit-independent regression-correlation values. It is these unit-independent values that are used to plot the two regression lines for comparison

Not only does multiple-regression analysis generate some surprises when it is used to sort out the relative contributions of mSAT and vSAT to GPA in biology majors at Freeman University, the picture is even more remarkable when the sexes are considered separately. For males, the vSAT : mSAT correlation is +0.62; for females it is −0.01; these are significantly different for the sample size involved, and we see the same discrepancy year after year. Or again, gender-specific analysis tells us that for males, the mSAT correlation with GPA is actually stronger than the one for vSAT (0.21 versus 0.17, a significant with each other; the X_1-variable regression against the dependent variable is plotted as a thin line, while the X_2-variable regression is shown as a thick line. The value of a in the regression-plane formula is listed as the Y-axis intercept.

Below these values are the F-statistics for various combinations of variables, with the associated df values. These statistics include the F-values for:

1. the true correlation of X_1 alone with Y, allowing for the correlation between X_1 and X_2;

2. the true correlation of X_2 alone with Y, allowing for the correlation between the independent variables;

3. the true correlation of X_1 and X_2 together with Y.

The tables of threshold F-values will allow you to check the P-value associated with each statistic. There are two tables because the X_1-alone and X_2-alone values require one table (the topmost of the two scrolling fields) while the value for the two together requires another (the bottommost of the two); use the df value below each statistic to locate the correct line of the table to consult.

The F-statistic for the combined variables will usually be smaller than one or both of the F-statistics for the individual variables. This somewhat counterintuitive result is a reflection of the statistical reality that we are comparing three rather than two variables. Thus we are increasing the chances of false positives, just as when we had to use ANOVAs to compare three or more parametric distributions.

Finally, *BioStats Basics Online* calculates the proportion of the variance accounted for by each variable and displays those values in the lower right. To get the variance still unaccounted for, add these two values and subtract their sum from 1.0.

difference when five successive years of data are combined), whereas for females the vSAT is a much better predictor (0.39 versus 0.18—also significant, and consistent from one year to the next). If an admissions committee wants to use SAT scores to help predict GPA in future biology majors, then treating males and females separately would allow them to improve their predictions somewhat (and reduce the variance). For instance, if the female-weighted SAT correlations are used, the predictive value of SATs rises from 0.13 to 0.19 for females, but falls to 0.08 for males.

We've not said much about the underlying math, and that's because it is complex. But to go just a bit further with our visual-intuitive approach, if you look back at Figure 12-4 you will see two lines where the multiple-regression plane intersects the plane rising vertically from each independent-variable axis. These lines are the true regressions: the line above the X_1 axis is the true linear-regression line of the independent variable X_1 against the dependent variable with the correlation (if any) between X_1 and X_2 factored in. The line above the X_2 axis is the true linear-regression line of the independent variable X_2 against the dependent variable with any correlation between X_1 and X_2 factored in. If the correlation between the independent variables is 1.0, the two regression lines would be the same, of course. If the correlation is 0.0, the two intersections would generate the same lines that simple linear regression would yield—that is, regression of one independent variable against the dependent variable without any consideration of shared pathways or correlations with other independent variables. For any correlation less than 1.0 there may be a tilt of the multiple-regression plane (as there is in Figure 12-4) between the two independent-variable axes.

12.5 Higher-Level Multiple Regression

If you wanted to add high school GPAs to this analysis, you would need to add a third independent variable (X_3), and take into account many more correlations: $X_1 : X_2$, $X_2 : X_3$, $X_1 : X_3$, $X_1 : Y$, $X_2 : Y$, and $X_3 : Y$. A fourth independent variable would increase the number of correlations to ten; a fifth would take it to fifteen; and so on. There is no theoretical limit to this iteration, but the multiple-axis "plane" becomes impossible for ordinary mortals to imagine, and the math spins out of control even faster. But serious computers running elaborate statistical software can perform the same sort of disentanglement of variables that mere two-variable multiple regression accomplishes.

We have never needed anything more than three-variable multiple regression, which is probably more a comment on the poverty of our imaginations—an inability to think of all the factors that might be involved in controlling the value of a dependent variable—than any inherent simplicity of nature. (*BioStats Basics Online* limits itself to the two-variable case; we simply could not fit all the necessary fields for input and display onto a single card.) But the tests are readily available should you ever need them, and the logic we've developed for the two-variable case is sufficient for understanding any nightmarish n-variable version of multiple regression.

Finally, we should mention a set of statistical techniques known as ANCOVA (analysis of covariance). Sometimes you may have two independent variables, one of which is continuous (like SAT scores) while the other is discrete (gender, for instance). This calls for a hybrid analysis in which one variable is regressed while the other is treated like the variables in ANOVA analysis. This remarkable body of tests, a testament to the cleverness of statisticians, is rarely given more than a mention in introductory texts because ANCOVA is complex and infrequently called into play; we are adopting the same "it's out there" approach, doing nothing more than alerting you to its existence.

Points to remember

✔ Cause-and-effect relationships become complicated when two or more causes (independent variables) help determine the value of an effect (dependent) variable. The complexity grows if the "independent" variables themselves are correlated (i.e., covary).

✔ Path analysis helps sort out all the possible interactions. By using the measurable correlations, it is often possible to infer the correlation values of the interactions that cannot be measured and thus compute the true magnitude of cause-and-effect relationships.

✔ Multiple regression is a statistically precise technique for analyzing and quantifying interactions with two or more covarying independent variables and a single dependent variable. It requires all of the variables to be parametrically distributed.

✔ Both path analysis and multiple regression sometimes demonstrate that apparently strong cause-and-effect interactions are actually illusions generated by correlations between the independent variables.

✔ Both path analysis and multiple regression are able to estimate the proportion of variance in the dependent variable that is unexplained by the independent variables.

Exercises

1. The outcomes of baseball games depend on some combination of pitching, hitting, fielding, and luck. There is very little difference between the quality of fielding from one team to another, but hitting (as measured by batting averages) and pitching (as measured by earned-run averages, or ERAs, a measure of the average number of runs given up by pitchers) do vary. Table A gives a partial list of data from the 1995 major league season. (Complete data for several seasons are in file 12—Baseball Data or 12-BBALL.dat.) To what extent do batting averages and ERAs account for a team's win : loss record? Which factor is more important? (Note: lower ERAs are better.)

2. To grow, plants need light, warmth, water, carbon dioxide, and minerals. Carbon dioxide is distributed nearly uniformly around the globe, and light correlates well with mean annual temperature. Soil fertility varies enormously, even over very short distances. In a global analysis of plant growth (or, in ecological jargon, "productivity"), we can look at the relative importance of two easily measured variables: temperature and

Table A   Partial list of baseball statistics

American League Eastern Division

Team	Win : loss record	ERA	Batting average
Boston	.615	4.40	.281
New York (Y)	.500	4.85	.277
Baltimore	.462	4.64	.259
Toronto	.422	5.03	.266
Detroit	.405	5.70	.257

American League Central Division

Team	Win : loss record	ERA	Batting average
Cleveland	.698	3.87	.290
Kansas City	.513	4.27	.261
Milwaukee	.496	4.73	.273
Chicago (WS)	.443	5.21	.286
Minnesota	.374	5.85	.278

precipitation. Table B is a partial list from 42 sites. (The complete data are in file 12—Productivity Data or 12-PLANT.dat.) Are both variables important? From the initial calculation of the independent-variable: dependent-variable correlation (before

allowing for any correlation between the independent variables), how much of the total variance in productivity is accounted for by precipitation and temperature? How much after multiple-regression analysis? Which factor is more important globally?

Table B   Partial list of plant-growth data

Mean annual precipitation (cm/year) at site	Mean annual temperature (°C) at site	Productivity (g/m²/year) at site
100	−12.5	250
400	−9	750
300	−1.5	250
400	−1	700
200	+1	500
300	+1	400
700	+2	1000
500	+2.5	700
500	+3	750
500	+3	850

3. The admissions committee at some universities quantify candidates on an academic "A" scale and a personal "P" scale running from 1 (best) to 5 (worst). For future biology majors, the A and P scales do not approximate a normal curve, but for subsequent history majors they do. Table C presents a partial list of A and P scores for 50 history students, along with their first-year GPAs. (The complete data are in file 12—Freshman Data or 12-FROSH.dat.) To what extent does the A-rating correlate with GPA?

Table C   Partial list of student rankings and GPAs

A-rank	P-rank	GPA
3	4	3.5
2	2	3.3
3	3	3.5
1	2	3.6
2	3	3.3
3	3	3.5
3	3	3.2
2	2	3.3
3	4	2.9
3	2	3.2

Does it do a better job than SAT scores (discussed in the text)? How do the *P*-ranks fit in: are they a way to boost the chances of academically poor students, or are they independent of *A*-rank, or do high *A*- and high *P*-ranks go together in well-rounded students?

More Than the Basics

Accounting for variance

In path analysis, we seek to explain the variance of a quantity in terms of the variances of contributing factors. In the simplest case, this takes the form of some number of measured factors, $X_1 \ldots X_k$ and the assumed unmeasured factor, U, combining to give a result, Y. This is analogous to saying

$$Y = b_1 X_1 + b_2 X_2 + \cdots b_k X_k + U$$

Because b_n is really a correlation statistic, and because of the formula for addition of variances given in More Than the Basics: Additivity Once More (Chapter 11), we can state the formula for variances as

$$V_Y = \sum_i p_{iY}^2 V_i + \sum_{i,j} V_{i,j} p_{iY} p_{jY} + r_{UY}^2$$

If all measurements are divided by their sets' SDs, forcing all the variances to equal 1, the final result is

$$1 = \sum_i p_{iY}^2 + \sum_{i,j} r_{ij} p_{iY} p_{jY} + r_{UY}^2$$

This is the essential theory of path analysis, and the root of the formula given in the text for simple path analysis. In this formula, the sum of correlations represents the variance directly due to the variables—the variance that would be measured if the variables were independent or set by the experimenter and no unknown factors were present. The sum of correlation products represents the variance due to covariance between the variables—the variance that would be subtracted out by controlling the variables independently. The final correlation is the influence of all factors not measured—measurement error, other influences, inaccuracy of the means due to small sample size, and any disregarded covariances.

Calculating multiple regression

To determine the slope (b_{Y_1}) of the $X_1 - Y$ regression line with X_2 held constant, use the formula

$$b'_{Y_1} = \frac{(r_{1_y} - r_{2_y}r_{1,2})}{}$$

$$b_{Y_1} = b'_{Y_1}\frac{s_Y}{s_{X_1}}$$

where b'_{Y_2} is the standardized partial regression coefficient, and

s is the indicated standard error.

To determine the slope (b_{Y_2}) of the $X_2 - Y$ regression line with X_1 held constant, use the formula

$$b'_{Y_2} = \frac{(r_{2_y} - r_{1_y}r_{1,2})}{}$$

$$b_{Y_2} = b'_{Y_2}\frac{s_Y}{s_{X_2}}$$

To determine the y-axis intercept (a) of the regression plane, use the formula

$$a = \bar{Y} - b_{Y_1}\bar{X}_1 - b_{Y_2}\bar{X}_2$$

To perform multiple-regression analysis, use the formula

$$\hat{Y}_i = a + b_{Y_1}X_{1,i} + b_{Y_2}X_{2,i}$$

$$SS_{total} = \sum(Y_i - \bar{Y})^2; \; df_{total} = n - 1$$

$$SS_{regression} = \sum(\hat{Y}_i - \bar{Y})^2; \; df_{regression} = 2$$

$$SS_{residual} = SS_{total} - SS_{regression};$$

$$df_{residual} = df_{total} - df_{regression}$$

$$F = \frac{SS_{regression}}{df_{regression}}\frac{df_{residual}}{SS_{residual}}$$

With degrees of freedom: $df_{numerator} = 2$; $df_{denominator} = n - 3$.

To test the hypothesis "X_1 has no independent effect on Y," use the formula

$$F = \frac{r_{2Y}b'_{Y_2}}{}\frac{(n - 3)}{}$$

With degrees of freedom: $df_{numerator} = 1$; $df_{denominator} = n - 3$.

To test the hypothesis "X_2 has no independent effect on Y," use the formula

$$F = \frac{r_{1Y}b'_{Y_1}}{}\frac{(n - 3)}{}$$

With degrees of freedom: $df_{numerator} = 1$; $df_{denominator} = n - 3$.

Formulas for multiple-regression analysis with more than two causative variables are beyond the scope of this book.

12 Relationships between Variables II

None of the Above $\lceil$ 13

What happens if you've done your research, got your data, and turned to the table of contents to find the right statistical test, only to discover that it's not there? It's not inconceivable, nor even comfortably uncommon. Or what if you have data for which you know the null distribution precisely but the distribution isn't normal. Are you really restricted to the relatively weak nonparametric tests, which assume that you know nothing about the null distribution?

A good first step is to create a path diagram of the problems, the steps in the process that yield the observed outcome, with the alternative choice points at each step. Sometimes this will make clear just what needs to be done—how, for example, subdividing the problem into two statistical analyses will unravel the difficulty. But if this mind-clearing exercise in path analysis doesn't do the trick, it's not the end of the world. Three methods exist for dealing with unusual data and special cases:

1. recast or transform your data into a form appropriate to a less powerful but more inclusive test;
2. consult one of the many dry, dusty tomes, erudite mathematicians, or convoluted computer programs on the subject;
3. make up your own test.

13.1 The Quick-but-Dirty Approach

The **quick-but-dirty approach** subjects data to a far less stringent but easily performed test to see if there is a trend. By far the most common down-market analysis used in this way is the chi-square test.

Though there may be no simple test that's perfectly suited to your analysis, sometimes there's a test that will do an adequate job, or at least give you some idea of whether it's worth proceeding further. The canonical method for this is to invent a null distribution of *categories* and use chi-square. Though it's seldom the appropriate test, it is always a useful second-best and good for ferreting out (or ruling out) trends.

For instance, here are some data we once collected in an experiment in which we had numbered all 8000 bees in a hive. We were attempting to test the dance-language hypothesis, the idea that honey bees have a symbolic language that allows them to communicate the distance and direction of food sources to one another by means of a dance that has components that correlate with each of these variables. We trained two small groups of foragers to sites in opposite directions (Figure 13-1) and then monitored as many of their dances as we could. Our idea was that if the direction information was being used, then the recruit bees attending the dances by north-station foragers should mainly turn up at the north station, while the recruit bees attending the dances by south-station foragers should mainly turn up at the south station. If the language is not used (the null hypothesis), there should be no correlation between dance and arrival directions.

We underestimated the difficulty of recording all the dance interactions, particularly when several dances were going on simultaneously. We also overestimated the recruitment rate: while six bees may actually attend a dance, perhaps one will be stimulated to leave the hive, and perhaps only 20% of these searching bees will find the small, highly artificial feeder used in the test. When we looked at the data, we saw that of 39 recruits captured at stations after having been observed attending a dance, 25 arrived in the direction predicted by the dance. The one-tailed binomial P-value for these data is 0.054—not quite significant.

But when we looked more closely at the data, we saw that we actually had two groups of bees: those that found a station

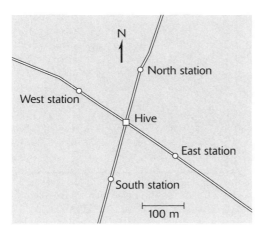

13-1 Experimental layout for a bee-language experiment. The hive was located at the intersection of two tracks in desert terrain, and foragers were trained to the north and south stations. Dancing was observed in the hive; the direction of the dance attended and the direction of arrival were recorded for individually numbered recruits. (The east and west stations were used to monitor the number of stray searching bees; there were none.)

quickly and those that took a long time. Those that arrived quickly almost always turned up at the station indicated by the dance; those that took a great deal of time arrived randomly (Figure 13-2). Looking even more carefully at our data, we found that when dance rates were low, and thus we could be sure of having seen every dancer–recruit interaction, then unsuccessful recruits—bees that attended a dance and then promptly left the hive, but then returned without having found a station and then attended another dance—searched only for between four and eight minutes before returning.

So we divided the data at eight minutes, considering separately the bees arriving within eight minutes and those that took longer (probably returning to the hive and attending,

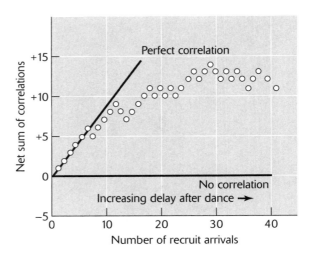

13-2 Recruits tended to arrive in the direction of the dance they had recently attended, but arrival direction of recruit bees that took longer times was uncorrelated with the direction of the observed dance.

unobserved, another dance to who-knows-which station). We ran the chi-square test for independence:

	Early-arriving	Late-arriving
Correct station	14	11
Wrong station	3	11

The picture changed. The chi-square analysis yields $P < 0.05$, indicating that right/wrong is not independent of early/late. We then used the chi-square goodness-of-fit test to compare the first-try recruits ($n = 17$) with the null hypothesis (8.5 recruits in each direction), finding a one-tailed $P < 0.005$; for the late arrivals, $P = 1.0$ ($n = 20$).

So does this mean our experiment yielded significant results? No, absolutely not: only our post hoc analysis did. But it does tell us how to run the next test; it also indicates the strength of the trend and allows us to compute the minimum sample size that such a trend will make significant ($n = 10$ early-arriving bees). This example clearly illustrates both the value and danger of post hoc analyses.

An unplanned chi-square analysis is so common as to be considered tame. On a more aggressive note, paired data can be treated as if unpaired (perhaps thus qualifying the data for a less powerful but more permissive test), and null hypotheses revised to stricter ones (thus holding the data to a stricter standard, but perhaps thereby qualifying it for a standard test); in both cases, you are using a more conservative test, insulating yourself from post hoc–based criticism. Finally, you can push existing tests to the limit based on their well-known (but theoretically inexplicable) properties; the most common example is force-feeding data with somewhat diverse SDs into ANOVA. Most statistical tests are robust enough to handle anything not outright dishonest. Tightly clustered circular data can be "unrolled" for parametric tests (that is, be treated as a linear distribution if the possibility of overlapping tails is remote), and continuous data treated as discrete (and occasionally vice versa). In the end, though, you may need simply to gather more data.

13.2 The Academic Approach

On the other hand, whatever your data, someone in the vast and varied field of statistics has probably figured out what can be done to them. Astonishingly often, an otherwise incomprehensible mathematician will volunteer useful information when asked, supplying either the name of a person to contact or a book to consult. Researchers who do a lot of analysis will almost always have a favorite statistician to whom they run with their problems and whom they can recommend to you. Many scientific fields have accumulated a suprisingly large lore on their own "pet" distributions; nobody knows more about the Poisson distribution than particle physicists, for instance. And it is surprising how often one can find a professor who has encountered a problem just like this a few years back and would just love to describe (often at great length) his or her clever and innovative technique to an interested student.

Most statistics texts, like this one, have bibliographies that provide a plethora of places to look. Their terminology is often somewhat difficult, and it seems to be a requirement of the field that no two statistics texts employ precisely the same words or symbols to refer to the same things (counterintuitive terms always being favored), but generally, perseverance will win the day.

There are, of course, vast hordes of computer programs available for statistical analysis, but the majority reveal little or no information on how to use their features, spitting out impossibly precise numbers without a word of explanation. Yet learning to use these programs, while often difficult, can be worth the effort if you expect to be doing a great deal of analysis.

13.3 The Hard Way: Monte Carlo Simulations

The **Monte Carlo method** simulates thousands of passes through the null-hypothesis contingencies of a statistical problem, thereby generating a null distribution that can serve as the basis of analysis against observed data.

Some problems, however, are so tricky as to defy all conventional analysis. One remarkably effective solution is the ***Monte Carlo method***, in which we create a null distribution by simulating the phenomenon over and over again, using the distribution of data thus generated as the null distribution and then comparing the actual observed distribution to it. For instance, the 10-sample equal-probability binomial null distribution we discussed in Chapter 3 could have been generated by actually flipping 10 coins thousands of times, or by consulting a table of random numbers and scoring the ratio of evens to odds in thousands of 10-digit groups if we had not had the binomial formula to use instead. (The term "Monte Carlo simulation" properly refers to computer-aided methods, first employed in the Second World War; without the benefit of a computer, such methods pose a hideously boring task. We will use the term to include all simulations, including coin flipping and iterative—i.e., endlessly repeated—pencil-and-paper computations.)

Let's look at two examples of how the Monte Carlo technique can be employed, one simple enough to solve with the product law if you have the patience, the second beyond all reach of pencil and paper. The first ("simple") example involves a problem posed in a popular book on probability: what are the odds of a basketball player with a 40% chance of hitting any single freethrow of getting exactly 11 of 20 such shots? Now sure you could do this with the product law, and you would get the right answer eventually (assuming you neglected to make some trivial error in math along the way; we find simple addition to be *our* biggest risk). Alternatively, you could write a simple Monte Carlo simulation with this generic structure:

Initialize: set "11-hits" total to zero
   Do the following for *n* times (where you input the *n*)
      Do the following for 20 times
         Select a random number from 0 to 100
         If the number is 40 or lower, add 1 to "Hits"
      If "Hits" total is 11, add one to "11-hits" total
      Set "Hits" total to zero
      Continue in this loop until you reach *n*
   Set "Proportion of 11-hit tries" to "11-hits" divided by *n*
   Display "Proportion of 11-hit tries"
Stop

A button on the Alternatives: Monte Carlo Simulations card (IV. F-1) in *BioStats Basics Online* will take you to a set of Monte Carlo simulations; one (card M-2) is for this 11-hit problem, but with some extra user-friendly features that allow it to serve as a demo. Note also that this program could be adapted to give you the distribution of 1-hit, 2-hit . . . 20-hit cases, or any minimum or maximum combinations of hits per run, or any run length other than 20, or calculate the numbers for a player with some hit rate other than 40%. A few key strokes in the basic program (shown on card M-3) can save hours of unnecessary product law computations to take into account some trivial change in the question or conditions.

Now for a more complicated example. A biology professor at our university asserted not too long ago that the choice of courses by students in his department was based entirely on how easily the courses were graded. This sort of suggestion, impugning as it did the honor of his departmental majors, had to

Box 13.1

BioStats Basics Online   **Monte Carlo analysis**

Monte Carlo analysis uses a computer program to generate a null distribution by simulating repeated trips through the web of contingencies in the phenomenon under analysis. The two Monte Carlo simulations described in this chapter are reached from a button on the Alternatives: Monte Carlo Simulations card (IV. F-1). One (card M-2 in the Monte Carlo stack) computes the outcomes for the basketball problem described in the text; the other (card M-4) computes the outcomes for the GPA problem. Neither is programmed to display the null distribution graphically, though Figure 13-3 shows what the GPA distribution looks like, and the table on

card M-5 gives the null-distribution numbers for the GPA problem, as well as the various means and *P*-value thresholds.

The emphasis in these two demonstrations is to help you visualize the individual passes through the contingency pathway. These visual aids are technically unnecessary in a Monte Carlo simulation, but they do provide an early indication of any errors in the programming. The actual program for the basketball problem (with the unnecessary display steps in italics, as well as the points at which the parameters of the problem could be readily changed) is shown on card M-3.

be met with rigorous statistical inspection, but the challenge was a hard one. Clearly, if students are choosing courses on the basis of GPA, they would wind up with higher in-department GPAs than chance would predict. But what *would* chance predict?

The null distribution of in-department GPAs is an incredibly complex creature to describe, for a number of reasons. Some courses have prerequisites, others are limited-enrollment, still others conflict with each other or with university requirements. A not-insignificant number of students took courses over the summer or at other institutions. The list of complications goes on and on.

So the seemingly innocuous question of the null distribution is, in the case of GPAs, entirely beyond the grasp of mathematics. Even if it turned out to be parametric, we have no way of knowing what the mean or SD ought to be. Something drastically new is required.

13.4 Computers: From Zero to Null in 14 Hours

Quite simply, the problem is that there is no null distribution. The solution is to generate one—not with math, however, but with brute force. We wrote a computer program (there is a user-friendly version that you can reach from the Alternatives: Monte Carlo Simulations card in *BioStats Basics Online*) that followed a random path through the department's program, sampling randomly from the actual grade distribution from each course, navigating the prerequisites, occasionally (at the average rate) finding the courses full or in conflict, and leaving enough space for university requirements. Each virtual student could complete a course of study in about two seconds of computer time. Then the program would dutifully start over and do it again. Best of all, it could do this while we were out of the office, and computers don't get paid overtime.

When we came back the next morning, the model student's 1,000,000 forays through the university hadn't quite generated a perfectly reliable null distribution yet, but it had made one thing clear: if our students base their class choices on grades, they're not very good at it. The random student managed an in-department GPA of 3.24, the real students 3.26 (Figure 13-3). A grade point difference of 0.02 is pretty small (and, as it turned out, insignificant for our sample size of real students, as shown on card M-5 of the Monte Carlo stack).

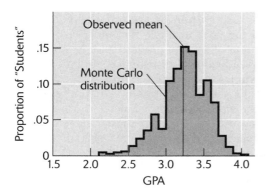

13-3 Here is the null distribution of GPAs based on 1,000,000 virtual students passing "randomly" through a biology department. The cyberstudents are constrained to complete the various departmental distribution requirements, avoid impossible numbers of spring- or fall-term courses, enroll in courses only if they have taken the prerequisites, and deal with limited-enrollment courses.

13 None of the Above

13.5 The Theory behind Statistics

There's even more power to this method: you don't have to stop with analyzing single occurrences one at a time. It's also possible to design entirely new statistical tests in this way, and this shouldn't surprise you at all: not all of those tables in the appendix were generated by abstract mathematics. Indeed, many statistical distributions were originally produced by stochastic (i.e., chance-based) methods like flipping a coin or consulting a table of random numbers, with the mathematicians following later to add the theory. Our experience shows that a statistical test must do two things: convert a batch of data into a test statistic, and then convert this statistic into a *P*-value by means of a table of critical values. And neither of those sounds very hard . . .

A **test statistic** is a bottom-line measure of the outcome of the set of contingencies that together contribute to a phenomenon.

Unfortunately, the first one sometimes is. A ***test statistic*** is a measure of extremity, a mathematical estimate of just how weird a set of data is. Although the statistic doesn't have to be the same as the *P*-value, it should definitely go up or go down—but not both—as the probability of a sample being extreme gets higher. In the case of grades, GPA was the test statistic.

If there is no obvious formula for a test statistic for your situation you will have to rely on your own cleverness. In general, a comprehensive knowledge of statistical tests and methods will allow you to guess a good formula, or at least a good-enough formula.

Monte Carlo analyses use the **distribution of the test statistic** to compute probabilities. As with all tests, the area under the null-hypothesis curve from the observed mean to the end of the tail, corrected for sample size and the number of tails in the analysis, generates the *P*-value.

Assuming that you've come up with a test statistic, you then have to determine the ***distribution of the test statistic*** under the null hypothesis; in the GPA example, the null prediction is that the actual distribution of student grades will have the same mean as the computer-derived null distribution. To get the null distribution, you program a computer (or find a savvy friend who is willing to do it for you) to generate a few hundred thousand data sets from the null hypothesis, calculate for each of these the corresponding test statistic, sort these values numerically from largest to smallest, and then find the test-statistic value (GPA in our case) for each *P*-value. (We did this in more detail in Chapter 4; this walk-through is just a reminder.)

In this departmental-GPA example, to get the critical value

for $P < 0.05$ in a one-tailed test situation, take the statistic one-twentieth of the way down the ordered list; this is the point at which the chance of drawing *one* grade at random from the null distribution that is this far from the mean in this particular *direction* falls below 5%. For instance, in Figure 13-3—or card M-5—you would begin at the high-GPA end (since the prediction is that students will choose the easily graded courses), and move left—down on the card M-5—until you have enclosed 5% of the null distribution. To repeat, this is the probability of *one* individual student achieving a GPA this high by chance. If you choose $P = 0.01$ as your threshold, you would locate the 1% point in the null-distribution tail.

Let's follow the rest of the analysis for the biology-student-GPA example, assuming a $P = 0.05$ threshold. Your next step is to correct for your actual sample size, n, by dividing the proportion of the null-curve area from the *null* mean to the *observed* mean by $\sqrt{n}$. (In our case, there were 49 students, so the scale factor was $\sqrt{49} = 7$). This determines the proportion of the area under the null curve from the null mean toward the predicted end that corresponds to the $P < 0.05$ threshold for your sample size. Now note equivalent GPA. This is the null-hypothesis threshold: a mean observed GPA higher than this is significant; a lower mean is consistent with the null hypothesis that students do not choose a course by its ease of grading.

Next, you need to determine the proportion of the area under the null-hypothesis curve from the *observed* GPA (not the *threshold* GPA of the previous paragraph) to the high end of the distribution; you are using the high end because the hypothesis specifically predicts that the observed GPA should be higher, not lower or just different. Subtract this value from 0.5 (the area from the mean to the end of the high-GPA tail) to get the proportion of the null curve from the *null* mean to the *observed* sample mean; if the value is negative, you can stop because the null hypothesis has not been contradicted regardless of the magnitude of the difference, since the observed mean fell in the wrong tail.

Finally, if the value is positive, you would compare the observed GPA with this threshold GPA: if the observed mean GPA falls between the null mean and the $P = 0.05$ threshold (as it does in this case), there is no significant difference; if the observed mean had been higher than the threshold value, the difference would have been significant.

In a two-tailed situation, you would follow these same steps *except* that you would use $P/2$ as the threshold (since the *P*-value is equally divided between the two ends of the distribution) and only consider the tail in which your sample mean falls; it is thus impossible to get the negative difference described in the one-tailed discussion above.

The two authors of this book differ in their preferred way to write Monte Carlo simulations. One of us (the elder) uses a high-level (i.e., easy) language like HyperTalk, which allows a clear visual presentation of the ongoing results and simple correction of programming errors; the other (younger) author prefers lower-level (i.e., difficult) languages like C, which are more elegant, flexible, and faster to write and to run, but which are harder to learn and debug. Part of the Monte Carlo stack in *BioStats Basics Online* (reached by clicking on Normality: A Hard-Core Example) displays a relatively simple program, written in the C programming language, that can serve as a skeleton for Monte Carlo simulations. It was used to generate the table used in *BioStats Basics Online* for the test for normality, using a test statistic from the Kolmogorov-Smirnov test of goodness-of-fit for continuous distributions. The generic program in Section 13.3, "The Hard Way: Monte Carlo Simulations," gives a pretty good idea of what the program in a higher-level language will look like, and card M-3 shows the program for the basketball example.

The point, then, is that if there is no "correct" test, your knowledge of what statistics is about and what the tests assume will almost always allow you to find an adequate test or write a simple program to create a null distribution. Remember that Fermat, Arbuthnott, Gosset, Rayleigh, Galton, Pearson, and many of the other major names in the history of statistics were not professional mathematicians; they encountered specific problems and devised solutions for them. Most of the tests we use today are the legacy of this specific-solution process that, incidentally, led to a generally useful new method. The test for normality in *BioStats Basics Online*, the most helpful method we have contributed, arose from exactly this same need-it-now situation, and we hope you will feel "empowered" to write your own test when the situation demands.

Points to remember

✔ Data—particularly data collected without the opportunity to institute experimental controls, or to establish which parameters were to be measured and in what way, or without the currently entertained (i.e., post hoc) hypothesis specifically in mind—are often difficult to analyze.

✔ Creating a path diagram of the variables can sometimes divide a seeming impossible problem into two or more statistically manageable pieces.

✔ The data can sometimes be recast into a statistically less demanding form; the most common technique is to sort continuous data into categories and analyze it with a chi-square test.

✔ Sometimes the problems with the form of the data or the hypothesis have been encountered before and been solved, often on a one-time basis; professional statisticians often know where to find the appropriate makeshift analysis.

✔ Analyses of data with no clear or easily calculated null distribution can often be made by generating a null distribution through brute force, usually by writing a computer program to run thousands of simulations of the set of null-hypothesis possibilities. This is the Monte Carlo approach.

✔ Monte Carlo distributions can usually be analyzed with great precision because they provide a precise null-hypothesis mean and a reliable distribution. Together these permit you to determine the exact proportion of the area in the null distribution between the null and the observed (sample) mean. Combined with the sample size contributing to the sample mean, the corresponding P-value can be readily calculated.

13 None of the Above

Exercises

1. Consider the following gambling system: you wager $1 on a 50 : 50 event—say, red versus black at roulette. If you win, you wager $1 again; if you lose, you wager $2. In the case of the $2 bet, if you win, you wager $1; if you lose, you wager $4. The same rule applies to all bets: you wager $1 if you have just won, and bet double the amount of the loss if you lose. The consequence of this system is that you turn a $1 net profit every time you win, no matter how long a losing streak you've had. Thus, if you lose the $1 and then the $2 bets but win the $4 wager, your net take is $1. What we want is the null distribution of money you would need to have with you to play this system—for example, $7 for the example above, in which you lost $1 and then $2 but then won the $4 bet. Create this null distribution with the product law. Then, to be more realistic, use the actual roulette odds of 18/38 (two of the numbers, 0 and 00, are neither red nor black). Now put yourself in the position of the operators of the roulette wheel: how can you prevent this system from working? The standard casino answer to this doubling ploy is to limit the range of bets at any given roulette table. If you can write programs, use the actual roulette odds and play the system 1000 times to a win on a table with a $1–10 range and a table with a $1–100 range. Does the system work under these circumstances, or is the casino's counterplay effective? If you cannot program, use coin flipping to play the system 25 times using the 50 : 50 odds for both ranges and determine your average net win or loss. In either case, when you hit the table maximum, wager the top amount until you win, then return to $1.

2. Suppose we want to analyze what factors contribute the most to a winning baseball team. To begin with, we need a null distribution that assumes that potentially important factors play no role. To construct such a distribution, let's assume that every hitter on a baseball team has a batting average of 0.250, every hit is a single (no doubles, triples, or home runs), no bases are stolen, no walks are pitched, there are no forced outs or double plays, and no fielding errors occur. If you are not familiar with baseball, you will not have worried about most of these potential complications in the first place. Just remember that (1) a batting average of 0.250 means the batter has a 25% chance of hitting a ball and making it safely to a base when he

is at bat; (2) if the batter does not hit a single in this simulation, he is "out"; (3) three outs end an inning for the team at bat, and any runners on the bases "die"; (4) a single puts a runner on first base (if there was a runner already there, he moves to second and a preexisting runner on second would advance to third—only if the bases are "loaded" with runners on first, second, and third when the next batter hits a single is a run scored, as the player on third moves to home plate.

Clearly if each batter has only a 25% chance of a hit, he has a 75% chance of winding up out; thus the probability of even one batter reaching first base in a given inning is small, and the odds of a score in an inning are tiny under the null-hypothesis conditions dictated by this question.

What is the distribution of scores for the team? Don't solve this with the product law. If you cannot program this on a computer, solve it with a coin for single innings (two heads means a hit, any other combination is an out). If you do program, solve it for nine-inning games.

If you want to take this further,

you could redistribute the hits as 77% singles, 14% doubles, 2% triples, and 7% home runs. Is the distribution of game scores more realistic now?

3. Imagine that you are on a game show and are asked to choose among three doors, A, B, and C. Behind one is a valuable prize; there is nothing behind the other two. Once you have selected a door, the host opens one of the other two doors to reveal that there is nothing behind it, and then asks you if you want to change your choice. Are you better off changing, sticking with the original, or does it not matter? This classic problem can be solved with the product law, but it is easy to make one or more errors in the calculation since the question involves conditional probability—that is, like the card-draw problem discussed in Chapter 3, the odds can change from the beginning of the choice sequence to the end. Although you are free to try a product law approach, also solve the problem with a Monte Carlo simulation (either by hand, beginning with a path diagram, or through programming).

Once Over Lightly

This chapter is a supersonic once-over of basic statistical techniques. It glosses over many important details, and so is best used as an overview or a quick review. As explained in the text, some of the terms here are informal or used in a somewhat special sense for purposes of clarity.

14.1 Distribution Types

The main purpose of statistics is to determine the chance that a given set of data—the sample distribution—could have been drawn by chance from another distribution. This second distribution—the one to which the sample distribution is being compared—may be

(a) a ***parent distribution***: the data set that completely and perfectly describes the second distribution. In theory, this means accurate measures of every member of the set ("population") under consideration. For the height of American males, for instance, it would include a measurement for every male in America. In reality, a very large unbiased sample does nearly as well.

(b) a ***null distribution***: a data set that completely and perfectly describes a theoretical expectation ("hypothetical distribution"). For animals choosing between two arms of a maze, for instance, the null distribution is exactly $50:50$.

(c) another ***sample distribution***: a set of data collected under circumstances that differ in some well-understood way from the first sample distribution.

14.2 Types of Data

The data in a distribution can be

(a) *continuous*: varying continuously with regard to some variable. For example, height varies continuously in whatever units of length you choose to measure it, whether centimeters or millimeters or microns or nanometers.

(b) *discrete*: varying in discrete units—the number of segments in an insect antenna, for example.

(c) *categorical*: existing in qualitatively different states—the green versus yellow character states of peas made famous by Mendel, for instance.

14.3 Characterizing Distributions

Many distributions can characterized by one or more of the following:

(a) *mean*: the arithmetical average of the data, the sum of all the data values divided by the number of data values. Only continuous and discrete data have means; the mean is a useful measure only for parametric data, though it is often presented (for no very good reason) with nonparametric data.

(b) *median*: the value of the middle datum when the data are arranged in order of the value of the variable being measured. Only continuous and discrete data have medians. The median is the best measure of the "average" of a nonparametric data set.

(c) *sample size*: n, the number of independent measurements of a particular variable that are included in a data set.

(d) *variance*: s^2, the difference of each data point's value from the mean is squared, the squared values summed, and the sum divided by the number of data points (the sample size, n). Only continuous and discrete data have variances.

(e) *standard deviation*: s or SD, the typical difference between a datum and the mean. In theory, the square root of the variance; in practice, $n - 1$ is substituted for n in the first step of computing the variance to compensate for the tendency of small sample sets to have misleadingly small standard deviations. For a parametric distribution, about 68% of the data are encompassed within one SD above and below the mean, and about 95% are found within two SDs; for nonparametric distributions, the standard deviation has little or (most often) no meaning.

(f) *standard error of the mean*: SEM, or just SE, the typical difference between the sample mean and the true mean for the sample size in question. The SE is the SD divided by the square root of n; the SD, therefore, is the SE for a sample size of one mean. About 68% of the sample means fall within one SE of the true mean, and about 95% fall within two SEs; the two-SE range is often referred to as the 95% confidence interval, which it roughly approximates.

14.4 Distribution Shapes

The distribution can be

(a) *parametric* (technically "normal" or "Gaussian" as used here)—a distribution that has a bell-curve shape described by the Gauss formula, and is completely summarized by two parameters, the mean and standard deviation. Parametric distributions arise from independent measurement errors or from the independent contribution of multiple factors to the variable being measured, or both; height and IQ are parametric.

(b) *nonparametric* (technically "nonnormal" as used here): any distribution not described by the Gauss formula, and thus which cannot be succinctly summarized by the two parameters of mean and SD. Following are several common sorts of nonparametic distributions:

skewed: most distributions that are asymmetrical, so that the mean and median are quite different. Adult human weight is usually skewed by a long tail of unusually heavy individuals; as a result, the mean value is heavier than the median.

peaked: most symmetrical distributions that have unusually small tails, and thus small standard deviations compared to a Gaussian curve. Because natural selection often selects against the extremes of a distribution of character values, the range of variation seen in many natural characters is truncated by selection against the individuals in the tails of the distribution.

flattened: most symmetrical distributions that have un-usually wide tails, and thus large standard deviations com-pared to a Gaussian curve. The win:loss records of the teams in a sports league generally form a flattened distribution be-cause some teams tend to win or lose more often than would be expected by chance, presumably because they are unusu-ally good or bad relative to the other teams.

multimodal: distributions with more than one peak; of-ten indicates that two different populations are mixed in one sample. The distribution of adult foot length, for instance, is bimodal, a result of the overlap of two parametric distribu-tions with distinctly different means: female foot length and male foot length.

(c) *binomial*: distributions in which there are two mutually exclusive states—head and tails, for instance—whose individual probabilities are known and sum to 1.0. For individual proba-bilities of 0.5 and 0.5, and when many individual outcomes are summed—the proportion of heads in 25 tosses, for instance—the distribution is essentially parametric.

(d) *circular*: distributions of periodic data, like the vanish-ing bearings of homing pigeons from a release site.

14.5 Statistical Tests

Statistical tests can be used to characterize

(a) *distribution shapes*: parametric or nonparametric, clas-sifications that can determine what sorts of further statistical procedures are appropriate.

(b) *distribution variances*: differences in standard devia-tion that may reflect a lack of independence in the data.

(c) *distribution means* (or medians): the most common sort of difference tested for. In analyses of parametric and binomial cases, the tests measure the relative area under the parent, null, or second-sample distribution from the measured value obtained from the sample distribution to the nearest end, scaled for sam-ple size, and then doubled. This is a "two-tailed" measure of the chance of obtaining a sample mean that is at least as extreme relative to the mean. When there is good reason to test for only one kind of extreme, only the area from the sample mean to the appropriate "tail" is computed. Nonparametric tests also have one-tailed and two-tailed values, though they are obtained less directly. (While these tests compare distributions, no nonparametric test actually uses mean or even median values; the paired *t*-test and ANOVA do not explicitly use mean values.)

(d) *relationships between variables*: correlations or putative cause-and-effect interactions between two or more parameters.

14.6 Which Test Is Appropriate?

The appropriate statistical test in any circumstance depends on the nature of the question being asked and the nature of the distribution(s) being examined. The purpose of all these statistical approaches, of course, is to help distinguish chance from causation in as broad a range of plausible cases as possible. The most common tests include:

(a) the *binomial test*: tests for a difference in *means* between a sample and the null distribution; the null distribution consists of two mutually exclusive events whose individual probabilities are known.

(b) the *Poisson test*: tests for a difference in *variance* between a parent and a sample binomial distribution that reflects a lack of independence of events in the sample distribution. This test is useful when the event of interest is relatively rare and the number of total events in the sample is large. When alternative events have the same probability, the *F*-test can be used to judge independence.

(c) the *Gauss test*: tests for a difference in *means* between a parent and a sample distribution. The parent distribution is parametric, composed of continuous data or of discrete data that encompass numerous different values, and its mean and SD are perfectly known.

(d) the *test of normality*: tests a sample *distribution* to see if it could have been drawn from a parametric parent distribution. This is a necessary preliminary for the *t*-tests.

(e) the *F-test*: tests for a difference in *variance* between two parametric distributions; this is a necessary preliminary for using the *t*-tests.

(f) the ***t-tests***: tests that compare the *means* of two distributions of continuous data or of discrete data that encompasses numerous different values. Perhaps the most commonly used statistical tests; the SD of the sample distribution is estimated from the sample data. Because of the sensitivity of the *t*-tests, nonparametric data are, where possible, generally *transformed* into a parametric distribution by one of several mathematical procedures that "normalize" the data; the same transformation must be applied to all the distributions being compared. There are four variations:

the ***single-sample t-test***: compares the *mean* of a parent distribution whose SD is unknown against a sample distribution, whose SD is estimated from the data in the sample set. The data must be parametric and there must be good reason to believe that the SD of the sample is similar to that of the SD of the parent distribution.

the ***paired-sample t-test***: compares the *means* of two data sets in which each datum in the first set can be matched with a corresponding datum in the second set. The data must be parametric and the SDs of the two samples must be similar; typically these data pairs consist of measurements of the same individual under two different conditions. This is the most sensitive of the two-sample *t*-tests.

the ***unpaired t-test***: compares the *mean* of one sample set against the mean of another. The data must be parametric and the SDs of the two samples must be similar.

the ***t-test for unequal SDs***: compares the *means* of two sample distributions against each other. The data must be parametric, but the SDs of the two samples need not be similar. This is the least sensitive variant of the *t*-test.

(g) the ***chi-square tests***: test for difference in *distribution* between two data sets consisting of categorical data. Because these tests are so simple and quick, they are often used on other kinds of data that are artificially categorized for the purpose, even though the proper test would be more sensitive. There are two versions:

the ***chi-square test for goodness of fit***: compares a sample *distribution* to a null distribution.

the ***chi-square test for independence***: compares two sample *distributions*. It is less sensitive than the goodness-of-fit test because it must allow for sampling error in both distributions, whereas a null distribution is perfectly known.

(h) the ***signed-rank test***: compares the *distribution* of paired, continuous or discrete nonparametric data.

(i) the ***rank-sum test***: compares the *distribution* of two sets of unpaired, continuous or discrete, nonparametric data. This is the least sensitive two-distribution test of all; for large sample sizes, the ***U-test*** is generally substituted.

(j) the ***circular tests***: the ***Rayleigh z-test*** determines the *mean* "direction" and degree of clustering—lack of *variance*—of periodic data. The ***Rayleigh u-test*** can determine if the mean direction of a sample is different from a predicted mean. The ***Watson-Williams test*** can determine if the mean direction differs between two samples.

(k) ***correlation analyses***: tests that look for any systematic change in the value of one parameter as another changes. No cause-and-effect sequence is presupposed; correlations can range from $+1.0$ to -1.0.

nonparametric correlation analysis: a correlation test that can use nonparametric data.

parametric correlation analysis: a correlation test in which both parameters must be from parametric distributions. If a cause-and-effect relationship is suspected, linear-regression analysis would ordinarily be used instead. This test is a necessary prerequisite to multiple-regression analysis, described below.

(l) ***linear regression analysis***: a test to measure the putative cause-and-effect relationship between an independent variable (the supposed cause) and a dependent variable (the effect). The dependent variable must be parametrically distributed, and the independent variable must be relatively precisely known; usually it is under the experimenter's control.

(m) ***path analysis***: a technique for apportioning the hypothetical cause-and-effect contributions to a measured variable from multiple sources that themselves interact. There are no well-defined ways to assign probability values to the contribution values.

14 Once Over Lightly

(n) *multiple-regression analysis*: a test that quantifies the relative apparent contribution of two or more correlated "independent" variables on the dependent variable; this is a special but very common case of path analysis, and one that yields conventional probability values. The relationship between the independent variables is first established using parametric correlation analysis; thus, unlike the situation necessary for linear regression analysis, the independent variables must also be parametrically distributed.

(o) *multiple-sample tests*: tests that deal with instances in which the *means* or *distributions* of more than two samples are being compared. These tests compensate for the chances of obtaining false positives when making multiple comparisons.

    analysis of variance (ANOVA): a multiple-distribution form of the t-test.

    (i) *one-way ANOVA*: this test determines whether one of the distributions is significantly different from the others; to identify which, use the *Tukey-Kramer* method.

    (ii) *two-way ANOVA*: this test allows grouping of multiple data sets to look at the influence of two variables, both on the measured dependent variable and on each other. A variety of higher-level ANOVAs exist to extend this approach.

    chi-square test for independence: the conventional form of this test can deal with multiple distributions.

    Friedman test: a multiple-sample form of the signed-rank test.

    Kruskal-Wallis test: a multiple-sample form of the rank-sum test.

(p) *multiple-pairwise comparisons*: corrections that deal with instances in which the *means* or *distributions* of more than one pair of samples have been collected. These corrections compensate for the chances of obtaining false positives when performing multiple tests. They take two forms:

    corrections: P-value adjustments, which allow for the chance of a false positive appearing anywhere among the set of paired comparisons; for studies that generate numerous paired results, the effect of these corrections can be devastating.

    average number of false positives: the number of false positives the set of paired comparisons is likely to generate;

for studies that produce many paired results that all test the same hypothesis, this approach is more appropriate and much more forgiving.

(q) *Monte Carlo simulation*: a technique for creating a null distribution and finding the probability of a particular outcome by collating the outcomes of thousands of simulated tests in which the null hypothesis is assumed to be correct.

Exercises

1. What is the difference between sample, parent, and null distributions?

2. What is the essential difference between parametric and nonparametric data?

3. What is the essential difference between a one-tailed and a two-tailed test?

4. What does a *P*-value of 0.05 actually mean?

5. What is the essential difference between continuous and categorical data? Where do discrete data fit in?

6. What is the essential difference between paired and unpaired data?

7. What is the essential difference between correlation and regression analysis? What limits the use of each?

8. What is the *t*-test good for? When would you use the Gauss test instead?

9. What is the chi-square test good for? Why are there two versions of it?

10. What is the signed-rank test good for? When would you use the rank-sum test instead?

11. What is ANOVA good for? Why not just use several *t*-tests? What are ANOVA's limitations?

12. Why do you need to use statistics anyway?

13. Is a sample size of one ever appropriate?

14. What are normalization techniques? Why might you use them?

15. What is multiple-regression analysis useful for?

16. Why would you use the test for normality?

17. Why would you use the *F*-test?

14 Once Over Lightly

Answers to Exercises

Chapter 2

1. The data are discrete. Because there are few categories, this must be plotted as a bar graph. With so little data, it is not obvious whether this distribution could approximate a parametric curve. The data look parametric, and so we could compute a mean (2.5 for the first 10 entries) and standard deviation (0.92). If you judge the distribution to be non-parametric, you would compute only the median, which is 3.

2. The data are continuous, though they have rounded to the nearest tenth of a point. There is no evidence from this limited sample that they are parametric: there is no obvious symmetry since the curve is skewed to the right. Thus the median (3.1) is the only relevant value. Later in the text, we will use parametric GPA distributions; we obtain these by considering only first-year or under-class grades; the skewing to the right evident here seems to result from taking upper-class courses after self-selection for departments in which the student has a special interest and thus does well in its courses.

3. There is an evident negative relationship between A-ranking and GPA (negative because lower A-ranks and higher GPAs are best); when we study correlation analysis we will see how to quantify this relationship and show that the trend is significant. There is no trend with the P-ranks.

4. These data are discrete, though they may appear to be continuous: all values are in units of 10. Nevertheless, there are so many possible values that you may treat them as continuous if you have enough data. With only 20 measurements in this case, lumping of values and plotting a bar graph seems the best course. The data appear to be parametric. Thus you can calculate from the first 10 values a mean of 630 and an SD of about 72.

5. The data are categorical. They may be plotted as a two-element bar graph.

6. The data are categorical. They may be plotted as a two-element bar graph.

7. The data are discrete. Because there are many possible values, the data could be plotted as a line graph if there were enough data. As it is, the best representation of the data is probably obtained by lumping values and drawing a bar graph. The data are clearly not parametric: there are no tails, but rather an abundance of small and large sizes.

8. The data are continuous. Because there are many possible values, the data could be plotted as a line graph if there were enough data. As it is, the best representation of the data is probably obtained by lumping values and drawing a bar graph. The data are not parametric because males and females have been combined; each group's height distribution on its own is parametric.

9. If there were no relationship between height and shoe size, we would expect the four category groups each to have about five students; instead, we see a shortfall in the small–tall and large–short categories. This result illustrates a quick way to look for trends; we will see how to quantify these categorical comparisons in a later chapter.

10. There is a fairly good correlation between increasing shoe size and increasing height. Note how much clearer the relationship is when the data are plotted in this way; much quantitative information is lost when data are categorized.

Chapter 3

1. The chance that two individuals will have the same birthday is 1/365. The odds that at least two in a group of four will have the same birthday is 6/365: individual A could share his birthday with individuals B, C, or D; individual B could share a birthday with C or D (the A/B case was included in the individual-A computation); individual C could share his birthday with C. Thus there are 6 possible ways to share birthdays, each with the same probability. Since no one result excludes any other—they could all have the same birthday—the individual probabilities can be added.

2. A twelve can be rolled only with a six on each of the dice; the chances of that occurring are $(1/6) \times (1/6)$, or $1/36$. A six can be rolled with a five and a one (on die A and die B, respectively), a four and a two, a three and a three, a two and a four, and a one and a five. Thus there are 5 ways, among the 36 different ways a pair of dice can be rolled: $5/36$. A seven can be rolled with a six and a one, a five and a two, a four and a three, a three and a four, a two and a five, and a one and a six; thus the odds are $6/36$. For the royal-flush problem, the logic is: the first card can be any of 20 high cards (one of the five highest from any of the four suits), and the odds of being dealt such a card are $20/52$; the second card must be one of the remaining four high cards in the suit of the first card deal, and thus the odds are $4/51$ (51 because after the first card is dealt there are only 51 options left); the next must be one of the three remaining high cards in the suit, or $3/50$; the next is one of the two remaining needed to fill out the flush, or $2/49$; and the last is the unique missing card, which has a probability of being dealt to you of $1/48$. Therefore the net probability is slightly better than one in a million: $(20 \times 4 \times 3 \times 2 \times 1)/(52 \times 51 \times 50 \times 49 \times 48) = 0.000001539$.

3. For the call versus silence test, $P < 0.00001$ (off the scale in *BioStats Basics Online*); the test is properly one-tailed since there is an a priori expectation. All other comparisons are two-tailed. The exact values you get depends on the test you use, since the two-choice test uses a shortcut that speeds its computations but lowers its precision when the results are not significant. The appropriate values using the binomial test are:

Test	P-value	Test	P-value	Test	P-value	Test	P-value
2	0.4762	6	0.6878	10	0.0019	14	0.0081
3	0.0002	7	0.0100	11	0.0000	15	0.0007
4	0.2198	8	0.0059	12	0.5620	16	0.0002
5	0.002	9	0.0001	13	0.2039		

4. Scaling the Test 14 results up from $72/120$ to $105/175$ yields a *P*-value of 0.0099; scaling down past $42/70$ will yield a

value greater than 0.1. The Test 6 results must be scaled up to 1224/2352 to achieve a P-value of about 0.050.

5. The values are:

Trials	P-value	Trials	P-value	Trials	P-value
100	0.2070	400	0.0095	800	0.0002
200	0.0693	600	0.0014	1000	0.0000

These are all two-tailed values because any deviation represents a bias.

6. The two-tailed value is 0.0265; if you had observations that preceded the collection of this data set to suggest male-offspring bias, you could have used a one-tailed test, for which $P = 0.0133$.

7. This is a two-tailed test because you have no reason to predict a particular outcome. The P-value for this outcome is about 0.0001; the null hypothesis can be rejected.

8. With no pre-test odds, you would reasonably assume that you have a 95% chance of having colon cancer if the test is positive. But if 999 out of 1000 people being tested do not have colon cancer, then it stands to reason that 5% of them (about 50) will get (false) positive test results while the one actual cancer victim will have a 95% chance of being diagnosed correctly. Thus there will be roughly a 50-to-1 excess of false positives; this must mean that a positive result is equal to approximately a 2% chance of actually having the cancer. Bayesian analysis reaches this conclusion by taking the pre-test odds as 1 : 1000 (which equals 0.001001); the likelihood ratio is true positives/false positives, which is 0.95/0.05 = 19. Post-test odds are pre-test odds × likelihood ratio, or (0.001 × 19) = 0.019. Converting these odds to a probability, we find odds/(1 + odds), or 0.019/1.019 = 0.0187—close to the 2% approximation we derived in the verbal argument above. So you probably don't have the disease but would be well advised to proceed with additional tests.

9. The CD (coefficient of dispersion) is 1.018, very close to the 1.0 predicted for independence. Thus we conclude that there are no particularly lax units, at least by this measure.

10. The CD is 1.307, which is unusually high; the arrivals appear to be clumped.

11. The CD is 0.786, which is quite low; the creosote bushes appear to inhibit the growth of nearby bushes.

Chapter 4

1. Both these distributions appear to be normal.

2. Neither of these distributions is normal. Heights are skewed to the right toward tall individuals, whereas weights (which are not normal in the general population) are oddly distributed.

3. Fielders average more than 3 inches taller than average males; with our sample size, the P-value is <0.0001.

4. Though the sample mean is higher than the parental mean, the difference is not significant at the level of $P<0.05$.

5. Even as a two-tailed test, the P-value is <0.0001; pitchers are lousy hitters.

6. The early nesters have lay significantly more eggs. As a two-tailed test, $P<0.02$; if we had been using Darwin's hypothesis as an experimental prediction, then the data are significant at $P<0.01$.

7. The survivors have significantly longer beaks $(P<0.01)$.

8. Even if you run this as a one-tailed test (which would be fair given the initial hypothesis), the difference is not significant at this sample size.

9. Each data set satisfies the normality test, though the sample size is quite small. The F-test tells you that the SDs are too different for the conventional unpaired t-test. Thus the proper test is the unpaired t-test for different SDs (described in Chapter 5). Even two-tailed, the difference in means yields a $P<0.01$, indicating that this highly potent mutagen can travel through the blood from the lungs to the cervix.

10. The mean number of chips is 1262.8, with a standard deviation of 116.1. A chip count of 999 is thus 263.8 chips below the mean. Dividing the SD into this difference yields 2.26; thus we want to know how much of the normal curve is more than

2.26 SDs below the mean. Interpolating between the 2.2 and 2.3 entries in Table 4-1, we get a value of 0.4882 for the area from the mean to the tail; thus the rest of the area is $0.5000 - 0.4882 = 0.0118$. Dividing this value into the total area under the curve (1.0), we get a value of 84.7. Thus, on average one in 85 bags should be one or more chips short.

Chapter 5

1. The mSAT data yield an F-ratio of about 2.6, which is too extreme for the sample size. Thus you cannot compare these data sets with any version of the t-test that assumes similar SDs.

2. The beak data yield an F-ratio of about 1.2, which is sufficiently small for this sample size to allow you to assume that the SDs are similar.

3. The data pass the parametric test (though the sample size is too small to be certain; larger samples are unambiguously parametric) and yield an F-ratio close to 1.0. Thus the proper test is the paired t-test. This test indicates that there is no significant difference between the display rates.

4. The data appear to be normally distributed. The F-ratio of about 2.1 is low enough with this sample size to use a standard t-test. The paired t-test is the correct one to use. The difference is significant at the $P < 0.01$ level, indicating that the chicks find hawk silhouettes innately frightening. After repeated exposure to the hawk model with no alarm calls, however, the freezing response becomes less marked, so there was something in what the critics had to say.

5. Because these data fail the F-test, you must use the t-test for unequal SDs. This test indicates that the difference in mean score on the mSAT is not significant for this sample size.

6. Because these data pass the F-test, we use the unpaired t-test, which yields a t-statistic that is significant at $P < 0.02$.

7. Both data sets are parametric, but the F-ratio of about 2.1 is too large for this sample size. Therefore you must use the t-test for unequal SDs. The difference between the two data sets is spectacular, but the table only goes down to $P < 0.01$.

8. Both sets of data are probably parametric, and pass the F-test for this sample size. Thus the appropriate test is the unpaired t-test. The difference is significant well beyond $P < 0.01$.

9. In each case, the SEs are a good guide to whether the calculated P-value is significant or not, and if significant, whether it is at the $P < 0.05$ or $P < 0.01$ level.

Chapter 6

1. The raw data yield a D-value of 0.115. The ln-transformed data have a D-value of 0.073; the square-root-transformed data have a D-value of 0.055; the sample-mean value (using samples of 10) is 0.086. Thus the best technique to use on these data is the square-root transform.

2. The raw data yield a D-value of 0.151. The ln-transformed data have a D-value of 0.112; the square-root-transformed data have a D-value of 0.130; the sample-mean value (with samples of 10) is 0.070. Thus the best technique to use on these data is the sample-mean transform; but given the limitations of making comparisons with sample-mean data, you may want to use the ln-technique instead.

3. The raw data yield a D-value of 0.146. The e^x-transformed data have a D-value of 0.111; the sample-mean value (with samples of 10) is 0.087. Thus the best technique to use on these data is the sample-mean transform; but given the limitations of making comparisons with sample-mean data, you might want to use the e^x technique instead.

4. The raw data yield a D-value of 0.113. The x^2 transform improves this to 0.098; sampling produces 0.082. Thus the best technique to use on these data is the sample-mean transform; but given the limitations of making comparisons with sample-mean data, you might want to use the x^2 technique instead.

Chapter 7

1. All three data sets appear to be parametric and pass the F-test. There is a significant difference in ANOVA variances which the Tukey-Kramer method isolates to $A.\ m.\ carnica$, the German

honey bee. There is no opportunity to use the two-way ANOVA on these data.

2. All four data sets appear to be parametric and pass the *F*-test. There is a significant difference in ANOVA variances which the Tukey-Kramer methods isolates to the male and female natural science salaries. The two-way ANOVA finds a significant effect of field (natural sciences versus humanities) on starting salaries, but though the female mean is higher, there is no significant difference; there is no interaction between the variables (field and gender).

3. There is no significant difference in variances among the subgroups, nor does anything turn up when we analyze the data with the two-way test.

Chapter 8

1. The expected null result is that no lizards will change color. Thus we sum the B >> B and B >> G categories to obtain the number of lizards that began in the brown state, and likewise we sum the G >> G and G >> B categories to find the number that started out green; these values are the expected numbers of brown and green lizards we will count at the end of the test if the stimulus has no effect. Thus we can set up six chi-square tables:

Test A: Darkness, 23°C

	Observed	Expected
Brown	8	8
Green	4	4

Test B: Darkness, 37°C

	Observed	Expected
Brown	1	14
Green	24	11

Test C: Halogen, 23°C

	Observed	Expected
Brown	33	15
Green	0	18

Test D: Halogen, 37°C

	Observed	Expected
Brown	12	24
Green	20	8

Test E: Ultraviolet, 23°C

	Observed	Expected
Brown	14	10
Green	5	9

Test F: Fluorescent, 23°C

	Observed	Expected
Brown	13	10
Green	6	9

The corresponding chi-square values are:

Test A $\chi^2 = 0$ $P > 0.1$
Test B $\chi^2 = 27.4$ $P < 0.005$
Test C $\chi^2 = 39.6$ $P < 0.005$
Test D $\chi^2 = 24.0$ $P < 0.005$
Test E $\chi^2 = 3.4$ $P < 0.1$
Test F $\chi^2 = 1.9$ $P > 0.1$

It seems clear that elevated temperatures tend to cause lizards to switch from brown to green, whereas something in halogen light not also found in weak UV or fluorescent lighting triggers a shift to brown.

2. The data comparing no silhouette to a model can be tested with the goodness-of-fit test since it is clear that in the absence of any silhouette, no crouching occurs (the null expectation). The tests comparing hawk responses to gull responses (Tests C and F) require the test for independence since there is no null prediction. The chi-square tables are as follows:

TESTED ALONE

Test A: Goose

	Observed	Expected
Crouch	0	10
No crouch	28	18

Test B: Hawk

	Observed	Expected
Crouch	0	12
No crouch	28	16

TESTED TOGETHER

Test D: Goose

	Observed	Expected
Crouch	0	5
No crouch	28	23

Test E: Hawk

	Observed	Expected
Crouch	0	13
No crouch	28	15

SILHOUETTES COMPARED

Test C

	Gull	Hawk
Crouch	10	12
No crouch	18	16

Test F

	Gull	Hawk
Crouch	5	13
No crouch	23	15

The results are:

Test A	$\chi^2 = 15.6$	$P < 0.005$
Test B	$\chi^2 = 21.0$	$P < 0.005$
Test C	$\chi^2 = 0.3$	$P > 0.1$
Test D	$\chi^2 = 6.1$	$P < 0.02$
Test E	$\chi^2 = 24.3$	$P < 0.005$
Test F	$\chi^2 = 5.2$	$P < 0.05$

It seems clear that chicks tend to respond to silhouettes, regardless of shape, though judging by the chi-square scores (Test A vs B and Test D vs E) the hawk shape is more effective, especially when birds are tested together. The test for independence shows a significant difference for hawk versus gull shapes for chicks tested together. These were the conditions and one of the scoring variables used by Lorenz and Tinbergen. Unsuccessful repetitions have tended to use isolated chicks and/or employed scoring variables that are insensitive under the testing conditions chosen.

3. Since you have a null prediction (60 : 40), you can use the chi-square goodness-of-fit test on the first part of the question. The table is:

	Observed	Expected	
Coke	23	13.6	$\chi^2 = 10.8$
Pepsi	11	20.4	$P < 0.005$

For the second part, we must use the test for independence (the expected values are in parentheses):

	Males	Females
Coke	15 (12.9)	8 (10.1)
Pepsi	4 (6.1)	7 (1.6)

Because one observed value is below 5.0, the test for independence will not work.

4. Both these comparisons require the test for independence:

	Males	Females
Dogs	18 (14.1)	8 (11.9)
Cats	1 (4.9)	8 (4.1)

Since chi square cannot deal with cases in which more than 20% of the values are below 5, this looks hopeless. You would need to use Fisher's Exact Test on these data.

The second comparison is highly problematic because there are several small rows:

Number of syllables	Males	Females
4	2 (1.1)	0 (0.9)
5	11 (6.0)	0 (5.0)
6	3 (4.3)	5 (3.7)
7	3 (5.4)	7 (4.6)
8	0 (1.1)	2 (0.9)
9	0 (1.1)	2 (0.9)

Combining the groups into pairs, we get:

Number of syllables	Males	Females
4–5	13 (7.1)	0 (5.9)
6–7	6 (7.8)	12 (8.2)
8–9	0 (2.2)	4 (1.8)

But this won't work either because there are two expected values below 5.0. How about combining into trios:

Number of syllables	Males	Females	
4–6	16 (11.4)	5 (7.6)	$\chi^2 = 10.5$
7–9	3 (7.6)	11 (6.4)	$P < 0.005$

The difference is highly significant.

5. The test for goodness-of-fit can be used to determine if imprinting is effective: if it isn't, we would expect a $50:50$ split between the alternative models (Test A). The same logic applies to testing for the effectiveness of sound (Test B) and color (Test C). Test D can be used with the goodness-of-fit test to judge whether sound is more or less important than color, since if they were equally important we would again expect a $50:50$ split. The tables are:

Test A: Correct model vs wrong color and sound

	Observed	Expected	
Follow	43	24	$\chi^2 = 30.1$
Don't follow	5	24	$P < 0.005$

Test B: Correct model vs wrong sound

	Observed	Expected	
Follow	37	24	$\chi^2 = 14.1$
Don't follow	11	24	$P < 0.005$

Test C: Correct model vs wrong color

	Observed	Expected	
Follow	32	24	$\chi^2 = 5.3$
Don't follow	16	24	$P < 0.05$

Test D: Wrong color vs wrong sound

	Observed	Expected	
Follow	30	24	$\chi^2 = 3.0$
Don't follow	18	24	$P < 0.1$

Clearly imprinting works, and both color and sound are important. The chi-square values (Test B vs Test C) suggest that sound is more important than color, but the effect is not significant at this sample size (Test D).

6. Because there is no expected distribution, the correct test is chi-square for independence. Given the reasonable worry about radiation acting either as a killer of cells or as a mutagenic influence, these data are best treated as a two-tailed comparison. As such, radiation treatment improves six-year survival by almost a factor of 2, which is significant at $P < 0.01$ and is very close to $P = 0.005$.

7. In each case there is a hypothesis but no expected distribution. Thus the correct test in chi-square for independence. The test should be two-tailed because these are the first data to suggest a hypothesis. The statistic for heart-related deaths is 0.1867, which is clearly not significant. The statistic for overall deaths is 4.0018, which (given the trend is in the expected direction based on previous work, and thus is one-tailed) is significant at $P < 0.025$.

8. The proper test in the absence of an expected distribution is chi-square for independence. The hypothesis is clear: predators should take more light-colored moths in Birmingham and more dark moths in Dorset; thus this is a one-tailed test. Both sets of results are significant with $P < 0.0025$.

Chapter 9

1. Since each male was tested under two conditions, we can use a paired analysis. The signed rank test yields a smaller sum of 135.5 with a sample size of 30; this corresponds to a (barely) significant $P < 0.05$.

2. This is grouped data because each toad was tested under the same range of different conditions, and thus the correct test is the Friedman test. Our eyes, plus column dropping, show big effects with Models B and E; these models either lack an outline or they are vertical. Model D also appears to be fairly different; its distinctive characteristic is that it is small. This experiment should have been run with each model alternated with a standard model or other control condition so that the data could have been analyzed as separate pairs.

3. Because these data are unpaired, the rank-sum test is appropriate. The A-rank comparison reveals that female biology majors have better (i.e., lower, where 1 is the best score) academic rankings than their male counterparts ($P < 0.02$). There is no significant difference in P-ranks, though the median male score was better. With regard to GPA, female GPAs are better ($P < 0.005$); these cumulative GPAs could have been normalized and an unpaired t-test performed, but the sample size is so large and the rank-sum test so easy that it is worth trying the nonparametric test first. At first glance the chi-square test for independence looks hopeless for the A-rank and P-rank data:

	A-rank			P-rank	
Score	Male	Female	Score	Male	Female
1	2	4	1	5	0
2	7	13	2	2	3
3	7	6	3	11	13
4	6	0	4	6	7
5	2	0	5	0	0

By combining categories (and thus discarding information—the difference in scores between individuals within a combined group) we can satisfy the test rules:

	A-rank			P-rank	
Score	Male	Female	Score	Male	Female
1–2	9	17	1–2	7	3
3–5	15	6	3	11	13
			4–5	6	7
	$\chi^2 = 6.3$			$\chi^2 = 1.8$	
	$P < 0.02$			$P > 0.1$	

Thus chi-square for independence yields about the same results as the rank-sum test. Either is a valid approach, but in this case the rank-sum was easier to apply given the form of the data.

4. It seems that parents select longer first and middle names for future female biology majors. Using the rank-sum test for this unpaired data, we obtain a P-value of < 0.005.

5. These paired data are appropriate for the signed-rank test. The comparison between a normal male and a male with a dorsal sword yields a smaller sum of 74, which, with a sample size of 18, is not significant. The comparison between a normal male and one with a ventral sword produces quite different results:

a smaller sum of 41, which, with a sample size of 22, yields $P < 0.01$.

6. The groups are very different: the H-statistic is 110.3 against a threshold value of 13.3 for $P < 0.01$. Through column dropping, we can show that sport-utility vehicles are the most important source of difference, with large cars close behind.

7. The "after" data are not parametric. (In the original publication, the data were compared using a paired t-test!) Since the data are paired, the proper test is the signed-rank test. The test is properly performed one-tailed since the hypothesis is that I3C will lead to estrogen breakdown. It's tempting to assume that since the two-tailed value is $0.10 > P > 0.05$, the one-tailed value will be $P < 0.05$. However, you should have noticed that the effect is in the wrong direction; thus the one-tailed value is $P < 0.95$. Maybe actually eating lots of broccoli isn't really as important as being the sort of person that does.

Chapter 10

1. The proper analysis is the z-test because there is no predicted mean time. The mean vector length is 0.08, with a z-statistic of 1.33, which yields a $P > 0.1$; thus the correlation claimed in the article vanishes under statistical scrutiny.

2. In both experiments, the foragers are well oriented with z-statistics of 19.5 and 29.6 (Tests A and B, respectively; $P < 0.001$). The u-test is appropriate since there are three predicted bearings. Unfortunately, they are so close together that each is consistent with the results. However, the u-statistic for the extrapolation hypothesis is consistently the highest:

	Sun's arc	Extrapolation	15°/hour
Test A	5.0	6.1	5.8
Test B	6.9	7.4	3.7

3. The brass-wearing groups were both well oriented ($P < 0.01$), though the match to the expected direction was not as good ($P < 0.05$). The magnet-wearing group at Site A was oriented ($P = 0.05$), but not in the expected direction; the Site B group

was not significantly oriented. Despite the apparently superior orientation of the brass-wearers, in neither case was there a significant difference (using the Watson-Williams test) between the brass and magnet groups.

4. All tests showed significant orientation in the predicted direction. However, since data were collected only from a limited range of angles, you might wonder if the same pattern might not be observed if the same number of bees went to each station. To test this, we can run an analysis with about the same number of bees evenly distributed (e.g., 10 at each station) and compare the z- and u-statistics:

Test	z-statistic	u-statistic	Mean bearing
35A actual	53.2	9.3	47°
35A random	52.1	9.0	53°
60 actual	52.0	9.2	52°
60 random	31.3	7.3	53°
35B actual	17.3	5.7	37°
35B random	15.6	4.9	53°
72 actual	14.9	5.0	65°
72 random	15.6	4.9	53°

As this comparison makes clear, circular analysis over a restricted range of possible directions may not be very useful. In theory, it is valid only if you sample over the entire 360° range, or can show that there were few if any data that fell outside the sample area (so that you can reasonably infer zero values for the unsampled arc). This situation usually obtains if you observe a bell-curve distribution that declines to zero at each side of the mean *within* your sampling array. This was the case in Question 2 because the collection stations were spread over about 90° and the ends of the distribution were evident; here, however, the range is only 63°, and in one case (35A) there is no reason to believe that stations farther to the right of the 85° site would not have captured recruits. Circular analysis in the other cases may be justified, but the u-test doesn't tell us much, as we've seen. Do not despair, however, because the Watson-Williams test is capable of distinguishing any two angular distributions if there are sufficient data. Indeed, comparing the distributions in pairs (ignoring 35A because it doesn't appear complete) with the Watson-Williams test does show that the distributions are significantly different.

5. Both groups of beetles are very well oriented; the F-statistic of the Watson-Williams test indicates that their mean directions are different at $P < 0.001$.

Chapter 11

1. The 1992 distribution is parametric; the 1993 distribution is marginal. Even if you opt for the parametric correlation analysis, there is no correlation (-0.0017; it's -0.016 by the nonparametric method). You might have chosen regression analysis, supposing that the 1992 record was the independent variable. While the idea that team quality in one year could be a plausible cause of the team record the next is reasonable, regression analysis is inappropriate because the value of the independent variable (the x-axis value) is not a precise measure of team quality; we saw in an earlier chapter that if that were so, the distribution of win : loss records would be flat, with the top team winning all its games.

2. Because altitude is the putative cause (independent variable) and height is parametrically distributed, these data can be subjected to regression analysis. The correlation of -0.96 indicates a very strong correlation between altitude and growth. In fact, these data come from greenhouse-reared plants germinated from seeds collected at these various sites; thus the growth differences observed are genetically controlled.

3. Nonparametric correlation analysis is appropriate here. The correlation is -0.97.

4. The slight positive correlation (0.16) between mSAT (the independent variable) and first-year GPA (the dependent variable) is not significant.

5. The regression correlation is -0.66, which is not significant with this sample size. Nevertheless, the correlation looks strong. If we transform the anomaly-strength data by taking the log of each value, the correlation improves to -0.88, with an F-value that is significant at $P < 0.01$.

6. Although shoe sizes are parametrically distributed, syllable number is not. The nonparametric correlation is −0.63, which is significant at $P < 0.01$. This may seem surprising, but it illustrates a secondary consequence. We saw in Chapter 9 (Question 4) that female biology majors have significantly more syllables in their names than males. We have also seen that male shoe sizes are significantly larger than those of females. The correlation we see here is generated by these gender differences.

7. These data can be analyzed by regression analysis since the independent variable (effective speed) is precisely known. Speed is clearly an important factor in pricing, though more so for laptops: for desktops the correlation is 0.746 (significant well beyond $P = 0.01$) whereas for laptops the correlation is 0.900 (significant well beyond $P = 0.01$). The idea that the very fastest are unusually expensive is not borne out: the fastest laptop and two of the three fastest desktops fall below the regression line. Laptops are certainly more expensive than desktops: the regression for desktops is $4.35/MHz whereas for laptops it is $8.16/MHz over the range of speeds available—roughly a twofold price premium for portables. Extrapolating this line back, the Y-intercept for desktops when X (speed) is zero is $789; for laptops the cost of a 0-MHz machine is in theory $1689. Presumably there is a base cost or base value independent of processor speed.

8. Alas, there is not much you can do with this intriguing data set. Only nonparametric correlation analysis is appropriate for the bimodal data (smoking and vitamin use) and weight. All are well correlated, positively or negatively, with exercise. Nonparametric correlation analysis is also necessary for the parametric data because the exercise variable is not known to be parametric. Regression analysis is inappropriate because the exercise variable is not precisely known; instead, the value is given as a (wide) range. And you cannot perform regression analysis to compare putative causes with colon cancer because the value of the cause variable is not known: it's just the average of data from 10,000 nurses in a particular exercise group. Thus the data are parsed in the wrong way for any interesting computations. This data set also illustrates the prevalence of correlated "causes": in many health analyses, an entire suite of behavioral and physiological variables track one another. Thus if you had taken data only on,

say, fiber intake and colon cancer, you would have found a strong correlation; but by looking for other associated variables, it's clear that a number of other putative causes account for the effect equally well. Perhaps only one is important; or maybe they work together; or it could be that they simply correlate with the actual cause, which is not being measured (e.g., race, or height, or socioeconomic status).

9. Because the independent variable (relatedness) is known exactly, and the dependent variable (egg failure) is parametrically distributed, these data can be subjected to regression analysis. The correlation is excellent (0.92) and significant at $P < 0.01$. Apparently inbreeding is bad for this species.

Chapter 12

1. There is a strong positive correlation of win : loss records and batting, and a strong negative correlation with pitching; both are highly significant. There is a modest correlation (0.22, $P < 0.05$) between batting and pitching, indicating that teams tend to be stronger or weaker in both. Batting averages account for about 32% of the variance in win : loss records; ERAs are responsible for about 42% of the variance. The remaining roughly 27% can be chalked up to chance.

2. There are strong correlations between precipitation and productivity as well as between temperature and productivity; together, they would seem to account for 119% of the variance. (We get these values by taking the squares of the independent-variable : dependent-variable correlations on the first card.) But there is also a strong link between temperature and precipitation, which means the initial correlations are overestimates. When the correlation between the independent variable is factored in (which occurs on the second card), precipitation accounts for 19% of the variance and temperature 28%; together they generate 47% of the variance, with soil fertility and other factors probably contributing the rest.

3. The A-ranking has a strong correlation with GPA among history majors, accounting for 45% of the variance in grades. (The correlation is negative because a good A-rank is low and a good GPA is high.) This is much more than the 13% accounted for by

SATs. The admissions committee, though its integration of SAT scores, APs, high school grades, and academic letters, is about half as successful as it could hope to be in predicting freshman academic performance. There is also a slight correlation between P-rank and GPA, but it is not significant. There is no correlation between A- and P-ranks; thus there is no reason to suppose the admissions committee, at least when dealing with potential history majors, is doing any sort of compensation.

Chapter 13

1. The distributions of amounts of money needed are

Bet	50:50 case	18:20 case
$ 1	50.0%	47.4%
3	25.0	24.9
7	12.5	13.1
15	6.8	6.9
31	3.3	3.6
63	1.7	1.9
127	0.8	1.0
255	0.4	0.5
511	0.2	0.3
1023	0.1	0.1
2047	0.1	0.1
4191	>0.1	>0.1

The typical long-term net change in your assets would be:

50:50 case		
No limit	**$1–10 table**	**$1–100 table**
+$1.00/bet	−$0.07/bet	−0.70/bet

18:20 case		
No limit	**$1–10 table**	**$1–100 table**
+$0.95/bet	−$0.08/bet	−0.74/bet

2. For single innings, the null distribution is:

96.2% no runs
 2.5% one run
 0.9% two runs
 0.3% three runs
 0.1% four runs

For full games, the null distribution is:

70.1% no runs
16.5% one run
 8.0% two runs
 3.1% three runs
 1.4% four runs
 0.6% five runs
 0.3% six runs
 0.1% seven runs

These theoretical values, which (except for rounding the batting average slightly to .250 for simplicity) are based on league averages. They yield scores that are well below the observed values, though the second set (which includes hits other than singles) is somewhat closer to reality. Given that scoreless games are rare, this null distribution demonstrates that other factors are at work that tend to "focus" hits into particular innings.

3. The three-door problem has confused any number of professional mathematicians. Intuitively, it seem obvious that the odds of having selected the correct door cannot possibly change once you have announced your choice. Nevertheless, by sticking with your original choice the odds of finding a prize behind it are 1 in 3, whereas by switching to the other unopened door, the odds rise to 2 in 3. Seems impossible, but the Monte Carlo simulation should make this evident. A summary of the correct conditional-probability pathways follows on the next page.

Strategy: Don't switch

Original choice	Actual prize	(Host door)*	Your choice	Outcome
A	A	(B or C)	A	win
A	B	(C)	A	lose
A	C	(B)	A	lose
B	A	(C)	B	lose
B	B	(A or C)	B	win
B	C	(B)	B	lose
C	A	(B)	C	lose
C	B	(C)	C	lose
C	C	(A or B)	C	win
			net win rate: 3/9	

*The choice by the host is automatic when your original choice is wrong, and arbitrary when your original choice is correct; only your two choices and the location of the prize contribute to the matrix of options. This is why in the three cases in which your first choice is correct, and your strategy is not to switch, the host responses are irrelevant and thus consolidated.

Strategy: Switch choice

Original choice	Actual prize	(Host door)*	Your choice	Outcome
A	A	(B or C)	C or B	lose
A	B	C	B	win
A	C	B	C	win
B	A	C	A	win
B	B	A or C	C or A	lose
B	C	B	C	win
C	A	B	A	win
C	B	C	B	win
C	C	A or B	B or C	lose
			net win rate: 6/9	

*The choice by the host is automatic when your original choice is wrong, and arbitrary when your original choice is correct; only your two choices and the location of the prize contribute to the matrix of options. This is why in the three cases in which your first choice is correct, and you are playing a switch strategy, the host responses and your automatic switch responses are consolidated.

Chapter 14

1. A null distribution is a perfectly known theoretical prediction. A parent distribution is a perfectly known distribution of real data; in theory it includes all the data there are, but in practice it merely requires a sample size so large that the uncertainty in the distribution is negligible. A sample distribution is a limited data set of real measurements.

2. Parametric data are distributed in the manner of a bell curve: the mean and median are equal, the distribution is symmetrical, 68.3% of the data lie within one standard deviation of the mean, 95% within two standard deviations, and so on. Nonparametric data are distributed in any of a number of other ways.

3. A one-tailed test presumes a well-founded initial hypothesis predicting a particular result of the experimental manipulation; a two-tailed test looks for any difference between the experimental distribution and the one it is being compared to.

4. A P-value of < 0.05 means that there is less than a 5% chance that the distribution you are testing was drawn at random from the distribution it is being compared against.

5. Categorical data exist as mutually exclusive outcomes, whereas continuous data exist as values along an axis with an infinite number of potential subdivisions. Discrete data consist of a set of possible values on an axis for which no intermediates are possible.

6. Paired data consist of pairs of values representing the same individual (or matched individuals) under two different conditions; unpaired data use data from different, unmatched individuals.

7. Regression analysis requires a plausible cause-and-effect relationship between the variables; in addition, the effect (dependent) variable must be parametrically distributed. Correlation analysis requires no cause-and-effect sequence, nor does nonparametric correlation analysis require that the variables be parametrically distributed.

8. The t-test compares a parametrically distributed sample distribution against a parent distribution or another sample distribution without requiring you to know the SD of either

distribution. The Gauss test compares a sample distribution against a parent or null distribution whose SD is perfectly known.

9. The chi-square test is for categorical data (though it can be applied to continuous data if the data are lumped into a small number of discrete bins). There are two versions because the goodness-of-fit test can compare only a sample categorical distribution against a parent or null distribution, whereas the less powerful independence test compares two or more sample distributions.

10. The signed-rank test is for nonparametric paired data; the rank-sum test is for nonparametric unpaired data.

11. ANOVA skirts the problem of false positives, which arises when more than one t-test comparison is made. However, ANOVA is not as sensitive and cannot by itself tell you which mean is significantly different from the others.

12. You need statistics to establish quantitatively the probability that your data could have been generated by chance rather than a causative effect. Statistics is also invaluable in guiding the initial design of experiments so as to minimize sample sizes and experimental effort.

13. A sample size of one can be significant if the datum is at least two standard deviations from the mean.

14. Normalization allows you to convert a nonparametric distribution into a parametric one, which then allows the many powerful parametric tests and techniques to be applied to your data. Among these techniques are mean sampling, squaring, taking the square root, raising the data to e^x, and taking the log.

15. Multiple-regression analysis allows you to sort out the relative effects of two independent variables on a dependent variable, even if the independent variables themselves interact.

16. The test for normality is useful to see if a data set is parametric and thus can be subjected to parametric tests. It is also useful to see if your attempt at a data transformation was successful.

17. The F-test can be used to test for independence in data sets or to determine whether two data sets have similar SDs. (The assumption of similar SDs underlies many tests.)

Statistics Labs

1. Distribution Lab I

The object of this lab is to develop data-collection techniques, to plot various distributions, and test for gender-specific differences using some basic statistical tests. Some of the data will be used (or reused) later in the course, so be sure to save it.

You need to collect some data about undergraduates, keeping track of individuals and their gender. This will permit you to use data later in the course to compare measurements from the same individual; thus, all the data from each individual needs to be recorded as a unit (i.e., all on the same line of the data-collection form you create).

The data to collect will include height, shoe size, number of siblings, height of siblings (only for sibs at least sixteen years old), and approximate distance to home.

Break up into pairs or trios and set off to interview students for about an hour; a sample size of 20 of each gender is about right. (Experience suggests that students are most responsive to groups containing at least one member of each gender. Some males you interview may not be sure of their sisters' heights; in such cases, try a taller-or-shorter-than-me approach to zero in on the correct value.)

When you return, plot the data, both by hand and using *BioStats Basics Online*. (Be sure to store the data using the appropriate button in *BioStats Basics Online* so that you will not need to enter your data a second time.) Plot the height and shoe-size data both separately by gender and together; note that you must add 1.0 to each male shoe size to convert it to female size. Each group and the class as a whole should compute a mean and standard deviation for each kind of data for which these parameters are appropriate. If you omit any data sets from these computations, explain why. Which of your data sets are continuous? Which are discrete? Which are categorical?

Your lab report should include a description of your methods (both in data collection and analysis), plus summary plots and statistical descriptions of your data. Include any appropriate remarks about the possible meaning of the results and potential problems with the data or its interpretation. The data should be in an appendix.

2. Distribution Lab II

In this lab you will analyze some of the data gathered in the previous lab.

For the sets that are appropriate, test your data for normality and report the results. What was your null hypothesis? Regardless of the results of this test, treat the data from each sex for height and shoe size as normally distributed for the rest of the analysis.

Test to see if the height and shoe-size data from the two sexes have similar SDs and report the results. What was your null hypothesis? Regardless of the outcome of this test, assume that the SDs are the same for the rest of your comparisons.

The mean and SD for male and female heights in the general population are 69.5 ± 2.5 and 63.5 ± 2.4 inches, respectively; you may treat these data as exact. Are your samples of male and/or female students significantly different from the parent distributions? What test did you use? Why? What was your null hypothesis?

The mean shoe size for males and females (in female units) for the general population are 9.7 and 7.2, respectively; you may treat these data as exact. Are your samples of male and/or female student shoe sizes significantly different from the parent distributions? What test did you use? Why? What was your null hypothesis?

Compare the heights of the males and females you interviewed. Are they significantly different? Which test did you use? Why? What was your null hypothesis?

Compare the heights of males and females in families using your data on siblings. What do you do with data from students with no sibs? What do you do with data from students with only same-sex sibs? What do you do if there are two or more opposite-sex sibs? What do you do if there are both same-sex and opposite-sex sibs in the same family? If your eventual sample size is too small (fewer than 10 comparisons), combine it with a set from another group. Is there a significant difference? What test did you use? Why? What was your null hypothesis?

Your lab report should include a description of your methods (both in data collection and analysis), summary plots and statistical descriptions of your data, and the results of each of your comparisons (with appropriate remarks about the possible meaning of the results and potential problems with the data or its interpretation). The data should be in an appendix.

3. ANOVA/Chi-Square Lab

In this lab you will gather and analyze data that require multiple comparisons. The data will be analyzed using analysis of variance and chi-square.

To use ANOVA, you must have parametric data. The variable will be height, which is parametrically distributed. The other factors to record for each person you interview are gender, race (at least three categories), whether the individual plays a varsity sport, preferred pet (dog or cat), preferred morning beverage (coffee, tea, or juice), whether the student is taller or shorter than the same-sex parent, and whether the student's father is taller or shorter than the student's mother. You will need to gather enough data to have at least 10 same-sex entries for the third-most-common race; it may be necessary to share data among groups to reach this threshold.

In looking for patterns of relationships between height and multiple other variables, you can analyze your data with ANOVA to see if any variable has a significant impact. Use the one-way ANOVA to examine height in the gender for which you have the most data for the third-most-common race. Examine the patterns of height against race. If your sample sizes are unequal, what did you do to make them the same? Why do you think this is an unbiased approach? Is there a significant effect according the ANOVA test? If so, can the Tukey-Kramer method isolate the unusual group? What was your null hypothesis?

Use the two-way ANOVA to examine the effect of two variables on height. In this case, use gender as one variable and varsity participation as the other. Justify your method of equalizing sample sizes. Is there a significant difference? If so, does the analysis isolate the important variable or variables, or is there an interaction? What was your null hypothesis?

Chi-square is used for categorical comparisons; the data are not parametric. Chi-square for goodness of fit is appropriate if you have a null distribution (usually the expectation of a null hypothesis); the test compares the observed distribution with the expected. Chi-square for independence compares two sets of data and has no null expectation: it simply asks if the distributions are different from each other.

Among parental couples with students here, is one sex taller than the other? Which test is appropriate? What is your null hypothesis?

Are student heights different from those of their same-sex parent? Which test is appropriate? What is your null hypothesis?

Do pet preferences differ between sexes? Which test is appropriate? What is your null hypothesis?

Do morning-beverage preferences differ between sexes? Which test is appropriate? What is your null hypothesis?

Do morning-beverage preferences differ between varsity athletes and other students? Which test is appropriate? What is your null hypothesis?

Your lab report should include a description of your methods (both in data collection and analysis), summary plots and statistical descriptions of your data, and the results of each of your comparisons (with appropriate remarks about the possible meaning of the results and potential problems with the data or its interpretation). The data should be in an appendix.

4. Nonparametric Lab

The object of this lab is to learn how to analyze nonparametric data. You will need to collect some data by interviewing fellow students. Keep track of each student's response and gender. A sample size of 20 of each sex should be about right.

The questions to ask:

Is the student a varsity athlete?

How many hours a week does the student estimate he or she spends studying?

How many hours a week does the student estimate he or she spends on "official" nonathletic extracurricular activities (e.g., clubs and organizations, publications, student government, etc.).

How many hours a week does the student estimate he or she spends on athletics?

How many hours a week does the student estimate he or she spends sleeping or resting during term? During exams? During vacation periods?

How many hours a week does the student estimate he or she spends socializing?

Test first to see whether students spend more time studying or participating in extracurricular activities (including athletics). Is there a significant difference? What test did you use? Why? What was your null hypothesis? Could you have analyzed these data with chi-square? If so, try it and compare the results of the two tests.

Next, test to see if there is any sex difference in time spent studying. Is there a significant difference? What test did you use? Why? What was your null hypothesis?

For nonathletes, analyze the distribution of time spent sleeping and resting during term, during exams, and during vacations. Is there a significant difference? What test did you use? Why? What was your null hypothesis?

Next, check to see if there is any gender difference in time distributions. To do this, enter the mean values for each activity in a row, with the other gender in the other row. For sleeping or resting time, just use the term-time value. Is there a significant difference? What test did you use? Why? What was your null hypothesis?

Now see if participation in varsity athletics has any impact. Again, use mean values for each activity, with athletes in one row and nonathletes in the other. Is there a significant difference? What test did you use? Why? What was your null hypothesis?

Your lab report should include a description of your methods (both in data collection and analysis), summary plots and statistical descriptions of your data, and the results of each of your comparisons (with appropriate remarks about the possible meaning of the results and potential problems with the data or its interpretation). The data should be in an appendix.

5. Correlation and Regression Lab

The object of this lab is to look for correlations and, where appropriate, plot regressions. You will need to interview fellow students and ask questions about themselves *and* their current/most recent boyfriend/girlfriend. The relevant data to gather are height, age (or class), and number of sibs; in addition, ask what the ideal opposite-sex height is for them. Record the gender of the student interviewed.

Most comparisons will be pairings of the same variable in each couple. Start with number of sibs. What is the appropriate test (i.e., nonparametric correlation, parametric correlation, or regression)? Is there a significant relationship? What was your null hypothesis?

How about age? What is the appropriate test (i.e., nonparametric correlation, parametric correlation, or regression)? Is there a significant relationship? What was your null hypothesis?

Now try height in couples. What is the appropriate test (i.e., nonparametric correlation, parametric correlation, or regression)? Is there a significant relationship? What was your null hypothesis?

Finally, look at ideal heights. Analyze the sexes separately, comparing each individual's own height with the ideal height the respondent specified for the opposite sex. (You may assume that the preference for opposite-sex height is normally distributed.) What is the appropriate test (i.e., nonparametric correlation, parametric correlation, or regression)? Is there a significant relationship? What was your null hypothesis? Is the correlation coefficient larger or smaller than in the previous height comparison? What might this mean?

Finally, compare the ideal and actual heights of opposite-sex partners. What is the appropriate test (i.e., nonparametric correlation, parametric correlation, or regression)? Is there a significant relationship? What was your null hypothesis? Is the correlation coefficient larger or smaller than in the other height comparisons? What might this mean?

Your lab report should include a description of your methods (both in data collection and analysis), summary plots and statistical descriptions of your data, and the results of each of your comparisons (with appropriate remarks about the possible meaning of the results and potential problems with the data or its interpretation). The data should be in an appendix.

6. Circular Statistics Lab

The object of this lab is to explore the reports that humans have an internal sense of direction, perhaps based on an ability to sense the earth's magnetic field. (The ability to use magnetic fields is found in many species of animals, including honey bees, monarch butterflies, cave-dwelling newts, salmon and tuna, homing pigeons and most species of migrating birds, dolphins, and field mice; it appears to be absent from many others.)

Robin Baker, of the University of Manchester, reported that college students transported blindfolded to distant sites could accurately point in the homeward direction; students wearing magnets (as opposed to brass weights) were disoriented. Baker also reported good results using blindfolded students in a darkened room: they were placed in a swivel chair, told which direction was north, blindfolded, rotated slowly, gently stopped while facing some other direction, and then asked to point north. Again, wearing magnets was said to have greatly reduced the accuracy of pointing.

Other workers have tried to repeat these various tests; some have been successful, most have not. Skeptics believe positive results depend on sloppy technique and the use of unintended cues; believers assume negative results are a consequence of sloppy technique and interference with relevant cues. You can decide for yourself.

This lab will use the rotating-chair protocol. Use one another as subjects; if the sample size is too small (i.e., fewer than 20), recruit some other students to participate. You will need to decide on a protocol and write it out; in particular, the instructions given to subjects should be read, not delivered off the cuff. For consistency, you will need to select one pair of students to run the test; the others will need to be outside the test room, unable to hear what is being said. You will need a blindfold, a pair of ear protectors (the kind sold in hardware store to be worn when operating noisy equipment), a swivel chair, a small magnet on a piece of tape, a brass object of about the same size similarly attached to tape, and a room with the blinds drawn. The testers should draw up a long list of turn numbers (e.g., 1.5, 2.0, 1.25, 1.75, 2.5, 1.5, 2.25, 1.75, 1.25, etc.; these number must be unknown to the subjects in advance, and should always be greater than one turn).

For calibration, the testers should begin by telling the unblindfolded subject which way is north. Then for calibration ask each subject to point in the direction of a well-known landmark—the main administration building, for instance. Record this data to the nearest 10° or 15°, along with the sex of the student. Next, have the subject sit cross-legged in the chair and allow him or her to put on the blindfold. Point the subject north and tell him or her that this is north. On even- or odd-numbered tests (the testers should decide and not tell the subjects) attach the magnet via the tape to the blindfold (as high as possible and roughly between the eyes) and rotate the subject slowly clockwise the number of turns specified by the first value on the list; ask which direction the subject is now facing. Rotate the subject the same number of turns counterclockwise and tell him or her that this is north again. Exchange the brass/magnet for the magnet/brass and repeat the test using the number of turns specified by the second value on the list; after getting the subject's estimate of direction, return to the original position by reversing the turning. Remove the brass/weight and allow the subject to put on the ear protectors. Repeat the test one last time with the third number of turns on the list. For the next subject, begin with the fourth value on the list, and so on; when you reach the bottom of the list, start again at the top.

In analyzing your data, you will want to use one of the circular statistics test, either with a predicted direction or just looking for consistent orientation. Comparisons between sexes requires the Watson-Williams test. The data can be entered on the appropriate card in the *BioStats Basics Online* program. You will need to decide how to split up the work among groups. The testers should be relieved of the need to enter data, though they will probably need to interpret the written numbers to the rest of the group.

The questions to ask:

Is there any ability to point in the direction of the campus landmark? If not, is there any clustering at all? Is there any difference between sexes—i.e., perhaps one sex was clustered and the other wasn't, or perhaps one was better oriented than the other?

Is there any ability to judge direction wearing a brass weight but no ear protectors? If not, is there any clustering at all? Is there any difference between sexes—i.e., perhaps one sex was clustered and the other wasn't, or perhaps one was better oriented than the other?

Is there any ability to judge direction wearing a magnet but no ear protectors? If not, is there any clustering at all? Is there any difference between sexes—i.e., perhaps one sex was clustered and the other wasn't, or perhaps one was better oriented than the other?

Is there any ability to judge direction wearing the ear protectors? If not, is there any clustering at all? Is there any difference between sexes—i.e., perhaps one sex was clustered and the other wasn't, or perhaps one was better oriented than the other?

Your lab report should include a description of your methods (both in data collection and analysis), summary plots and statistical descriptions of your data, and the results of each of your comparisons (with appropriate remarks about the possible meaning of the results and potential problems with the data or its interpretation). The data should be in an appendix.

Selected Statistical Tables

A. Areas of the normal curve   *371*

B. Critical values of the *F*-distribution   *372*

C. Critical values of Student's *t*-distribution   *386*

D. Critical values of the chi-square distribution   *387*

E. Critical values of the Wilcoxon rank sum   *389*

F. Critical values of the Friedman χ_r^2 distribution   *392*

G. Critical values of *U*, the Mann-Whitney statistic   *393*

H. Critical values of the Kruskal-Wallis *H*-distribution   *397*

I. Critical values of Rayleigh's *u*   *398*

J. Critical values of Rayleigh's *z*   *399*

K. Critical values of the Spearman rank correlation coefficient, r_s *400*

L. Critical values of the correlation coefficient, *r*   *402*

Table A Areas of the normal curve

Standard deviation units	0.	0.01	0.02	0.03	0.04	0.05	0.06	0.07	0.08	0.09	Standard deviation units
0.0	.0000	.0040	.0080	.0120	.0160	.0199	.0239	.0279	.0319	.0359	0.0
0.1	.0398	.0438	.0478	.0517	.0557	.0596	.0636	.0675	.0714	.0753	0.1
0.2	.0793	.0832	.0871	.0910	.0948	.0987	.1026	.1064	.1103	.1141	0.2
0.3	.1179	.1217	.1255	.1293	.1331	.1368	.1406	.1443	.1480	.1517	0.3
0.4	.1554	.1591	.1628	.1664	.1700	.1736	.1772	.1808	.1844	.1879	0.4
0.5	.1915	.1950	.1985	.2019	.2054	.2088	.2123	.2157	.2190	.2224	0.5
0.6	.2257	.2291	.2324	.2357	.2389	.2422	.2454	.2486	.2517	.2549	0.6
0.7	.2580	.2611	.2642	.2673	.2704	.2734	.2764	.2794	.2823	.2852	0.7
0.8	.2881	.2910	.2939	.2967	.2995	.3023	.3051	.3078	.3106	.3133	0.8
0.9	.3159	.3186	.3212	.3238	.3264	.3289	.3315	.3340	.3365	.3389	0.9
1.0	.3413	.3438	.3461	.3485	.3508	.3531	.3554	.3577	.3599	.3621	1.0
1.1	.3643	.3665	.3686	.3708	.3729	.3749	.3770	.3790	.3810	.3830	1.1
1.2	.3849	.3869	.3888	.3907	.3925	.3944	.3962	.3980	.3997	.4015	1.2
1.3	.4032	.4049	.4066	.4082	.4099	.4115	.4131	.4147	.4162	.4177	1.3
1.4	.4192	.4207	.4222	.4236	.4251	.4265	.4279	.4292	.4306	.4319	1.4
1.5	.4332	.4345	.4357	.4370	.4382	.4394	.4406	.4418	.4429	.4441	1.5
1.6	.4452	.4463	.4474	.4484	.4495	.4505	.4515	.4525	.4535	.4545	1.6
1.7	.4554	.4564	.4573	.4582	.4591	.4599	.4608	.4616	.4625	.4633	1.7
1.8	.4641	.4649	.4656	.4664	.4671	.4678	.4686	.4693	.4699	.4706	1.8
1.9	.4713	.4719	.4726	.4732	.4738	.4744	.4750	.4756	.4761	.4767	1.9
2.0	.4772	.4778	.4783	.4788	.4793	.4798	.4803	.4808	.4812	.4817	2.0
2.1	.4821	.4826	.4830	.4834	.4838	.4842	.4846	.4850	.4854	.4857	2.1
2.2	.4861	.4864	.4868	.4871	.4875	.4878	.4881	.4884	.4887	.4890	2.2
2.3	.4893	.4896	.4898	.4901	.4904	.4906	.4909	.4911	.4913	.4916	2.3
2.4	.4918	.4920	.4922	.4925	.4927	.4929	.4931	.4932	.4934	.4936	2.4
2.5	.4938	.4940	.4941	.4943	.4945	.4946	.4948	.4949	.4951	.4952	2.5
2.6	.4953	.4955	.4956	.4957	.4959	.4960	.4961	.4962	.4963	.4964	2.6
2.7	.4965	.4966	.4967	.4968	.4969	.4970	.4971	.4972	.4973	.4974	2.7
2.8	.4974	.4975	.4976	.4977	.4977	.4978	.4979	.4979	.4980	.4981	2.8
2.9	.4981	.4982	.4982	.4983	.4984	.4984	.4985	.4985	.4986	.4986	2.9
3.0	.4987	.4987	.4987	.4988	.4988	.4989	.4989	.4989	.4990	.4990	3.0
3.1	.4990	.4991	.4991	.4991	.4992	.4992	.4992	.4992	.4993	.4993	3.1
3.2	.4993	.4993	.4994	.4994	.4994	.4994	.4994	.4995	.4995	.4995	3.2
3.3	.4995	.4995	.4995	.4996	.4996	.4996	.4996	.4996	.4996	.4997	3.3
3.4	.4997	.4997	.4997	.4997	.4997	.4997	.4997	.4997	.4997	.4998	3.4
3.5	.499767										
3.6	.499841										
3.7	.499892										
3.8	.499928										
3.9	.499952										
4.0	.499968										
4.1	.499979										
4.2	.499987										
4.3	.499991										
4.4	.499995										
4.5	.499997										
4.6	.499998										
4.7	.499999										
4.8	.499999										
4.9	.500000										

Table B   Critical values of the *F*-distribution

v_1
(degrees of freedom of numerator mean squares)

	α	1	2	3	4	5	6	7	8	9	10	11	12
	.75	.172	.389	.494	.553	.591	.617	.636	.650	.661	.670	.680	.684
	.50	1.00	1.50	1.71	1.82	1.89	1.94	1.98	2.00	2.03	2.04	2.06	2.07
	.25	5.83	7.50	8.20	8.58	8.82	8.98	9.10	9.19	9.26	9.32	9.37	9.41
	.10	39.9	49.5	53.6	55.8	57.2	58.2	58.9	59.4	59.9	60.2	60.5	60.7
1	.05	161	199	216	225	230	234	237	239	241	241	243	244
	.025	648	800	864	900	922	937	948	957	963	969	973	977
	.01	4050	5000	5400	5620	5760	5860	5930	5980	6020	6060	6080	6110
	.005	16200	20000	21600	22500	23100	23400	23700	23900	24100	24200	24300	24400
	.001	405300	500000	540400	562500	576400	585900	592900	598100	602300	605600	608400	610700
	.75	.133	.333	.439	.500	.540	.568	.588	.604	.616	.626	.633	.641
	.50	.667	1.00	1.13	1.21	1.25	1.28	1.30	1.32	1.33	1.35	1.36	1.36
	.25	2.57	3.00	3.15	3.23	3.28	3.31	3.34	3.35	3.37	3.38	3.39	3.39
	.10	8.53	9.00	9.16	9.24	9.29	9.33	9.35	9.37	9.38	9.39	9.40	9.41
2	.05	18.5	19.0	19.2	19.2	19.3	19.3	19.4	19.4	19.4	19.4	19.4	19.4
	.025	38.5	39.0	39.2	39.2	39.3	39.3	39.4	39.4	39.4	39.4	39.4	39.4
	.01	98.5	99.0	99.2	99.2	99.3	99.3	99.4	99.4	99.4	99.4	99.4	99.4
	.005	198	199	199	199	199	199	199	199	199	199	199	199
	.001	999	999	999	999	999	999	999	999	999	999	999	999
	.75	.122	.317	.424	.489	.531	.561	.581	.600	.613	.624	.633	.641
	.50	.585	.881	1.00	1.06	1.10	1.13	1.15	1.16	1.17	1.18	1.19	1.20
	.25	2.02	2.28	2.36	2.39	2.41	2.42	2.43	2.44	2.44	2.44	2.45	2.45
	.10	5.54	5.46	5.39	5.34	5.31	5.28	5.27	5.25	5.24	5.23	5.22	5.22
3	.05	10.1	9.55	9.28	9.12	9.01	8.94	8.89	8.85	8.81	8.79	8.76	8.74
	.025	17.4	16.0	15.4	15.1	14.9	14.7	14.6	14.5	14.5	14.4	14.3	14.3
	.01	34.1	30.8	29.5	28.7	28.2	27.9	27.7	27.5	27.3	27.2	27.1	27.1
	.005	55.6	49.8	47.5	46.2	45.4	44.8	44.4	44.1	43.9	43.7	43.5	43.4
	.001	167	149	141	137	135	133	132	131	130	129	128	128
	.75	.117	.309	.418	.484	.528	.560	.583	.601	.615	.627	.637	.645
	.50	.549	.828	.941	1.00	1.04	1.06	1.08	1.09	1.10	1.11	1.12	1.13
	.25	1.81	2.00	2.05	2.06	2.07	2.08	2.08	2.08	2.08	2.08	2.08	2.08
	.10	4.54	4.32	4.19	4.11	4.05	4.01	3.98	3.95	3.94	3.92	3.91	3.90
4	.05	7.71	6.94	6.59	6.39	6.26	6.16	6.09	6.04	6.00	5.96	5.93	5.91
	.025	12.2	10.6	9.98	9.60	9.36	9.20	9.07	8.98	8.90	8.84	8.79	8.75
	.01	21.2	18.0	16.7	16.0	15.5	15.2	15.0	14.8	14.7	14.5	14.4	14.4
	.005	31.3	26.3	24.3	23.2	22.4	22.0	21.6	21.4	21.1	21.0	20.8	20.7
	.001	74.1	61.3	56.2	53.4	51.7	50.5	49.7	49.0	48.5	48.1	47.7	47.4
	.75	.113	.305	.415	.483	.528	.560	.584	.604	.618	.631	.641	.650
	.50	.528	.799	.907	.965	1.00	1.02	1.04	1.05	1.06	1.07	1.08	1.09
	.25	1.69	1.85	1.88	1.89	1.89	1.89	1.89	1.89	1.89	1.89	1.89	1.89
	.10	4.06	3.78	3.62	3.52	3.45	3.40	3.37	3.34	3.32	3.30	3.28	3.27
5	.05	6.61	5.79	5.41	5.19	5.05	4.95	4.88	4.82	4.77	4.74	4.71	4.68
	.025	10.0	8.43	7.76	7.39	7.15	6.98	6.85	6.76	6.68	6.62	6.57	6.52
	.01	16.3	13.3	12.1	11.4	11.0	10.7	10.5	10.3	10.2	10.1	9.99	9.89
	.005	22.8	18.3	16.5	15.6	14.9	14.5	14.2	14.0	13.8	13.6	13.5	13.4
	.001	47.2	37.1	33.2	31.1	29.7	28.8	28.2	27.6	27.2	26.9	26.6	26.4

v_2 (degrees of freedom of denominator mean squares)

ν_1
(degrees of freedom of numerator mean squares)

ν_2	α	1	2	3	4	5	6	7	8	9	10	11	12
6	.75	.111	.302	.413	.481	.524	.561	.586	.606	.621	.635	.645	.654
	.50	.515	.780	.886	.942	.977	1.00	1.02	1.03	1.04	1.05	1.06	1.06
	.25	1.62	1.76	1.78	1.79	1.79	1.78	1.78	1.78	1.77	1.77	1.77	1.77
	.10	3.78	3.46	3.29	3.18	3.11	3.05	3.01	2.98	2.96	2.94	2.92	2.90
	.05	5.99	5.14	4.76	4.53	4.39	4.28	4.21	4.15	4.10	4.06	4.03	4.00
	.025	8.81	7.26	6.60	6.23	5.99	5.82	5.70	5.60	5.52	5.46	5.41	5.37
	.01	13.7	10.9	9.78	9.15	8.75	8.47	8.26	8.10	7.98	7.87	7.79	7.72
	.005	18.6	14.5	12.9	12.0	11.5	11.1	10.8	10.6	10.4	10.3	10.1	10.0
	.001	35.5	27.0	23.7	21.9	20.8	20.0	19.5	19.0	18.7	18.4	18.2	18.0
7	.75	.110	.300	.412	.481	.528	.562	.588	.608	.624	.637	.649	.658
	.50	.506	.767	.871	.926	.960	.983	1.00	1.01	1.02	1.03	1.04	1.04
	.25	1.57	1.70	1.72	1.72	1.71	1.71	1.70	1.70	1.69	1.69	1.68	1.68
	.10	3.59	3.26	3.07	2.96	2.88	2.83	2.78	2.75	2.72	2.70	2.68	3.67
	.05	5.59	4.74	4.35	4.12	3.97	3.87	3.77	3.73	3.68	3.64	3.60	3.57
	.025	8.07	6.54	5.89	5.52	5.29	5.12	4.99	4.89	4.82	4.76	4.71	4.67
	.01	12.2	9.55	8.45	7.85	7.46	7.19	6.99	6.84	6.72	6.62	6.54	6.47
	.005	16.2	12.4	10.9	10.1	9.52	9.16	8.89	8.68	8.52	8.38	8.27	8.18
	.001	29.3	21.7	18.8	17.2	16.2	15.5	15.0	14.6	14.3	14.1	13.9	13.7
8	.75	.109	.298	.411	.481	.529	.563	.589	.610	.627	.640	.654	.661
	.50	.499	.757	.860	.915	.948	.971	.988	1.00	1.01	1.02	1.03	1.03
	.25	1.54	1.66	1.67	1.66	1.66	1.65	1.64	1.64	1.63	1.63	1.63	1.625
	.10	3.46	3.11	2.92	2.81	2.73	2.67	2.62	2.59	2.56	2.54	2.52	2.50
	.05	5.32	4.46	4.07	3.84	3.69	3.58	3.50	3.44	3.39	3.35	3.31	3.28
	.025	7.57	6.06	5.42	5.05	4.82	4.65	4.53	4.43	4.36	4.30	4.25	4.20
	.01	11.3	8.65	7.59	7.01	6.63	6.37	6.18	6.03	5.91	5.81	5.73	5.67
	.005	14.7	11.0	9.60	8.81	8.30	7.95	7.69	7.50	7.34	7.21	7.10	7.01
	.001	25.4	18.5	15.8	14.4	13.5	12.9	12.4	12.0	11.8	11.5	11.3	11.2
9	.75	.108	.297	.410	.480	.529	.564	.591	.612	.629	.643	.654	.664
	.50	.494	.749	.852	.906	.939	.962	.978	.990	1.00	1.01	1.02	1.02
	.25	1.51	1.62	1.63	1.63	1.62	1.61	1.60	1.60	1.59	1.59	1.58	1.58
	.10	3.36	3.01	2.81	2.69	2.61	2.55	2.51	2.47	2.44	2.42	2.40	2.38
	.05	5.12	4.26	3.86	3.63	3.48	3.37	3.29	3.23	3.18	3.14	3.10	3.07
	.025	7.21	5.71	5.08	4.72	4.48	4.32	4.20	4.10	4.03	3.96	3.91	3.87
	.01	10.6	8.02	6.99	6.42	6.06	5.80	5.61	5.47	5.35	5.26	5.18	5.11
	.005	13.6	10.1	8.72	7.96	7.47	7.13	6.88	6.69	6.54	6.42	6.32	6.23
	.001	22.9	16.4	13.9	12.6	11.7	11.1	10.7	10.4	10.1	9.79	9.72	9.57
10	.75	.107	.296	.409	.480	.529	.565	.592	.613	.631	.645	.657	.667
	.50	.490	.743	.845	.899	.932	.954	.971	.983	.992	1.00	1.01	1.01
	.25	1.49	1.60	1.60	1.59	1.59	1.58	1.57	1.56	1.56	1.55	1.54	1.54
	.10	3.29	2.92	2.73	2.61	2.52	2.46	2.41	2.38	2.35	2.32	2.30	2.28
	.05	4.96	4.10	3.71	3.48	3.33	3.22	3.14	3.07	3.02	2.98	2.94	2.91
	.025	6.94	5.46	4.83	4.47	4.24	4.07	3.95	3.85	3.78	3.72	3.67	3.62
	.01	10.0	7.56	6.55	5.99	5.64	5.39	5.20	5.06	4.94	4.85	4.77	4.71
	.005	12.8	9.43	8.08	7.34	6.87	6.54	6.30	6.12	5.97	5.85	5.75	5.66
	.001	21.0	14.9	12.5	11.3	10.5	9.92	9.52	9.20	8.96	8.75	8.59	8.45

ν_2 (degrees of freedom of denominator mean squares)

(continued)

Table B Critical values of the *F*-distribution (*continued*)

ν_1
(degrees of freedom of numerator mean squares)

	α	1	2	3	4	5	6	7	8	9	10	11	12
	.75	.107	.295	.408	.481	.529	.565	.592	.614	.633	.645	.658	.667
	.50	.486	.739	.840	.893	.926	.948	.964	.977	.986	.994	1.00	1.01
	.25	1.47	1.58	1.58	1.57	1.56	1.55	1.54	1.53	1.53	1.52	1.51	1.51
	.10	3.23	2.86	2.66	2.54	2.45	2.39	2.34	2.30	2.27	2.25	2.23	2.21
11	.05	4.84	3.98	3.59	3.36	3.20	3.09	3.01	2.95	2.90	2.85	2.82	2.79
	.025	6.72	5.26	4.63	4.28	4.04	3.88	3.76	3.66	3.59	3.53	3.48	3.43
	.01	9.65	7.21	6.22	5.67	5.32	5.07	4.89	4.74	4.63	4.54	4.46	4.40
	.005	12.2	8.91	7.60	6.88	6.42	6.10	5.86	5.68	5.54	5.42	5.32	5.24
	.001	19.7	13.8	11.6	10.3	9.58	9.05	8.66	8.35	8.12	7.92	7.76	7.63
	.75	.106	.295	.408	.480	.530	.566	.594	.616	.633	.649	.661	.671
	.50	.484	.735	.835	.888	.921	.943	.959	.972	.981	.989	.995	1.00
	.25	1.46	1.56	1.56	1.55	1.54	1.53	1.52	1.51	1.51	1.50	1.50	1.49
	.10	3.18	2.81	2.61	2.48	2.39	2.33	2.28	2.24	2.21	2.19	2.17	2.15
12	.05	4.75	3.89	3.49	3.26	3.11	3.00	2.91	2.85	2.80	2.75	2.72	2.69
	.025	6.55	5.10	4.47	4.12	3.89	3.73	3.61	3.51	3.44	3.37	3.32	3.28
	.01	9.33	6.93	5.95	5.41	5.06	4.82	4.64	4.50	4.39	4.30	4.22	4.16
	.005	11.8	8.51	7.23	6.52	6.07	5.76	5.52	5.35	5.20	5.09	4.99	4.91
	.001	18.6	13.0	10.8	9.63	8.89	8.38	8.00	7.71	7.48	7.29	7.14	7.00
	.75	.106	.294	.408	.480	.530	.567	.595	.617	.635	.650	.662	.673
	.50	.481	.731	.832	.885	.917	.939	.955	.967	.977	.984	.990	.996
	.25	1.45	1.55	1.55	1.53	1.52	1.51	1.50	1.49	1.49	1.48	1.48	1.47
	.10	3.14	2.76	2.56	2.43	2.35	2.28	2.23	2.20	2.16	2.14	2.12	2.10
13	.05	4.67	3.81	3.41	3.18	3.03	2.92	2.83	2.77	2.71	2.67	2.64	2.60
	.025	6.41	4.97	4.35	4.00	3.77	3.60	3.48	3.39	3.31	3.25	3.20	3.15
	.01	9.07	6.70	5.74	4.21	4.86	4.62	4.44	4.30	4.19	4.10	4.03	3.96
	.005	11.4	8.19	6.93	6.23	5.79	5.48	5.25	5.08	4.94	4.82	4.73	4.64
	.001	17.8	12.3	10.2	9.07	8.35	7.86	7.49	7.21	6.98	6.80	6.65	6.52
	.75	.106	.294	.408	.480	.530	.567	.595	.618	.636	.651	.664	.674
	.50	.479	.729	.828	.881	.914	.936	.952	.964	.973	.981	.987	.992
	.25	1.44	1.53	1.53	1.52	1.51	1.50	1.49	1.48	1.47	1.46	1.46	1.45
	.10	3.10	2.73	2.52	2.39	2.31	2.24	2.19	2.45	1.42	2.10	2.07	2.05
14	.05	4.60	3.74	3.34	3.11	2.96	2.85	2.76	2.70	2.65	2.60	2.57	2.53
	.025	6.30	4.86	4.24	3.89	3.66	3.50	3.38	3.29	3.20	3.15	3.10	3.05
	.01	8.86	6.51	5.56	5.04	4.69	4.46	4.28	4.14	4.03	3.94	3.86	3.80
	.005	11.1	7.92	6.68	6.00	5.53	5.26	5.03	4.86	4.72	4.60	4.51	4.43
	.001	17.1	11.8	9.73	8.62	7.92	7.43	7.08	6.80	6.58	6.40	6.26	6.13
	.75	.105	.293	.407	.480	.530	.568	.596	.618	.637	.652	.667	.676
	.50	.478	.726	.826	.878	.911	.933	.948	.960	.970	.977	.983	.989
	.25	1.43	1.52	1.52	1.51	1.49	1.48	1.47	1.46	1.46	1.45	1.44	1.44
	.10	3.07	2.70	2.49	2.36	2.27	2.21	2.16	2.12	2.09	2.06	2.04	2.02
15	.05	4.54	3.68	3.29	3.06	2.90	2.79	2.71	2.64	2.59	2.54	2.51	2.48
	.025	6.20	4.77	4.15	3.80	3.58	3.41	3.29	3.20	3.12	3.06	3.01	2.96
	.01	8.68	6.36	5.42	4.89	4.56	4.32	4.14	4.00	3.89	3.80	3.73	3.67
	.005	10.8	7.70	6.48	5.80	5.37	5.07	4.85	4.67	4.54	4.42	4.33	4.25
	.001	16.6	11.3	9.34	8.25	7.57	7.09	6.74	6.47	6.26	6.08	5.94	5.81

ν_2 (degrees of freedom of denominator mean squares)

v_1
(degrees of freedom of numerator mean squares)

v_2		α	1	2	3	4	5	6	7	8	9	10	11	12
	16	.75	.105	.293	.407	.480	.531	.568	.597	.619	.638	.653	.666	.677
		.50	.476	.724	.823	.876	.908	.930	.946	.958	.967	.975	.981	.986
		.25	1.42	1.51	1.51	1.50	1.48	1.47	1.46	1.45	1.44	1.44	1.43	1.43
		.10	3.05	2.67	2.46	2.33	2.24	2.18	2.13	2.09	2.06	2.03	2.01	1.99
		.05	4.49	3.63	3.24	3.01	2.85	2.74	2.66	2.59	2.54	2.49	2.46	2.42
		.025	6.12	4.69	4.08	3.73	3.50	3.34	3.22	3.12	3.05	2.99	2.93	2.89
		.01	8.53	6.23	5.29	4.77	4.44	4.20	4.03	3.89	3.78	3.69	3.62	3.55
		.005	10.6	7.51	6.30	5.64	5.21	4.91	4.69	4.52	5.38	4.27	4.18	4.10
		.001	16.1	11.0	9.00	7.94	7.27	6.81	6.46	6.19	5.98	5.81	5.67	5.55
	17	.75	.105	.292	.407	.480	.531	.568	.597	.620	.639	.654	.667	.678
		.50	.475	.722	.821	.874	.906	.928	.943	.955	.965	.972	.978	.983
		.25	1.42	1.51	1.50	1.49	1.47	1.46	1.45	1.44	1.43	1.43	1.42	1.41
		.10	3.03	2.64	2.44	2.31	2.22	2.15	2.10	2.06	2.03	2.00	1.98	1.96
		.05	4.45	3.59	3.20	2.96	2.91	2.70	3.61	2.55	2.49	2.45	2.41	2.38
		.025	6.04	4.62	4.01	3.66	3.44	3.28	3.16	3.06	2.98	2.92	2.87	2.82
		.01	8.40	6.11	5.19	4.67	4.34	4.10	3.93	3.79	3.68	3.59	3.52	3.46
		.005	10.4	7.35	6.16	5.50	5.07	4.78	4.56	4.39	4.25	4.14	4.05	3.97
		.001	15.7	10.7	8.73	7.68	7.02	6.56	6.22	5.96	5.75	5.58	5.44	5.32
	18	.75	.105	.292	.407	.480	.531	.569	.598	.621	.639	.655	.668	.679
		.50	.474	.721	.819	.872	.904	.926	.941	.953	.962	.970	.976	.981
		.25	1.41	1.50	1.49	1.48	1.46	1.45	1.44	1.43	1.42	1.42	1.41	1.40
		.10	3.01	2.62	2.42	2.29	2.20	2.13	2.08	2.04	2.00	1.98	1.95	1.93
		.05	4.41	3.55	3.16	2.93	2.77	2.66	2.58	2.51	2.46	2.41	2.37	2.34
		.025	5.98	4.56	3.95	3.61	3.38	3.22	3.10	3.01	2.93	2.87	2.81	2.77
		.01	8.28	6.01	5.09	4.58	4.25	4.01	3.84	3.71	3.60	3.51	3.43	3.37
		.005	10.2	7.21	6.03	5.37	4.96	4.66	4.44	4.28	4.14	4.03	3.94	3.86
		.001	15.4	10.4	8.49	7.46	6.81	6.35	6.02	5.76	5.56	5.39	5.25	5.13
	19	.75	.104	.292	.407	.480	.531	.569	.598	.621	.640	.656	.669	.680
		.50	.473	.719	.818	.870	.902	.924	.939	.951	.961	.968	.974	.979
		.25	1.41	1.49	1.49	1.47	1.46	1.44	1.43	1.42	1.41	1.41	1.40	1.40
		.10	2.99	2.61	2.40	2.27	2.18	2.11	2.06	2.02	1.98	1.96	1.93	1.91
		.05	4.38	3.52	3.13	2.90	2.74	2.63	2.54	2.48	2.42	2.38	2.34	2.31
		.025	5.92	4.51	3.90	3.56	3.33	3.17	3.05	2.76	2.88	2.82	2.76	2.72
		.01	8.19	5.93	5.01	4.50	4.17	3.94	3.77	3.63	3.52	3.43	3.56	3.30
		.005	10.1	7.09	5.92	5.27	4.85	4.56	4.34	4.18	4.04	3.93	3.84	3.76
		.001	15.1	10.2	8.28	7.26	6.62	6.18	5.85	5.59	5.39	5.22	5.08	4.97
	20	.75	.104	.292	.407	.480	.531	.569	.598	.622	.641	.656	.671	.681
		.50	.472	.718	.816	.868	.900	.922	.938	.950	.959	.966	.972	.977
		.25	1.40	1.49	1.48	1.47	1.45	1.44	1.43	1.42	1.41	1.40	1.39	1.39
		.10	2.97	2.59	2.38	2.25	2.16	2.09	2.04	2.00	1.96	1.94	1.91	1.89
		.05	4.35	3.49	3.10	2.87	2.71	2.60	2.51	2.45	2.39	2.35	2.31	2.28
		.025	5.87	4.46	3.86	3.51	3.29	3.13	3.01	2.91	2.84	2.77	2.72	2.68
		.01	8.10	5.85	4.94	4.43	4.10	3.87	3.70	3.56	3.46	3.37	3.29	3.23
		.005	9.94	6.99	5.82	5.17	4.76	4.47	4.26	4.09	3.96	3.85	3.76	3.68
		.001	14.8	9.95	8.10	7.10	6.46	6.02	5.69	5.44	5.24	5.08	4.94	4.82

v_2 (degrees of freedom of denominator mean squares)

(continued)

Table B Critical values of the *F*-distribution (*continued*)

ν_1
(degrees of freedom of numerator mean squares)

	α	1	2	3	4	5	6	7	8	9	10	11	12
	.75	.104	.292	.407	.480	.532	.570	.599	.622	.641	.657	.671	.682
	.50	.471	.717	.815	.867	.899	.921	.936	.948	.957	.965	.971	.976
	.25	1.40	1.48	1.48	1.46	1.44	1.43	1.42	1.41	1.40	1.39	1.39	1.38
	.10	2.96	2.57	2.36	2.23	2.14	2.08	2.02	1.98	1.95	1.92	1.90	1.87
21	.05	4.32	3.47	3.07	2.84	2.68	2.57	2.49	2.42	2.37	2.32	2.28	2.25
	.025	5.83	4.42	3.82	3.48	3.25	3.09	2.97	2.87	2.80	2.73	2.68	2.64
	.01	8.02	5.75	4.87	4.37	4.04	3.81	3.64	3.51	3.40	3.31	3.24	3.17
	.005	9.83	6.89	5.73	5.09	4.99	4.39	4.18	4.01	3.88	3.77	3.68	3.60
	.001	14.6	9.77	7.94	6.95	6.32	5.88	5.56	5.31	5.11	4.95	4.81	4.70
	.75	.104	.292	.407	.481	.532	.570	.599	.623	.642	.658	.671	.683
	.50	.470	.715	.814	.866	.898	.919	.935	.947	.956	.963	.969	.974
	.25	1.40	1.48	1.47	1.45	1.44	1.42	1.41	1.40	1.39	1.39	1.38	1.37
	.10	2.95	2.56	2.35	2.22	2.13	2.06	2.01	1.97	1.93	1.90	1.88	1.86
22	.05	4.30	3.44	3.05	2.82	2.66	2.55	2.46	2.40	2.39	2.30	2.26	2.23
	.025	5.79	4.38	3.78	3.44	3.22	3.05	2.93	2.84	2.76	2.70	2.65	2.60
	.01	7.95	5.72	4.82	4.31	3.99	3.76	3.59	3.45	3.35	3.26	3.18	3.12
	.005	6.73	6.81	5.65	5.02	4.61	4.32	4.11	3.94	6.81	3.70	3.61	3.53
	.001	14.4	9.61	7.80	6.81	6.19	5.76	5.44	5.19	4.99	4.83	4.70	4.58
	.75	.104	.291	.406	.481	.532	.570	.600	.623	.642	.658	.672	.684
	.50	.470	.714	.813	.864	.896	.918	.934	.945	.955	.962	.968	.973
	.25	1.39	1.47	1.47	1.45	1.43	1.42	1.41	1.40	1.39	1.38	1.37	1.37
	.10	2.94	2.55	2.31	2.25	2.11	1.05	1.99	1.95	1.92	1.89	1.87	1.85
23	.05	4.28	3.42	3.03	2.80	2.64	2.53	2.44	2.37	2.32	2.27	2.24	2.20
	.025	5.75	4.35	3.75	3.41	3.18	3.02	2.90	2.81	2.73	2.67	2.62	2.57
	.01	7.88	5.66	4.76	4.26	3.94	3.71	3.54	3.41	3.30	3.21	3.14	3.07
	.005	9.63	6.73	5.58	4.95	4.54	4.26	4.05	3.88	3.75	3.64	3.55	3.47
	.001	14.2	9.47	7.67	6.69	6.08	5.65	5.33	5.09	4.89	4.73	4.59	4.48
	.75	.104	.291	.406	.480	.532	.570	.600	.623	.643	.659	.671	.684
	.50	.469	.714	.812	.863	.895	.917	.932	.944	.953	.961	.967	.972
	.25	1.39	1.47	1.46	1.44	1.43	1.41	1.40	1.39	1.38	1.37	1.37	1.36
	.10	2.93	2.54	2.33	2.19	2.10	2.04	1.98	1.94	1.91	1.88	1.85	1.83
24	.05	4.26	3.40	3.01	2.78	2.62	2.51	2.42	2.36	2.30	2.25	2.22	2.18
	.025	5.72	4.32	3.72	3.38	3.15	2.99	2.87	2.78	2.70	2.64	2.59	2.54
	.01	7.82	5.61	4.72	4.22	3.90	3.67	3.50	3.36	3.26	3.17	3.09	3.03
	.005	9.55	6.66	5.52	4.89	4.49	4.20	3.99	3.83	3.69	3.59	3.50	3.42
	.001	14.0	9.34	7.55	6.59	5.98	5.55	5.23	4.99	4.80	4.64	4.50	4.39
	.75	.104	.292	.407	.482	.532	.571	.600	.624	.643	.659	.673	.685
	.50	.468	.713	.811	.862	.984	.916	.931	.943	.952	.960	.966	.971
	.25	1.39	1.47	1.46	1.44	1.42	1.41	1.40	1.39	1.38	1.37	1.36	1.36
	.10	2.92	2.53	2.32	2.18	2.09	2.02	1.97	1.93	1.89	1.87	1.84	1.82
25	.05	4.24	3.39	2.99	2.76	2.60	2.49	2.40	2.34	2.28	2.24	2.20	2.16
	.025	5.69	4.29	3.69	3.35	3.13	2.97	2.85	2.75	2.68	2.61	2.56	2.51
	.01	7.77	5.57	4.68	4.18	3.86	3.63	3.46	3.32	3.22	3.13	3.06	2.99
	.005	9.48	6.60	5.46	4.84	4.43	4.15	3.94	3.78	3.64	3.54	3.44	3.37
	.001	13.9	9.22	7.45	6.49	5.88	5.46	5.15	4.91	4.71	4.56	4.42	4.31

ν_2 (degrees of freedom of denominator mean squares)

ν_1
(degrees of freedom of numerator mean squares)

	α	1	2	3	4	5	6	7	8	9	10	11	12
26	.75	.104	.291	.406	.480	.532	.571	.600	.624	.644	.660	.674	.686
	.50	.468	.712	.810	.861	.893	.915	.930	.942	.951	.959	.965	.970
	.25	1.38	1.46	1.45	1.44	1.42	1.41	1.39	1.38	1.37	1.37	1.36	1.35
	.10	2.91	2.52	2.31	2.17	2.08	2.01	1.96	1.92	1.88	1.86	1.83	1.81
	.05	4.23	3.37	2.98	2.74	2.59	2.47	2.39	2.32	2.27	2.22	2.18	2.15
	.025	5.66	4.27	3.67	3.33	3.10	2.94	2.82	2.73	2.65	2.59	2.54	2.49
	.01	7.72	5.53	4.64	4.14	3.82	3.59	3.42	3.29	3.18	3.09	3.02	2.96
	.005	9.41	6.54	5.41	4.79	4.38	4.10	3.89	3.73	3.60	3.49	340	3.33
	.001	13.7	9.12	7.36	6.41	5.80	5.38	5.07	4.83	4.64	4.48	4.35	4.24
27	.75	.104	.291	.406	.480	.532	.571	.601	.624	.644	.660	.674	.686
	.50	.467	.711	.809	.861	.892	.914	.930	.941	.950	.958	.964	.969
	.25	1.38	1.46	1.45	1.43	1.42	1.40	1.39	1.38	1.37	1.36	1.35	1.35
	.10	2.90	2.51	2.30	2.17	2.07	2.00	1.95	1.91	1.87	1.85	1.82	1.80
	.05	4.21	3.35	2.96	2.73	2.57	2.46	2.37	2.31	2.25	2.20	2.17	2.13
	.025	5.63	4.24	3.65	3.31	3.08	2.92	2.80	2.71	2.63	2.57	2.51	2.47
	.01	7.68	5.49	4.60	4.11	.3.78	3.56	3.39	3.26	3.45	3.06	2.99	2.93
	.005	9.34	6.49	5.36	4.74	4.34	4.06	3.85	3.69	3.56	3.45	3.36	3.28
	.001	13.6	9.02	7.27	6.33	5.73	5.31	5.00	4.76	4.57	4.41	4.28	4.17
28	.75	.103	.290	.406	.480	.532	.571	.601	.625	.644	.661	.675	.687
	.50	.467	.711	.808	.860	.892	.913	.929	.940	.950	.957	.963	.968
	.25	1.38	1.46	1.45	1.43	1.41	1.40	1.39	1.38	1.37	1.36	1.35	1.34
	.10	2.89	2.50	2.29	2.16	2.06	2.00	1.94	1.90	1.87	1.84	1.81	1.79
	.05	4.20	3.34	2.95	2.71	2.56	2.45	2.36	2.29	2.24	2.19	2.15	2.12
	.025	5.61	4.22	3.63	3.29	3.06	2.90	2.78	2.69	2.61	2.55	2.49	2.45
	.01	7.64	5.45	4.57	4.07	3.75	3.53	3.36	3.23	3.12	3.03	2.96	2.90
	.005	9.28	6.44	5.32	4.70	4.30	4.02	3.81	3.65	3.52	3.41	3.32	3.25
	.001	13.5	8.93	7.19	6.25	5.66	5.24	4.93	4.69	4.50	4.35	4.22	4.11
29	.75	.103	.290	.406	.480	.532	.571	.601	.625	.645	.661	.675	.687
	.50	.467	.710	.808	.859	.891	.912	.928	.940	.949	.956	.962	.967
	.25	1.38	1.45	1.45	1.43	1.41	1.40	1.38	1.37	1.36	1.35	1.35	1.34
	.10	2.89	2.50	2.28	2.15	2.06	1.99	1.93	1.89	2.86	1.83	1.80	1.78
	.05	4.18	3.33	2.93	2.70	2.55	2.43	2.35	2.28	2.22	2.18	2.14	2.10
	.025	5.59	4.20	3.60	3.27	3.04	2.88	2.76	2.67	2.59	2.51	2.47	2.43
	.01	7.60	5.42	4.54	4.04	3.73	3.50	3.33	3.20	3.09	3.00	2.93	2.87
	.005	9.23	6.40	5.28	4.66	4.26	3.98	3.77	3.61	3.48	3.38	3.29	3.21
	.001	13.4	8.85	7.12	6.19	5.59	5.18	4.87	4.64	4.45	4.29	4.16	4.05
30	.75	.103	.290	.406	.480	.532	.571	.601	.625	.645	.661	.676	.688
	.50	.466	.709	.807	.858	.890	.912	.927	.939	.948	.955	.961	.966
	.25	1.38	1.45	1.44	1.42	1.41	1.39	1.38	1.37	1.36	1.35	1.34	1.34
	.10	2.88	2.49	2.28	2.14	2.05	1.98	1.93	1.88	1.85	1.82	1.79	1.77
	.05	4.17	3.32	2.92	2.69	2.53	2.42	2.33	2.27	2.21	2.16	2.13	2.09
	.025	5.57	4.18	3.59	3.25	3.03	2.87	2.75	2.65	2.57	2.51	2.46	2.41
	.01	7.56	5.39	4.51	4.02	3.70	3.47	3.30	3.17	3.07	2.98	2.90	2.84
	.005	9.18	6.35	5.24	4.62	4.23	3.95	3.74	3.58	3.45	3.34	3.25	3.18
	.001	13.3	8.77	7.05	6.12	5.53	5.12	4.82	4.58	4.39	4.24	4.11	4.00

ν_2 (degrees of freedom of denominator mean squares)

(*continued*)

Table B   Critical values of the *F*-distribution   (*continued*)

v_1
(degrees of freedom of numerator mean squares)

	α	1	2	3	4	5	6	7	8	9	10	11	12
40	.75	.103	.289	.404	.480	.533	.572	.603	.627	.647	.662	.679	.689
	.50	.463	.705	.802	.854	.885	.907	.922	.934	.943	.950	.956	.961
	.25	1.36	1.44	1.42	1.40	1.39	1.37	1.36	1.35	1.34	1.33	1.32	1.31
	.10	2.84	2.44	2.23	2.09	2.00	1.93	1.87	1.83	1.79	1.76	1.74	1.74
	.05	4.08	3.23	2.84	2.61	2.45	2.34	2.25	2.18	2.12	2.08	2.04	2.04
	.025	5.42	4.05	3.46	3.13	2.90	2.74	2.62	2.53	2.45	2.39	2.33	2.29
	.01	7.31	5.18	4.31	3.83	3.51	3.29	3.12	2.99	2.89	2.80	2.73	2.66
	.005	8.83	6.07	4.98	4.37	3.99	3.71	3.51	3.35	3.22	3.12	3.03	2.95
	.001	12.6	8.25	6.60	5.70	5.13	4.73	4.44	4.21	4.02	3.87	3.74	3.64
60	.75	.102	.289	.405	.480	.534	.573	.604	.629	.650	.667	.682	.695
	.50	.461	.701	.798	.849	.880	.901	.917	.928	.937	.945	.951	.956
	.25	1.35	1.42	1.41	1.38	1.37	1.35	1.33	1.32	1.31	1.30	1.29	1.29
	.10	2.79	2.39	2.18	2.04	1.95	1.87	1.82	1.77	1.74	1.71	1.68	1.66
	.05	4.00	3.15	2.76	2.53	2.37	2.25	2.17	2.10	2.04	1.99	1.95	1.92
	.025	5.29	3.93	3.34	3.01	2.79	2.63	2.51	2.41	2.33	2.27	2.22	2.17
	.01	7.08	4.98	4.13	3.65	3.34	3.12	2.95	2.82	2.72	2.63	2.56	2.50
	.005	8.49	5.79	4.73	4.14	3.76	3.49	3.29	3.13	3.01	2.90	2.81	2.74
	.001	12.0	7.76	6.17	5.31	4.76	4.37	4.09	3.87	3.69	3.54	3.41	3.31
120	.75	.102	.288	.405	.481	.534	.574	.606	.631	.652	.670	.686	.699
	.50	.458	.697	.793	.844	.875	.896	.912	.923	.932	.939	.945	.950
	.25	1.34	1.40	1.39	1.37	1.35	1.33	1.31	1.30	1.29	1.28	1.27	1.26
	.10	2.75	2.35	2.13	1.99	1.90	1.82	1.77	1.72	1.68	1.65	1.63	1.60
	.05	3.92	3.07	2.68	2.45	2.29	2.17	2.09	2.02	1.96	1.91	1.87	1.83
	.025	5.15	3.80	3.23	2.89	2.67	2.52	2.39	2.30	2.22	2.16	2.10	2.05
	.01	6.85	4.79	3.95	3.48	3.17	2.96	2.79	2.66	2.56	2.47	2.40	2.34
	.005	8.18	5.54	4.50	3.92	3.55	3.28	3.09	2.93	2.81	2.71	2.62	2.54
	.001	11.4	7.32	5.79	4.95	4.42	4.04	3.77	3.55	3.38	3.24	3.11	3.02
∞	.75	.102	.288	.404	.481	.535	.576	.608	.634	.655	.674	.690	.703
	.50	.455	.693	.789	.839	.870	.891	.907	.918	.927	.934	.939	.945
	.25	1.32	1.39	1.37	1.35	1.33	1.31	1.29	1.28	1.27	1.25	1.24	1.24
	.10	2.71	2.30	2.08	1.94	1.85	1.77	1.72	1.67	1.63	1.60	1.57	1.55
	.05	3.84	3.00	2.60	2.37	2.21	2.10	2.01	1.94	1.88	1.83	1.79	1.75
	.025	5.02	3.69	3.11	2.79	2.57	2.41	2.29	2.19	2.11	2.05	1.99	1.94
	.01	6.63	4.61	3.78	3.32	3.02	2.80	2.64	2.51	2.41	2.32	2.25	2.18
	.005	7.88	5.30	4.28	3.72	3.35	3.09	2.90	2.74	2.62	2.52	2.43	2.36
	.001	10.8	6.91	5.42	4.62	4.10	3.74	3.47	3.27	3.10	2.96	2.84	2.74

v_2 (degrees of freedom of denominator mean squares)

$$\nu_1$$
(degrees of freedom of numerator mean squares)

	α	15	20	24	30	40	50	60	120	∞
	.75	.698	.712	.719	.727	.734	.738	.741	.749	.756
	.50	2.09	2.12	2.13	2.15	2.16	2.17	2.17	2.18	2.20
	.25	9.49	9.58	9.63	9.67	9.71	9.74	9.76	9.80	9.85
	.10	61.2	61.7	62.0	62.3	62.5	62.7	62.8	63.1	63.3
1	.05	246	248	249	250	251	252	252	253	254
	.025	985	993	997	1000	1010	1010	1010	1010	1020
	.01	6160	6210	6230	6260	6290	6300	6310	6340	6370
	.005	24630	24836	24940	25440	25148	25211	25253	25359	25465
	.001	615800	620900	623500	626100	628700	630300	631300	634000	636600
	.75	.657	.672	.680	.689	.697	.702	.705	.713	.721
	.50	1.38	1.39	1.40	1.41	1.42	1.43	1.43	1.43	1.44
	.25	3.41	3.43	3.43	3.44	3.45	3.46	3.46	3.47	3.48
	.10	9.42	9.44	9.45	9.46	9.47	9.47	9.47	9.48	9.49
2	.05	19.4	19.4	19.5	19.5	19.5	19.5	19.5	19.5	19.5
	.025	39.4	39.4	39.5	39.5	39.5	39.5	39.5	39.5	39.5
	.01	99.4	99.4	99.5	99.5	99.5	99.5	99.5	99.5	99.5
	.005	199	199	199	199	199	199	199	199	200
	.001	999	999	1000	1000	1000	1000	1000	1000	1000
	.75	.658	.675	.684	.694	.702	.708	.711	.721	.730
	.50	1.21	1.23	1.23	1.24	1.25	1.25	1.25	1.26	1.27
	.25	2.46	2.46	2.46	2.47	2.47	2.47	2.47	2.47	2.47
	.10	5.20	5.18	5.18	5.17	5.16	5.15	5.15	5.14	5.13
3	.05	8.70	8.66	8.64	8.62	8.59	8.58	8.57	8.55	8.53
	.025	14.3	14.2	14.1	14.1	14.0	14.0	14.0	13.9	13.9
	.01	26.9	26.7	26.6	26.5	26.4	26.3	26.3	26.2	26.1
	.005	43.1	42.8	42.6	42.5	42.3	42.2	42.1	42.0	41.8
	.001	127	126	126	125	125	125	124	124	124
	.75	.664	.683	.692	.702	.712	.718	.722	.733	.743
	.50	1.14	1.15	1.16	1.16	1.17	1.18	1.18	1.18	1.19
	.25	2.08	2.08	2.08	2.08	2.08	2.08	2.08	2.08	2.08
	.10	3.87	3.84	3.83	3.82	3.80	3.79	3.79	3.78	3.76
4	.05	5.86	5.80	5.77	5.75	5.72	5.70	5.69	5.66	5.63
	.025	8.66	8.56	8.51	8.46	8.41	8.38	8.36	8.31	8.26
	.01	14.2	14.0	13.9	13.8	13.7	13.7	13.7	13.6	13.5
	.005	20.4	20.2	20.0	19.9	19.8	19.7	19.6	19.5	19.3
	.001	46.8	46.1	45.8	45.4	45.1	44.9	44.8	44.4	44.0
	.75	.669	.690	.700	.711	.722	.728	.732	.743	.755
	.50	1.10	1.11	1.12	1.12	1.13	1.14	1.14	1.14	1.15
	.25	1.89	1.88	1.88	1.88	1.88	1.87	1.87	1.87	1.87
	.10	3.24	3.21	3.19	3.17	3.16	3.15	3.14	3.12	3.10
5	.05	4.62	4.56	4.53	4.50	4.46	4.44	4.43	4.40	4.36
	.025	6.43	6.33	6.28	6.23	6.18	6.14	6.12	6.07	6.02
	.01	9.72	9.55	9.47	9.38	9.29	9.24	9.20	9.11	9.02
	.005	13.1	12.9	12.8	12.7	12.5	12.4	12.4	12.3	12.1
	.001	25.9	25.4	25.1	24.9	24.6	24.4	24.3	24.1	23.7

ν_2 (degrees of freedom of denominator mean squares)

(*continued*)

Table B   Critical values of the *F*-distribution   (*continued*)

ν_1
(degrees of freedom of numerator mean squares)

ν_2	α	15	20	24	30	40	50	60	120	∞
6	.75	.675	.696	.707	.718	.729	.736	.741	.753	.765
	.50	1.07	1.08	1.09	1.10	1.10	1.11	1.11	1.12	1.12
	.25	1.76	1.76	1.75	1.75	1.75	1.74	1.74	1.74	1.74
	.10	2.87	2.84	2.82	2.80	2.78	2.77	2.76	2.74	2.72
	.05	3.94	3.87	3.84	3.81	3.77	3.75	3.74	3.70	3.67
	.025	5.27	5.17	5.12	5.07	5.01	4.98	4.96	4.90	4.85
	.01	7.56	7.40	7.31	7.23	7.14	7.09	7.06	6.97	6.88
	.005	9.81	9.59	9.47	9.36	9.24	9.17	9.12	9.00	8.88
	.001	17.6	17.1	16.9	16.7	16.4	16.3	16.2	16.0	15.8
7	.75	.679	.702	.713	.725	.737	.745	.749	.762	.775
	.50	1.05	1.07	1.07	1.08	1.08	1.09	1.09	1.10	1.10
	.25	1.68	1.67	1.67	1.66	1.66	1.65	1.65	1.65	1.65
	.10	2.63	2.59	2.58	2.56	2.54	2.52	2.51	2.49	2.47
	.05	3.51	3.44	3.41	3.38	3.34	3.32	3.30	3.27	3.23
	.025	4.57	4.47	4.42	4.36	4.31	4.27	4.25	4.20	4.14
	.01	6.31	6.16	6.07	5.99	5.91	5.86	5.82	5.74	5.65
	.005	7.97	7.75	7.65	7.53	7.42	7.35	7.31	7.19	7.08
	.001	13.3	12.9	12.7	12.5	12.3	12.2	12.1	11.9	11.7
8	.75	.684	.707	.718	.730	.743	.751	.756	.769	.783
	.50	1.04	1.05	1.06	1.07	1.08	1.07	1.08	1.08	1.09
	.25	1.62	1.61	1.60	1.60	1.59	1.59	1.59	1.58	1.58
	.10	2.46	2.42	2.40	2.38	2.36	2.35	2.34	2.32	2.29
	.05	3.22	3.15	3.12	3.08	3.04	3.02	3.01	2.97	2.93
	.025	4.10	4.00	3.95	3.89	3.84	3.80	3.78	3.73	3.67
	.01	5.52	5.36	5.28	5.20	5.12	5.07	5.03	4.95	4.86
	.005	6.81	6.61	6.50	6.40	6.29	6.22	6.18	6.06	5.95
	.001	10.8	10.5	10.3	10.1	9.9	9.8	9.7	9.5	9.3
9	.75	.687	.711	.723	.736	.749	.757	.762	.776	.791
	.50	1.03	1.04	1.05	1.05	1.06	1.07	1.07	1.07	1.08
	.25	1.57	1.56	1.56	1.55	1.55	1.54	1.54	1.53	1.53
	.10	2.34	2.30	2.28	2.25	2.23	2.22	2.21	2.18	2.16
	.05	3.01	2.94	2.90	2.86	2.83	2.81	2.79	2.75	2.71
	.025	3.77	3.67	3.61	3.56	3.51	3.47	3.45	3.39	3.33
	.01	4.96	4.81	4.73	4.65	4.57	4.52	4.48	4.40	4.31
	.005	6.03	5.83	5.73	5.62	5.52	5.45	5.41	5.30	5.19
	.001	9.24	8.90	8.72	8.55	8.37	8.26	8.19	8.00	7.81
10	.75	.691	.714	.727	.740	.754	.762	.767	.782	.797
	.50	1.02	1.03	1.04	1.05	1.05	1.06	1.06	1.06	1.07
	.25	1.53	1.52	1.52	1.51	1.51	1.50	1.50	1.49	1.48
	.10	2.24	2.20	2.18	2.16	2.13	2.12	2.11	2.08	2.06
	.05	2.85	2.77	2.74	2.70	2.66	2.64	2.62	2.58	2.54
	.025	3.52	3.42	3.37	3.31	3.26	3.22	3.20	3.14	3.08
	.01	4.56	4.41	4.33	4.25	4.17	4.12	4.08	4.00	3.91
	.005	5.47	5.27	5.17	5.07	4.97	4.90	4.86	4.75	4.64
	.001	8.13	7.80	7.64	7.47	7.30	7.19	7.12	6.94	6.76

ν_2 (degrees of freedom of denominator mean squares)

ν_1
(degrees of freedom of numerator mean squares)

	α	15	20	24	30	40	50	60	120	∞
	.75	.694	.719	.730	.744	.758	.767	.773	.788	.803
	.50	1.02	1.03	1.03	1.04	1.05	1.05	1.05	1.06	1.06
	.25	1.50	1.49	1.49	1.48	1.47	1.47	1.47	1.46	1.45
	.10	2.17	2.12	2.10	2.08	2.05	2.04	2.03	2.00	1.97
11	.05	2.72	2.65	2.61	2.57	2.53	2.51	2.49	2.45	2.40
	.025	3.33	3.23	3.17	3.12	3.06	3.02	3.00	2.94	2.88
	.01	4.25	4.10	4.02	3.94	3.86	3.81	3.78	3.69	3.60
	.005	5.05	4.86	4.76	4.65	4.55	4.49	4.45	4.34	4.23
	.001	7.32	7.01	6.85	6.68	6.52	6.42	6.35	6.17	6.00
	.75	.695	.721	.734	.748	.762	.771	.777	.792	.808
	.50	1.01	1.02	1.03	1.03	1.04	1.04	1.05	1.05	1.06
	.25	1.48	1.47	1.46	1.45	1.45	1.44	1.44	1.43	1.42
	.10	2.11	2.06	2.04	2.01	1.99	1.97	1.96	1.93	1.90
12	.05	2.62	2.54	2.51	2.47	2.43	2.40	2.38	2.34	2.30
	.025	3.18	3.07	3.02	2.96	2.91	2.87	2.85	2.79	2.72
	.01	4.01	3.86	3.78	3.70	3.62	3.57	3.54	3.45	3.36
	.005	4.72	4.53	4.43	4.33	4.23	4.16	4.12	4.01	3.90
	.001	6.71	6.40	6.25	6.09	5.93	5.83	5.76	5.59	5.42
	.75	.697	.723	.737	.751	.766	.775	.781	.797	.813
	.50	1.01	1.02	1.02	1.03	1.04	1.04	1.04	1.05	1.05
	.25	1.46	1.45	1.44	1.43	1.42	1.42	1.42	1.41	1.40
	.10	2.05	2.01	1.98	1.96	1.93	1.92	1.90	1.88	1.85
13	.05	2.53	2.46	2.42	2.38	2.34	2.31	2.30	2.25	2.21
	.025	3.05	2.95	2.89	2.84	2.78	2.74	2.72	2.66	2.60
	.01	3.82	3.66	3.59	3.51	3.43	3.37	3.34	3.25	3.17
	.005	4.46	4.27	4.17	4.07	3.97	3.91	3.87	3.76	2.65
	.001	6.23	5.93	5.78	5.63	5.47	5.37	5.30	5.14	4.97
	.75	.699	.726	.740	.754	.769	.778	.785	.801	.818
	.50	1.00	1.01	1.02	1.03	1.03	1.04	1.04	1.04	1.05
	.25	1.44	1.43	1.42	1.41	1.41	1.40	1.40	1.39	1.38
	.10	2.01	1.96	1.94	1.91	1.89	1.87	1.86	1.83	1.80
14	.05	2.46	2.39	2.35	2.31	2.27	2.24	2.22	2.18	2.13
	.025	2.95	2.84	2.79	2.73	2.67	2.64	2.61	2.55	2.49
	.01	3.66	3.51	3.43	3.35	3.27	3.21	3.18	3.09	3.00
	.005	4.25	4.06	3.96	3.86	3.76	3.70	3.66	3.55	3.44
	.001	5.85	5.56	5.41	5.25	5.10	5.00	4.94	4.77	4.60
	.75	.701	.728	.742	.757	.772	.782	.788	.805	.822
	.50	1.00	1.01	1.02	1.02	1.03	1.03	1.03	1.04	1.05
	.25	1.43	1.41	1.41	1.40	1.39	1.38	1.38	1.37	1.36
	.10	1.97	1.92	1.90	1.87	1.85	1.83	1.82	1.79	1.76
15	.05	2.40	2.33	2.29	2.25	2.20	2.18	2.16	2.11	2.07
	.025	2.86	2.76	2.70	2.64	2.59	2.55	2.52	2.46	2.40
	.01	3.52	3.37	3.29	3.21	3.12	3.08	3.05	2.96	2.87
	.005	4.07	3.88	3.79	3.69	3.59	3.52	3.48	3.37	3.26
	.001	5.54	5.25	5.10	4.95	4.80	4.70	4.64	4.47	4.31

ν_2 (degrees of freedom of denominator mean squares)

(continued)

Table B Critical values of the *F*-distribution (*continued*)

ν_1
(degrees of freedom of numerator mean squares)

	α	15	20	24	30	40	50	60	120	∞
16	.75	.703	.730	.744	.759	.775	.785	.791	.808	.826
	.50	.997	1.01	1.01	1.02	1.03	1.03	1.03	1.04	1.04
	.25	1.41	1.40	1.39	1.38	1.37	1.37	1.36	1.35	1.34
	.10	1.94	1.89	1.87	1.84	1.81	1.79	1.78	1.75	1.72
	.05	2.35	2.28	2.24	2.19	2.15	2.12	2.11	2.06	2.01
	.025	2.79	2.68	2.63	2.57	2.51	2.47	2.45	2.38	2.32
	.01	3.41	3.26	3.18	3.10	3.02	2.97	2.93	2.84	2.75
	.005	3.92	3.73	3.64	3.54	3.44	3.37	3.33	3.22	3.11
	.001	5.27	4.99	4.85	4.70	4.54	4.45	4.39	4.23	4.06
17	.75	.704	.732	.746	.762	.777	.787	.794	.811	.830
	.50	.995	1.01	1.01	1.02	1.02	1.03	1.03	1.03	1.04
	.25	1.40	1.39	1.38	1.37	1.36	1.36	1.35	1.34	1.33
	.10	1.91	1.86	1.84	1.81	1.78	1.76	1.75	1.72	1.69
	.05	2.31	2.23	2.19	2.15	2.10	2.08	2.06	2.01	1.96
	.025	2.72	2.62	2.56	2.50	2.44	2.41	2.38	2.32	2.25
	.01	3.31	3.16	3.08	3.00	2.92	2.87	2.83	2.75	2.65
	.005	3.79	3.61	3.51	3.41	3.31	3.25	3.21	3.10	2.98
	.001	5.05	4.78	4.63	4.48	4.33	4.24	4.18	4.02	3.85
18	.75	.706	.733	.748	.764	.780	.790	.797	.814	.833
	.50	.992	1.00	1.01	1.02	1.02	1.02	1.03	1.03	1.04
	.25	1.39	1.38	1.37	1.36	1.35	1.34	1.34	1.33	1.32
	.10	1.89	1.84	1.81	1.78	1.75	1.74	1.72	1.69	1.66
	.05	2.27	2.19	2.15	2.11	2.06	2.04	2.02	1.97	1.92
	.025	2.67	2.56	2.50	2.44	2.38	2.35	2.32	2.26	2.19
	.01	3.23	3.08	3.00	2.92	2.84	2.78	2.75	2.66	2.57
	.005	3.68	3.50	3.40	3.30	3.20	3.14	3.10	2.99	2.87
	.001	4.87	4.59	4.45	4.30	4.15	4.06	4.00	3.84	3.67
19	.75	.707	.735	.750	.766	.782	.792	.799	.817	.836
	.50	.990	1.00	1.01	1.01	1.02	1.02	1.02	1.03	1.04
	.25	1.38	1.37	1.36	1.35	1.34	1.33	1.33	1.32	1.30
	.10	1.86	1.81	1.79	1.76	1.73	1.71	1.70	1.67	1.63
	.05	2.23	2.16	2.11	2.07	2.03	2.00	1.98	1.93	1.88
	.025	2.62	2.51	2.45	2.39	2.33	2.30	2.27	2.20	2.13
	.01	3.15	3.00	2.92	2.84	2.76	2.71	2.67	2.58	2.49
	.005	3.59	3.40	3.31	3.21	3.11	3.04	3.00	2.89	2.78
	.001	4.70	4.43	4.29	4.14	3.99	3.90	3.84	3.68	3.51
20	.75	.708	.736	.751	.767	.784	.794	.801	.820	.840
	.50	.989	1.00	1.01	1.01	1.02	1.02	1.02	1.03	1.03
	.25	1.37	1.36	1.35	1.34	1.33	1.33	1.32	1.31	1.29
	.10	1.84	1.79	1.77	1.74	1.71	1.69	1.68	1.64	1.61
	.05	2.20	2.12	2.08	2.04	1.99	1.97	1.95	1.90	1.84
	.025	2.57	2.46	2.41	2.35	2.29	2.25	2.22	2.16	2.09
	.01	3.09	2.94	2.86	2.78	2.69	2.64	2.61	2.52	2.42
	.005	3.50	3.32	3.22	3.12	3.02	2.96	2.92	2.81	2.69
	.001	4.56	4.29	4.15	4.00	3.86	3.76	3.70	3.54	3.38

ν_2 (degrees of freedom of denominator mean squares)

ν_1
(degrees of freedom of numerator mean squares)

ν_2 (degrees of freedom of denominator mean squares)

	α	15	20	24	30	40	50	60	120	∞
21	.75	.709	.738	.753	.769	.786	.796	.803	.822	.842
	.50	.987	.998	1.00	1.01	1.02	1.02	1.02	1.03	1.03
	.25	1.37	1.35	1.34	1.33	1.32	1.32	1.31	1.30	1.28
	.10	1.83	1.78	1.75	1.72	1.69	1.67	1.66	1.62	1.59
	.05	2.18	2.10	2.05	2.01	1.96	1.94	1.92	1.87	1.81
	.025	2.53	2.42	2.37	2.31	2.25	2.21	2.18	2.11	2.04
	.01	3.03	2.88	2.80	2.72	2.64	2.58	2.55	2.46	2.36
	.005	3.43	3.24	3.15	3.05	2.95	2.88	2.84	2.73	2.61
	.001	4.44	4.17	4.03	3.88	3.74	3.64	3.58	3.42	3.26
22	.75	.710	.739	.754	.770	.787	.798	.805	.824	.845
	.50	.986	.997	1.00	1.01	1.01	1.02	1.02	1.03	1.03
	.25	1.36	1.34	1.33	1.32	1.31	1.31	1.30	1.29	1.28
	.10	1.81	1.76	1.73	1.70	1.67	1.65	1.64	1.60	1.57
	.05	2.15	2.07	2.03	1.98	1.94	1.91	1.89	1.84	1.78
	.025	2.50	2.39	2.33	2.27	2.21	2.17	2.14	2.08	2.00
	.01	2.98	2.83	2.75	2.67	2.58	2.53	2.50	2.40	2.31
	.005	3.36	3.18	3.08	2.98	2.88	2.82	2.77	2.66	2.55
	.001	4.33	4.06	3.92	3.78	3.63	3.54	3.48	3.32	3.15
23	.75	.711	.740	.756	.772	.789	.800	.807	.827	.847
	.50	.984	.996	1.00	1.01	1.01	1.02	1.02	1.02	1.03
	.25	1.35	1.34	1.33	1.32	1.31	1.30	1.30	1.28	1.27
	.10	1.80	1.74	1.72	1.69	1.66	1.64	1.62	1.59	1.55
	.05	2.13	2.05	2.01	1.96	1.91	1.89	1.86	1.81	1.76
	.025	2.47	2.36	2.30	2.24	2.18	2.14	2.11	2.04	1.97
	.01	2.93	2.78	2.70	2.62	2.54	2.48	2.45	2.35	2.26
	.005	3.30	3.12	3.02	2.92	2.82	2.76	2.71	2.60	2.48
	.001	4.23	3.96	3.82	3.68	3.53	3.44	3.38	3.22	3.05
24	.75	.712	.741	.757	.773	.791	.802	.809	.829	.850
	.50	.983	.994	1.00	1.01	1.01	1.02	1.02	1.02	1.03
	.25	1.35	1.33	1.32	1.31	1.30	1.29	1.29	1.28	1.26
	.10	1.78	1.73	1.70	1.67	1.64	1.62	1.61	1.57	1.53
	.05	2.11	2.03	1.98	1.94	1.89	1.86	1.84	1.79	1.73
	.025	2.44	2.33	2.27	2.21	2.15	2.11	2.08	2.01	1.94
	.01	2.89	2.74	2.66	2.58	2.49	2.44	2.40	2.31	2.21
	.005	3.25	3.06	2.97	2.87	2.77	2.70	2.66	2.55	2.43
	.001	4.14	3.87	3.74	3.59	3.45	3.36	3.29	3.14	2.97
25	.75	.712	.742	.758	.775	.792	.803	.811	.831	.852
	.50	.982	.993	.999	1.00	1.01	1.01	1.02	1.02	1.03
	.25	1.34	1.33	1.32	1.31	1.29	1.29	1.28	1.27	1.25
	.10	1.77	1.72	1.69	1.66	1.63	1.61	1.59	1.56	1.52
	.05	2.09	2.01	1.96	1.92	1.87	1.84	1.82	1.77	1.71
	.025	2.41	2.30	2.24	2.18	2.12	2.08	2.05	1.98	1.91
	.01	2.85	2.70	2.62	2.54	2.45	2.40	2.36	2.27	2.17
	.005	3.20	3.01	2.92	2.82	2.72	2.65	2.61	2.50	2.38
	.001	4.06	3.97	3.66	3.52	3.37	3.28	3.22	3.06	2.89

(continued)

Table B Critical values of the *F*-distribution (*continued*)

v_1
(degrees of freedom of numerator mean squares)

	α	15	20	24	30	40	50	60	120	∞
26	.75	.713	.743	.759	.776	.793	.805	.812	.832	.854
	.50	.981	.992	.998	1.00	1.01	1.01	1.01	1.02	1.03
	.25	1.34	1.32	1.31	1.30	1.29	1.28	1.28	1.26	1.25
	.10	1.76	1.71	1.68	1.65	1.61	1.59	1.58	1.54	1.50
	.05	2.07	1.99	1.95	1.90	1.85	1.82	1.80	1.75	1.69
	.025	2.39	2.28	2.22	2.16	2.09	2.05	2.03	1.95	1.88
	.01	2.81	2.66	2.58	2.50	2.42	2.36	2.33	2.23	2.13
	.005	3.15	2.97	2.87	2.77	2.67	2.61	2.56	2.45	2.33
	.001	3.99	3.72	3.59	3.44	3.30	3.21	3.15	2.99	2.82
27	.75	.714	.744	.760	.777	.795	.806	.814	.834	.856
	.50	.980	.991	.997	1.00	1.01	1.01	1.01	1.02	1.03
	.25	1.33	1.32	1.31	1.30	1.28	1.28	1.27	1.26	1.24
	.10	1.75	1.70	1.67	1.64	1.60	1.58	1.57	1.53	1.49
	.05	2.06	1.97	1.93	1.88	1.84	1.81	1.79	1.73	1.67
	.025	2.36	2.25	2.19	2.13	2.07	2.03	2.00	1.93	1.85
	.01	2.78	2.63	2.55	2.47	2.38	2.33	2.29	2.20	2.10
	.005	3.11	2.93	2.83	2.73	2.63	2.57	2.52	2.41	2.29
	.001	3.92	3.66	3.52	3.38	3.23	3.14	3.08	2.92	2.75
28	.75	.714	.745	.761	.778	.796	.807	.815	.856	.858
	.50	.979	.990	.996	1.00	1.01	1.01	1.01	1.02	1.02
	.25	1.33	1.31	1.30	1.29	1.28	1.27	1.27	1.25	1.24
	.10	1.74	1.69	1.66	1.63	1.59	1.57	1.56	1.52	1.48
	.05	2.04	1.96	1.91	1.87	1.82	1.79	1.77	1.71	1.65
	.025	2.34	2.23	2.17	2.11	2.05	2.01	1.98	1.91	1.83
	.01	2.75	2.60	2.52	2.44	2.35	2.30	2.26	2.17	2.06
	.005	3.07	2.89	2.79	2.69	2.59	2.53	2.48	2.37	2.25
	.001	3.86	3.60	3.46	3.32	3.18	3.09	3.02	2.86	2.69
29	.75	.715	.745	.762	.779	.797	.809	.816	.837	.860
	.50	.979	.990	.996	1.00	1.01	1.01	1.01	1.02	1.02
	.25	1.32	1.31	1.30	1.29	1.27	1.27	1.26	1.25	1.23
	.10	1.73	1.68	1.65	1.62	1.58	1.56	1.55	1.51	1.47
	.05	2.03	1.94	1.90	1.85	1.81	1.78	1.75	1.70	1.64
	.025	2.32	2.21	2.15	2.09	2.03	1.99	1.96	1.89	1.81
	.01	2.73	2.57	2.49	2.41	2.33	2.27	2.23	2.14	2.03
	.005	3.04	2.86	2.76	2.66	2.56	2.49	2.45	2.33	2.21
	.001	3.80	3.54	3.41	3.27	3.12	3.03	2.97	2.81	2.64
30	.75	.716	.746	.763	.780	.798	.810	.818	.839	.862
	.50	.978	.989	.994	1.00	1.01	1.01	1.01	1.02	1.02
	.25	1.32	1.30	1.29	1.28	1.27	1.26	1.26	1.24	1.23
	.10	1.72	1.67	1.64	1.61	1.57	1.55	1.54	1.50	1.46
	.05	2.01	1.93	1.89	1.84	1.79	1.76	1.74	1.68	1.62
	.025	2.31	2.20	2.14	2.07	2.01	1.97	1.94	1.87	1.79
	.01	2.70	2.55	2.47	2.39	2.30	2.25	2.21	2.11	2.01
	.005	3.01	2.82	2.73	2.63	2.52	2.46	2.42	2.30	2.18
	.001	3.75	3.49	3.36	3.22	3.07	2.98	2.92	2.76	2.59

v_2 (degrees of freedom of denominator mean squares)

ν_1
(degrees of freedom of numerator mean squares)

	α	15	20	24	30	40	50	60	120	∞
40	.75	.720	.752	.769	.787	.806	.819	.828	.851	.877
	.50	.972	.983	.989	.994	1.00	1.00	1.01	1.01	1.02
	.25	1.30	1.28	1.26	1.25	1.24	1.23	1.22	1.21	1.19
	.10	1.66	1.61	1.57	1.54	1.51	1.48	1.47	1.42	1.38
	.05	1.92	1.84	1.79	1.74	1.69	1.66	1.64	1.58	1.51
	.025	2.18	2.07	2.01	1.94	1.88	1.83	1.80	1.72	1.64
	.01	2.52	2.37	2.29	2.20	2.11	2.06	2.02	1.92	1.80
	.005	2.78	2.60	2.50	2.40	2.30	2.23	2.18	2.06	1.93
	.001	3.40	3.15	3.01	2.87	2.73	2.64	2.57	2.41	2.23
60	.75	.725	.758	.776	.796	.816	.830	.840	.865	.896
	.50	.967	.978	.983	.989	.994	.998	1.00	1.01	1.01
	.25	1.27	1.25	1.24	1.22	1.21	1.20	1.19	1.17	1.15
	.10	1.60	1.54	1.51	1.48	1.44	1.41	1.40	1.35	1.29
	.05	1.84	1.75	1.70	1.65	1.59	1.56	1.53	1.47	1.39
	.025	2.06	1.94	1.88	1.82	1.74	1.70	1.67	1.58	1.48
	.01	2.35	2.20	2.12	2.03	1.94	1.88	1.84	1.73	1.60
	.005	2.57	2.39	2.29	2.19	2.08	2.01	1.96	1.83	1.69
	.001	3.08	2.83	2.69	2.55	2.41	2.32	2.25	2.08	1.89
120	.75	.730	.765	.784	.805	.828	.843	.853	.884	.923
	.50	.961	.972	.978	.983	.989	.992	.994	1.00	1.01
	.25	1.24	1.22	1.21	1.19	1.18	1.17	1.16	1.13	1.10
	.10	1.55	1.48	1.45	1.41	1.37	1.34	1.32	1.26	1.19
	.05	1.75	1.66	1.61	1.55	1.50	1.46	1.43	1.35	1.25
	.025	1.95	1.82	1.76	1.69	1.61	1.56	1.53	1.43	1.31
	.01	2.19	2.03	1.95	1.86	1.76	1.70	1.66	1.53	1.38
	.005	2.37	2.19	2.09	1.98	1.87	1.80	1.75	1.61	1.43
	.001	2.78	2.53	2.40	2.26	2.11	2.02	1.95	1.76	1.54
∞	.75	.736	.773	.793	.816	.842	.860	.872	.910	1.00
	.50	.956	.967	.972	.978	.983	.987	.989	.994	1.00
	.25	1.22	1.19	1.18	1.16	1.14	1.13	1.12	1.08	1.00
	.10	1.49	1.42	1.38	1.34	1.30	1.26	1.24	1.17	1.00
	.05	1.67	1.57	1.52	1.46	1.39	1.35	1.32	1.22	1.00
	.025	1.83	1.71	1.64	1.57	1.48	1.43	1.39	1.27	1.00
	.01	2.04	1.88	1.79	1.70	1.59	1.52	1.47	1.32	1.00
	.005	2.19	2.00	1.90	1.79	1.67	1.59	1.53	1.36	1.00
	.001	2.51	2.27	2.13	1.99	1.84	1.73	1.66	1.45	1.00

ν_2 (degrees of freedom of denominator mean squares)

Table C Critical values of Student's *t*-distribution

ν \ α	0.9	0.5	0.4	0.2	0.1	0.05	0.02	0.01	0.001	α / ν
1	.158	1.000	1.376	3.078	6.314	12.706	31.821	63.657	636.619	1
2	.142	.816	1.061	1.886	2.920	4.303	6.965	9.925	31.598	2
3	.137	.765	.978	1.638	2.353	3.182	4.541	5.841	12.924	3
4	.134	.741	.941	1.533	2.132	2.776	3.747	4.604	8.610	4
5	.132	.727	.920	1.476	2.015	2.571	3.365	4.032	6.869	5
6	.131	.718	.906	1.440	1.943	2.447	3.143	3.707	5.959	6
7	.130	.711	.896	1.415	1.895	2.365	2.998	3.499	5.408	7
8	.130	.706	.889	1.397	1.860	2.306	2.896	3.355	5.041	8
9	.129	.703	.883	1.383	1.833	2.262	2.821	3.250	4.781	9
10	.129	.700	.879	1.372	1.812	2.228	2.764	3.169	4.587	10
11	.129	.697	.876	1.363	1.796	2.201	2.718	3.106	4.437	11
12	.128	.695	.873	1.356	1.782	2.179	2.681	3.055	4.318	12
13	.128	.694	.870	1.350	1.771	2.160	2.650	3.012	4.221	13
14	.128	.692	.868	1.345	1.761	2.145	2.624	2.977	4.140	14
15	.128	.691	.866	1.341	1.753	2.131	2.602	2.947	4.073	15
16	.128	.690	.865	1.337	1.746	2.120	2.583	2.921	4.015	16
17	.128	.689	.863	1.333	1.740	2.110	2.567	2.898	3.965	17
18	.127	.688	.862	1.330	1.734	2.101	2.552	2.878	3.922	18
19	.127	.688	.861	1.328	1.729	2.093	2.539	2.861	3.883	19
20	.127	.687	.860	1.325	1.725	2.086	2.528	2.845	3.850	20
21	.127	.686	.859	1.323	1.721	2.080	2.518	2.831	3.819	21
22	.127	.686	.858	1.321	1.717	2.074	2.508	2.819	3.792	22
23	.127	.685	.858	1.319	1.714	2.069	2.500	2.807	3.767	23
24	.127	.685	.857	1.318	1.711	2.064	2.492	2.797	3.745	24
25	.127	.684	.856	1.316	1.708	2.060	2.485	2.787	3.725	25
26	.127	.684	.856	1.315	1.706	2.056	2.479	2.779	3.707	26
27	.127	.684	.855	1.314	1.703	2.052	2.473	2.771	3.690	27
28	.127	.683	.855	1.313	1.701	2.048	2.467	2.763	3.674	28
29	.127	.683	.854	1.311	1.699	2.045	2.462	2.756	3.659	29
30	.127	.683	.854	1.310	1.697	2.042	2.457	2.750	3.646	30
40	.126	.681	.851	1.303	1.684	2.021	2.423	2.704	3.551	40
60	.126	.679	.848	1.296	1.671	2.000	2.390	2.660	3.460	60
120	.126	.677	.845	1.289	1.658	1.980	2.358	2.617	3.373	120
∞	.126	.674	.842	1.282	1.645	1.960	2.326	2.576	3.291	∞

Table D   Critical values of the chi-square distribution

ν \ α	.995	.975	.9	.5	.1	.05	.025	.01	.005	.001	α / ν
1	0.000	0.000	0.016	0.455	2.706	3.841	5.024	6.635	7.879	10.828	1
2	0.010	0.051	0.211	1.386	4.605	5.991	7.378	9.210	10.597	13.816	2
3	0.072	0.216	0.584	2.366	6.251	7.815	9.348	11.345	12.838	16.266	3
4	0.207	0.484	1.064	3.357	7.779	9.488	11.143	13.277	14.860	18.467	4
5	0.412	0.831	1.610	4.351	9.236	11.070	12.832	15.086	16.750	20.515	5
6	0.676	1.237	2.204	5.348	10.645	12.592	14.449	16.812	18.548	22.458	6
7	0.989	1.690	2.833	6.346	12.017	14.067	16.013	18.475	20.278	24.322	7
8	1.344	2.180	3.490	7.344	13.362	15.507	17.535	20.090	21.955	26.124	8
9	1.735	2.700	4.168	8.343	14.684	16.919	19.023	21.666	23.589	27.877	9
10	2.156	3.247	4.865	9.342	15.987	18.307	20.483	23.209	25.188	29.588	10
11	2.603	3.816	5.578	10.341	17.275	19.675	21.920	24.725	26.757	31.264	11
12	3.074	4.404	6.304	11.340	18.549	21.026	23.337	26.217	28.300	32.910	12
13	3.565	5.009	7.042	12.340	19.812	22.362	24.736	27.688	29.819	34.528	13
14	4.075	5.629	7.790	13.339	21.064	23.685	26.119	29.141	31.319	36.123	14
15	4.601	6.262	8.547	14.339	22.307	24.996	27.488	30.578	32.801	37.697	15
16	5.142	6.908	9.312	15.338	23.542	26.296	28.845	32.000	34.267	39.252	16
17	5.697	7.564	10.085	16.338	24.769	27.587	30.191	33.409	35.718	40.790	17
18	6.265	8.231	10.865	17.338	25.989	28.869	31.526	34.805	37.156	42.312	18
19	6.844	8.907	11.651	18.338	27.204	30.144	32.852	36.191	38.582	43.820	19
20	7.434	9.591	12.443	19.337	28.412	31.410	34.170	37.566	39.997	45.315	20
21	8.034	10.283	13.240	20.337	29.615	32.670	35.479	38.932	41.401	46.797	21
22	8.643	10.982	14.042	21.337	30.813	33.924	36.781	40.289	42.796	48.268	22
23	9.260	11.688	14.848	22.337	32.007	35.172	38.076	41.638	44.181	49.728	23
24	9.886	12.401	15.659	23.337	33.196	36.415	39.364	42.980	45.558	51.179	24
25	10.520	13.120	16.473	24.337	34.382	37.652	40.646	44.314	46.928	52.620	25
26	11.160	13.844	17.292	25.336	35.563	38.885	41.923	45.642	48.290	54.052	26
27	11.808	14.573	18.114	26.336	36.741	40.113	43.194	46.963	49.645	55.476	27
28	12.461	15.308	18.939	27.336	37.916	41.337	44.461	48.278	50.993	56.892	28
29	13.121	16.047	19.768	28.336	39.088	42.557	45.722	49.588	52.336	58.301	29
30	13.787	16.791	20.599	29.336	40.256	43.773	46.979	50.892	53.672	59.703	30
31	14.458	17.539	21.434	30.336	41.422	44.985	48.232	52.191	55.003	61.098	31
32	15.134	18.291	22.271	31.336	42.585	46.194	49.480	53.486	56.329	62.487	32
33	15.815	19.047	23.110	32.336	43.745	47.400	50.725	54.776	57.649	63.870	33
34	16.501	19.806	23.952	33.336	44.903	48.602	51.966	56.061	58.964	65.247	34
35	17.192	20.569	24.797	34.336	46.059	49.802	53.203	57.342	60.275	66.619	35
36	17.887	21.336	25.643	35.336	47.212	50.998	54.437	58.619	61.582	67.985	36
37	18.586	22.106	26.492	36.335	48.363	52.192	55.668	59.892	62.884	69.346	37
38	19.289	22.878	27.343	37.335	49.513	53.384	56.896	61.162	64.182	70.703	38
39	19.996	23.654	28.196	38.335	50.660	54.572	58.120	62.428	65.476	72.055	39
40	20.707	24.433	29.051	39.335	51.805	55.758	59.342	63.691	66.766	73.402	40
41	21.421	25.215	29.907	40.335	52.949	56.942	60.561	64.950	68.053	74.745	41
42	22.138	25.999	30.765	41.335	54.090	58.124	61.777	66.206	69.336	76.084	42
43	22.859	26.785	31.625	42.335	55.230	59.304	62.990	67.459	70.616	77.419	43
44	23.584	27.575	32.487	43.335	56.369	60.481	64.202	68.710	71.893	78.750	44
45	24.311	28.366	33.350	44.335	57.505	61.656	65.410	69.957	73.166	80.777	45
46	25.042	29.160	34.215	45.335	58.641	62.830	66.617	71.201	74.437	81.400	46
47	25.775	29.956	35.081	46.335	59.774	64.001	67.821	72.443	75.704	82.720	47
48	26.511	30.755	35.949	47.335	60.907	65.171	69.023	73.683	76.969	84.037	48
49	27.249	31.555	36.818	48.335	62.038	66.339	70.222	74.919	78.231	85.351	49
50	27.991	32.357	37.689	49.335	63.167	67.505	71.420	76.154	79.490	86.661	50

(*continued*)

Table D   Critical values of the chi-square distribution   (*continued*)

ν \ α	.995	.975	.9	.5	.1	.05	.025	.01	.005	.001 α / ν
51	28.735	33.162	38.560	50.335	64.295	68.669	72.616	77.386	80.747	87.968 51
52	29.481	33.968	39.433	51.335	65.422	69.832	73.810	78.616	82.001	89.272 52
53	30.230	34.776	40.308	52.335	66.548	70.993	75.002	79.843	83.253	90.573 53
54	30.981	35.586	41.183	53.335	67.673	72.153	76.192	81.069	84.502	91.872 54
55	31.735	36.398	42.060	54.335	68.796	73.311	77.380	82.292	85.749	93.168 55
56	32.490	37.212	42.937	55.335	69.918	74.468	78.567	83.513	86.994	94.460 56
57	33.248	38.027	43.816	56.335	71.040	75.624	79.752	84.733	88.237	95.751 57
58	34.008	38.844	44.696	57.335	72.160	76.778	80.936	85.950	89.477	97.039 58
59	34.770	39.662	45.577	58.335	73.279	77.931	82.117	87.166	90.715	98.324 59
60	35.534	40.482	46.459	59.335	74.397	79.082	83.298	88.379	91.952	99.607 60
61	36.300	41.303	47.342	60.335	75.514	80.232	84.476	89.591	93.186	100.888 61
62	37.068	42.126	48.226	61.335	76.630	81.381	85.654	90.802	94.419	102.166 62
63	37.838	42.950	49.111	62.335	77.745	82.529	86.830	92.010	95.649	103.442 63
64	38.610	43.776	49.996	63.335	78.860	83.675	88.004	93.217	96.878	104.716 64
65	39.383	44.603	50.883	64.335	79.973	84.821	89.177	94.422	98.105	105.988 65
66	40.158	45.431	51.770	65.335	81.085	85.965	90.349	95.626	99.331	107.258 66
67	40.935	46.261	52.659	66.335	82.197	87.108	91.519	96.828	100.55	108.526 67
68	41.713	47.092	53.548	67.334	83.308	88.250	92.689	98.028	101.78	109.791 68
69	42.494	47.924	54.438	68.334	84.418	89.391	93.856	99.228	103.00	111.055 69
70	43.275	48.758	55.329	69.334	85.527	90.531	95.023	100.43	104.21	112.317 70
71	44.058	49.592	56.221	70.334	86.635	91.670	96.189	101.62	105.43	113.577 71
72	44.843	50.428	57.113	71.334	87.743	92.808	97.353	102.82	106.65	114.835 72
73	45.629	51.265	58.006	72.334	88.850	93.945	98.516	104.01	107.86	116.092 73
74	46.417	52.103	58.900	73.334	89.956	95.081	99.678	105.20	109.07	117.346 74
75	47.206	52.942	59.795	74.334	91.061	96.217	100.84	106.39	110.29	118.599 75
76	47.997	53.782	60.690	75.334	92.166	97.351	102.00	107.58	111.50	119.850 76
77	48.788	54.623	61.586	76.334	93.270	98.484	103.16	108.77	112.70	121.100 77
78	49.582	55.466	62.483	77.334	94.373	99.617	104.32	109.96	113.91	122.348 78
79	50.376	56.309	63.380	78.334	95.476	100.75	105.47	111.14	115.12	123.594 79
80	51.172	57.153	64.278	79.334	96.578	101.88	106.63	112.33	116.32	124.839 80
81	51.969	57.998	65.176	80.334	97.680	103.01	107.78	113.51	117.52	126.082 81
82	52.767	58.845	66.076	81.334	98.780	104.14	108.94	114.69	118.73	127.324 82
83	53.567	59.692	66.976	82.334	99.880	105.27	110.09	115.88	119.93	128.565 83
84	54.368	60.540	67.876	83.334	100.98	106.39	111.24	117.06	121.13	129.804 84
85	55.170	61.389	68.777	84.334	102.08	107.52	112.39	118.24	122.32	131.041 85
86	55.973	62.239	69.679	85.334	103.18	108.65	113.54	119.41	123.52	132.277 86
87	56.777	63.089	70.581	86.334	104.28	109.77	114.69	120.59	124.72	133.512 87
88	57.582	63.941	71.484	87.334	105.37	110.90	115.84	121.77	125.91	134.745 88
89	58.389	64.793	72.387	88.334	106.47	112.02	116.99	122.94	127.11	135.978 89
90	59.196	65.647	73.291	89.334	107.56	113.15	118.14	124.12	128.30	137.208 90
91	60.005	66.501	74.196	90.334	108.66	114.27	119.28	125.29	129.49	138.438 91
92	60.815	67.356	75.101	91.334	109.76	115.39	120.43	126.46	130.68	139.666 92
93	61.625	68.211	76.006	92.334	110.85	116.51	121.57	127.63	131.87	140.893 93
94	62.437	69.068	76.912	93.334	111.94	117.63	122.72	128.80	133.06	142.119 94
95	63.250	69.925	77.818	94.334	113.04	118.75	123.86	129.97	134.25	143.344 95
96	64.063	70.783	78.725	95.334	114.13	119.87	125.00	131.14	135.43	144.567 96
97	64.878	71.642	79.633	96.334	115.22	120.99	126.14	132.31	136.62	145.789 97
98	65.694	72.501	80.541	97.334	116.32	122.11	127.28	133.48	137.80	147.010 98
99	66.510	73.361	81.449	98.334	117.41	123.23	128.42	134.64	138.99	148.230 99
100	67.328	74.222	82.358	99.334	118.50	124.34	129.56	135.81	140.17	149.449 100

Table E Critical values of the Wilcoxon rank sum

Nominal α

	0.05		0.025		0.01		0.005	
n	T	α	T	α	T	α	T	α
5	0	.0312						
	1	.0625						
6	2	.0469	0	.0156				
	3	.0781	1	.0312				
7	3	.0391	2	.0234	0	.0078		
	4	.0547	3	.0391	1	.0156		
8	5	.0391	3	.0195	1	.0078	0	.0039
	6	.0547	4	.0273	2	.0117	1	.0078
9	8	.0488	5	.0195	3	.0098	1	.0039
	9	.0645	6	.0273	4	.0137	2	.0059
10	10	.0420	8	.0244	5	.0098	3	.0049
	11	.0527	9	.0322	6	.0137	4	.0068
11	13	.0415	10	.0210	7	.0093	5	.0049
	14	.0508	11	.0269	8	.0122	6	.0068
12	17	.0461	13	.0212	9	.0081	7	.0046
	18	.0549	14	.0261	10	.0105	8	.0061
13	21	.0471	17	.0239	12	.0085	9	.0040
	22	.0549	18	.0287	13	.0107	10	.0052
14	25	.0453	21	.0247	15	.0083	12	.0043
	26	.0520	22	.0290	16	.0101	13	.0054
15	30	.0473	25	.0240	19	.0090	15	.0042
	31	.0535	26	.0277	20	.0108	16	.0051
16	35	.0467	29	.0222	23	.0091	19	.0046
	36	.0523	30	.0253	24	.0107	20	.0055
17	41	.0492	34	.0224	27	.0087	23	.0047
	42	.0544	35	.0253	28	.0101	24	.0055
18	47	.0494	40	.0241	32	.0091	27	.0045
	48	.0542	41	.0269	33	.0104	28	.0052
19	53	.0478	46	.0247	37	.0090	32	.0047
	54	.0521	47	.0273	38	.0102	33	.0054
20	60	.0487	52	.0242	43	.0096	37	.0047
	61	.0527	53	.0266	44	.0107	38	.0053
21	67	.0479	58	.0230	49	.0097	42	.0045
	68	.0516	59	.0251	50	.0108	43	.0051
22	75	.0492	65	.0231	55	.0095	48	.0046
	76	.0527	66	.0250	56	.0104	49	.0052

(*continued*)

Table E   Critical values of the Wilcoxon rank sum (*continued*)

	Nominal α							
	0.05		0.025		0.01		0.005	
n	T	α	T	α	T	α	T	α
23	83	.0490	73	.0242	62	.0098	54	.0046
	84	.0523	74	.0261	63	.0107	55	.0051
24	91	.0475	81	.0245	69	.0097	61	.0048
	92	.0505	82	.0263	70	.0106	62	.0053
25	100	.0479	89	.0241	76	.0094	68	.0048
	101	.0507	90	.0258	77	.0101	69	.0053
26	110	.0497	98	.0247	84	.0095	75	.0047
	111	.0524	99	.0263	85	.0102	76	.0051
27	119	.0477	107	.0246	92	.0093	83	.0048
	120	.0502	108	.0260	93	.0100	84	.0052
28	130	.0496	116	.0239	101	.0096	91	.0048
	131	.0521	117	.0252	102	.0102	92	.0051
29	140	.0482	126	.0240	110	.0095	100	.0049
	141	.0504	127	.0253	111	.0101	101	.0053
30	151	.0481	137	.0249	120	.0098	109	.0050
	152	.0502	138	.0261	121	.0104	110	.0053
31	163	.0491	147	.0239	130	.0099	118	.0049
	164	.0512	148	.0251	131	.0105	119	.0052
32	175	.0492	159	.0249	140	.0097	128	.0050
	176	.0512	160	.0260	141	.0103	129	.0053
33	187	.0485	170	.0242	151	.0099	138	.0049
	188	.0503	171	.0253	152	.0104	139	.0052
34	200	.0488	182	.0242	162	.0098	148	.0048
	201	.0506	183	.0252	163	.0103	149	.0051
35	213	.0484	195	.0247	173	.0096	159	.0048
	214	.0501	196	.0257	174	.0100	160	.0051
36	227	.0489	208	.0248	185	.0096	171	.0050
	228	.0505	209	.0258	186	.0100	172	.0052
37	241	.0487	221	.0245	198	.0099	182	.0048
	242	.0503	222	.0254	199	.0103	183	.0050
38	256	.0493	235	.0247	211	.0099	194	.0048
	257	.0509	236	.0256	212	.0104	195	.0050
39	271	.0493	249	.0246	224	.0099	207	.0049
	272	.0507	250	.0254	225	.0103	208	.0051

Nominal α

	0.05		0.025		0.01		0.005	
n	T	α	T	α	T	α	T	α
40	286	.0486	264	.0249	238	.0100	220	.0049
	287	.0500	265	.0257	239	.0104	221	.0051
41	302	.0488	279	.0248	252	.0100	233	.0048
	303	.0501	280	.0256	253	.0103	234	.0050
42	319	.0496	294	.0245	266	.0098	247	.0049
	320	.0509	295	.0252	267	.0102	248	.0051
43	336	.0498	310	.0245	281	.0098	261	.0048
	337	.0511	311	.0252	282	.0102	262	.0050
44	353	.0495	327	.0250	296	.0097	276	.0049
	354	.0507	328	.0257	297	.0101	277	.0051
45	371	.0498	343	.0244	312	.0098	291	.0049
	372	.0510	344	.0251	313	.0101	292	.0051
46	389	.0497	361	.0249	328	.0098	307	.0050
	390	.0508	362	.0256	329	.0101	308	.0052
47	407	.0490	378	.0245	345	.0099	322	.0048
	408	.0501	379	.0251	346	.0102	323	.0050
48	426	.0490	396	.0244	362	.0099	339	.0050
	427	.0500	397	.0251	363	.0102	340	.0051
49	446	.0495	415	.0247	379	.0098	355	.0049
	447	.0505	416	.0253	380	.0100	356	.0050
50	466	.0495	434	.0247	397	.0098	373	.0050
	467	.0506	435	.0253	398	.0101	374	.0051

Table F Critical values of the Friedman χ_r^2 distribution

a (n)	b (M)*	α: 0.50	0.20	0.10	0.05	0.02	0.01	0.005	0.002	0.001
3	2	3.000	4.000							
3	3	2.667	4.667	(6.000)	6.000					
3	4	2.000	4.500	6.000	6.500	(8.000)	(8.000)	8.000		
3	5	2.800	3.600	5.200	6.400	(8.400)	8.400	(10.000)	(10.000)	10.000
3	6	2.330	4.000	5.330	7.000	8.330	9.000	(10.330)	10.330	12.000
3	7	2.000	3.714	5.429	7.143	8.000	8.857	10.286	11.143	12.286
3	8	2.250	4.000	5.250	6.250	7.750	9.000	9.750	12.000	12.250
3	9	2.000	3.556	5.556	6.222	8.000	9.556	10.667	11.556	12.667
3	10	1.800	3.800	5.000	6.200	7.800	9.600	10.400	12.200	12.600
3	11	4.636	3.818	4.909	6.545	7.818	9.455	10.364	11.636	13.273
3	12	1.500	3.500	5.167	6.167	8.000	9.500	10.167	12.167	12.500
3	13	1.846	3.846	4.769	6.000	8.000	9.385	10.308	11.538	12.923
3	14	1.714	3.571	5.143	6.143	8.143	9.000	10.429	12.000	13.286
3	15	1.733	3.600	4.933	6.400	8.133	8.933	10.000	12.133	12.933
4	2	3.600	5.400	(6.000)	6.000					
4	3	3.400	5.400	6.600	7.400	8.200	(9.000)	(9.000)	9.000	
4	4	3.000	4.800	6.300	7.800	8.400	9.600	(10.200)	10.200	11.100
4	5	3.000	5.160	6.360	7.800	9.240	9.960	10.920	11.640	12.600
4	6	3.000	4.800	6.400	7.600	9.400	10.200	11.400	12.200	12.800
4	7	2.829	4.886	6.429	7.800	9.343	10.371	11.400	12.771	13.800
4	8	2.550	4.800	6.300	7.650	9.450	10.350	11.850	12.900	13.800
4	9			6.467	7.800	9.133	10.867	12.067		14.467
4	10			6.360	7.800	9.120	10.800	12.000		14.640
4	11			6.382	7.909	9.327	11.073	12.273		14.891
4	12			6.400	7.900	9.200	11.100	12.300		15.000
4	13			6.415	7.985	7.369	11.123	12.323		15.277
4	14			6.343	7.886	9.343	11.143	12.514		15.257
4	15			6.440	8.040	9.400	11.240	12.520		15.400
5	2			7.200	7.600	8.000	8.000			
5	3			7.467	8.533	9.600	10.133	10.667		11.467
5	4			7.600	8.800	9.800	11.200	12.000		13.200
5	5			7.680	8.960	10.240	11.680	12.480		14.400
5	6			7.733	9.067	10.400	11.867	13.067		15.200
5	7			7.771	9.143	10.514	12.114	13.257		15.657
5	8			7.800	9.300	10.600	12.300	13.500		16.000
5	9			7.733	9.244	10.667	12.444	13.689		16.356
5	10			7.760	9.280	10.720	12.480	13.840		16.480
6	2			8.286	9.143	9.429	9.714	10.000		
6	3			8.714	9.857	10.810	11.762	12.524		13.286
6	4			9.000	10.286	11.429	12.714	13.571		15.286
6	5			9.000	10.486	11.743	13.229	14.257		16.429
6	6			9.048	10.571	12.000	13.619	14.762		17.048
6	7			9.122	10.674	12.061	13.857	15.000		17.612
6	8			9.143	10.714	12.214	14.000	15.286		18.000
6	9			9.127	10.778	12.302	14.143	15.476		18.270
6	10			9.143	10.800	12.343	14.299	15.600		18.514

*For Kendall's coefficient of concordance, W, use the column headings in parentheses.

Source: D. B. Owen, *Handbook of Statistical Tables.* © 1962 U.S. Department of Energy. Published by Addison-Wesley, Reading, MA, Table 14.1, pp. 408–409.

Table G Critical values of *U*, the Mann-Whitney statistic

n_1	n_2	0.10	0.05	0.025	0.01	0.005	0.001
				α			
3	2	6					
	3	8	9				
4	2	8					
	3	11	12				
	4	13	15	16			
5	2	9	10				
	3	13	14	15			
	4	16	18	19	20		
	5	20	21	23	24	25	
6	2	11	12				
	3	15	16	17			
	4	19	21	22	23	24	
	5	23	25	27	28	29	
	6	27	29	31	33	34	
7	2	13	14				
	3	17	19	20	21		
	4	22	24	25	27	28	
	5	27	29	30	32	34	
	6	31	34	36	38	39	42
	7	36	38	41	43	45	48
8	2	14	15	16			
	3	19	21	22	24		
	4	25	27	28	30	31	
	5	30	32	34	36	38	40
	6	35	38	40	42	44	47
	7	40	43	46	49	50	54
	8	45	49	51	55	57	60
9	1	9					
	2	16	17	18			
	3	22	23	25	26	27	
	4	27	30	32	33	35	
	5	33	36	38	40	42	44
	6	39	42	44	47	49	52
	7	45	48	51	54	56	60
	8	50	54	57	61	63	67
	9	56	60	64	67	70	74
10	1	10					
	2	17	19	20			
	3	24	26	27	29	30	
	4	30	33	35	37	38	40
	5	37	39	42	44	46	49
	6	43	46	49	52	54	57
	7	49	53	56	59	61	65
	8	56	60	63	67	69	74
	9	62	66	70	74	77	82
	10	68	73	77	81	84	90

(continued)

Table G   Critical values of U, the Mann-Whitney statistic (*continued*)

n_1	n_2	0.10	0.05	0.025	α 0.01	0.005	0.001
11	1	11					
	2	19	21	22			
	3	26	28	30	32	33	
	4	33	36	38	40	42	44
	5	40	43	46	48	50	53
	6	47	50	53	57	59	62
	7	54	58	61	65	67	71
	8	61	65	69	73	75	80
	9	68	72	76	81	83	89
	10	74	79	84	88	92	98
	11	81	87	91	96	100	106
12	1	12					
	2	20	22	23			
	3	28	31	32	34	35	
	4	36	39	41	42	45	48
	5	43	47	49	52	54	58
	6	51	55	58	61	63	68
	7	58	63	66	70	72	77
	8	66	70	74	79	81	87
	9	73	78	82	87	90	96
	10	81	86	91	96	99	106
	11	88	94	99	104	108	115
	12	95	102	107	113	117	124
13	1	13					
	2	22	24	25	26		
	3	30	33	35	37	38	
	4	39	42	44	47	49	51
	5	47	50	53	56	58	62
	6	55	59	62	66	68	73
	7	63	67	71	75	78	83
	8	71	76	80	84	87	93
	9	79	84	89	94	97	103
	10	87	93	97	103	106	113
	11	95	101	106	112	116	123
	12	103	109	115	121	125	133
	13	111	118	124	130	135	143
14	1	14					
	2	24	25	27	28		
	3	32	35	37	40	41	
	4	41	45	47	50	52	55
	5	50	54	57	60	63	67
	6	59	63	67	71	73	78
	7	67	72	76	81	83	89
	8	76	81	86	90	94	100
	9	85	90	95	100	104	111
	10	93	99	104	110	114	121
	11	102	108	114	120	124	132
	12	110	117	123	130	134	143
	13	119	126	132	139	144	153
	14	127	135	141	149	154	164

n_1	n_2	0.10	0.05	0.025	α 0.01	0.005	0.001
15	1	15					
	2	25	27	29	30		
	3	35	38	40	42	43	
	4	44	48	50	53	55	59
	5	53	57	61	64	67	71
	6	63	67	71	75	78	83
	7	72	77	81	86	89	95
	8	81	87	91	96	100	106
	9	90	96	101	107	111	118
	10	99	106	111	117	121	129
	11	108	115	121	128	132	141
	12	117	125	131	138	143	152
	13	127	134	141	148	153	163
	14	136	144	151	159	164	174
	15	145	153	161	169	174	185
16	1	16					
	2	27	29	31	32		
	3	37	40	42	45	46	
	4	47	50	53	57	59	62
	5	57	61	65	68	71	75
	6	67	71	75	80	83	88
	7	76	82	86	91	94	101
	8	86	92	97	102	106	113
	9	96	102	107	113	117	125
	10	106	112	118	124	129	137
	11	115	122	129	135	140	149
	12	125	132	139	146	151	161
	13	134	143	149	157	163	173
	14	144	153	160	168	174	185
	15	154	163	170	179	185	197
	16	163	173	181	190	196	208
17	1	17					
	2	28	31	32	34		
	3	39	42	45	47	49	51
	4	50	53	57	60	62	66
	5	60	65	68	72	75	80
	6	71	76	80	84	87	93
	7	81	86	91	96	100	106
	8	91	97	102	108	112	119
	9	101	108	114	120	124	132
	10	112	119	125	132	136	145
	11	122	130	136	143	148	158
	12	132	140	147	155	160	170
	13	142	151	158	166	172	183
	14	153	161	169	178	184	195
	15	163	172	180	189	195	208
	16	173	183	191	201	207	220
	17	183	193	202	212	219	232

(continued)

Table G Critical values of U, the Mann-Whitney statistic (*continued*)

		α					
n_1	n_2	0.10	0.05	0.025	0.01	0.005	0.001
18	1	18					
	2	30	32	34	36		
	3	41	45	47	50	52	54
	4	52	56	60	63	66	69
	5	63	68	72	76	79	84
	6	74	80	84	89	92	98
	7	85	91	96	102	105	112
	8	96	103	108	114	118	126
	9	107	114	120	126	131	139
	10	118	125	132	139	143	153
	11	129	137	143	151	156	166
	12	139	148	155	163	169	179
	13	150	159	167	175	181	192
	14	161	170	178	187	194	206
	15	172	182	190	200	206	219
	16	182	193	202	212	218	232
	17	193	204	213	224	231	245
	18	204	215	225	236	243	258
19	1	18	19				
	2	31	34	36	37	38	
	3	43	47	50	53	54	57
	4	55	59	63	67	69	73
	5	67	72	76	80	83	88
	6	78	84	89	94	97	103
	7	90	96	101	107	111	118
	8	101	108	114	120	124	132
	9	113	120	126	133	138	146
	10	124	132	138	146	151	161
	11	136	144	151	159	164	175
	12	147	156	163	172	177	188
	13	158	167	175	184	190	202
	14	169	179	188	197	203	216
	15	181	191	200	210	216	230
	16	192	203	212	222	230	244
	17	203	214	224	235	242	257
	18	214	226	236	248	255	271
	19	226	238	248	260	268	284
20	1	19	20				
	2	33	36	38	39	40	
	3	45	49	52	55	57	60
	4	58	62	66	70	72	77
	5	70	75	80	84	87	93
	6	82	88	93	98	102	108
	7	94	101	106	112	116	124
	8	106	113	119	126	130	139
	9	118	126	132	140	144	154
	10	130	138	145	153	158	168
	11	142	151	158	167	172	183
	12	154	163	171	180	186	198
	13	166	176	184	193	200	212
	14	178	188	197	207	213	226
	15	190	200	210	220	227	241
	16	201	213	222	233	241	255
	17	213	225	235	247	254	270
	18	225	237	248	260	268	284
	19	237	250	261	273	281	298
	20	249	262	273	286	295	312

Table H Critical values of the Kruskal-Wallis *H*-distribution

n_1	n_2	n_3	α: 0.10	0.05	0.002	0.01	0.005	0.002	0.001
2	2	2	4.571						
3	2	1	4.286						
3	2	2	4.500	4.714					
3	3	1	4.571	5.143					
3	3	2	4.556	5.361	6.250				
3	3	3	4.622	5.600	6.489	(7.200)	7.200		
4	2	1	4.500						
4	2	2	4.458	5.333	6.000				
4	3	1	4.056	5.208					
4	3	2	4.511	5.444	6.144	6.444	7.000		
4	3	3	4.709	5.791	6.564	6.745	7.318	8.018	
4	4	1	4.167	4.967	(6.667)	6.667			
4	4	2	4.555	5.455	6.600	7.036	7.282	7.855	
4	4	3	4.545	5.598	6.712	7.144	7.598	8.227	8.909
4	4	4	4.654	5.692	6.962	7.654	8.000	8.654	9.269
5	2	1	4.200	5.000					
5	2	2	4.373	5.160	6.000	6.533			
5	3	1	4.018	4.960	6.044				
5	3	2	4.651	5.251	6.124	6.909	7.182		
5	3	3	4.533	5.648	6.533	7.079	7.636	8.048	8.727
5	4	1	3.987	4.985	6.431	6.955	7.364		
5	4	2	4.541	5.273	6.505	7.205	7.573	8.114	8.591
5	4	3	4.549	5.656	6.676	7.445	7.927	8.481	8.795
5	4	4	4.619	5.657	6.953	7.760	8.189	8.868	9.168
5	5	1	4.109	5.127	6.145	7.309	8.182		
5	5	2	4.623	5.338	6.446	7.338	8.131	6.446	7.338
5	5	3	4.545	5.705	6.866	7.578	8.316	8.809	9.521
5	5	4	4.523	5.666	7.000	7.823	8.523	9.163	9.606
5	5	5	4.940	5.780	7.220	8.000	8.780	9.620	9.920
6	1	1	—						
6	2	1	4.200	4.822					
6	2	2	4.545	5.345	6.182	6.982			
6	3	1	3.909	4.855	6.236				
6	3	2	4.682	5.348	6.227	6.970	7.515	8.182	
6	3	3	4.538	5.615	6.590	7.410	7.872	8.628	9.346
6	4	1	4.038	4.947	6.174	7.106	7.614		
6	4	2	4.494	5.340	6.571	7.340	7.846	8.494	8.827
6	4	3	4.604	5.610	6.725	7.500	8.033	8.918	9.170
6	4	4	4.595	5.681	6.900	7.795	8.381	9.167	9.861
6	5	1	4.128	4.990	6.138	7.182	8.077	8.515	
6	5	2	4.596	5.338	6.585	7.376	8.196	8.967	9.189
6	5	3	4.535	5.602	6.829	7.590	8.314	9.150	9.669
6	5	4	4.522	5.661	7.018	7.936	8.643	9.458	9.960
6	5	5	4.547	5.729	7.110	8.028	8.859	9.771	10.271
6	6	1	4.000	4.945	6.286	7.121	8.165	9.077	9.692
6	6	2	4.438	5.410	6.667	7.467	8.210	9.219	9.752
6	6	3	4.558	5.625	6.900	7.725	8.458	9.458	10.150
6	6	4	4.548	5.724	7.107	8.000	8.754	9.662	10.342
6	6	5	4.542	5.765	7.152	8.124	8.987	9.948	10.524
6	6	6	4.643	5.801	7.240	8.222	9.170	10.187	10.889
7	7	7	4.594	5.819	7.332	8.378	9.373	10.516	11.310
8	8	8	4.595	5.805	7.355	8.465	9.495	10.805	11.705

Source: Selected Tables in Mathematical Statistics, Vol. III, pp. 320–384, by permission of the American Mathematical Society. © 1975 by the American Mathematical Society.

Table I Critical values of Rayleigh's *u*

n	α: 0.25	0.10	0.05	0.025	0.01	0.005	0.0025	0.001	0.0005
8	0.688	1.296	1.649	1.947	2.280	2.498	2.691	2.916	3.066
9	0.687	1.294	1.649	1.948	2.286	2.507	2.705	2.937	3.094
10	0.685	1.293	1.648	1.950	2.290	2.514	2.716	2.954	3.115
11	0.684	1.292	1.648	1.950	2.293	2.520	2.725	2.967	3.133
12	0.684	1.291	1.648	1.951	2.296	2.525	2.732	2.978	3.147
13	0.683	1.290	1.647	1.952	2.299	2.529	2.738	2.987	3.159
14	0.682	1.290	1.647	1.953	2.301	2.532	2.743	2.995	3.169
15	0.682	1.289	1.647	1.953	2.302	2.535	2.748	3.002	3.177
16	0.681	1.289	1.647	1.953	2.304	2.538	2.751	3.008	3.185
17	0.681	1.288	1.647	1.954	2.305	2.540	2.755	3.013	3.191
18	0.681	1.288	1.647	1.954	2.306	2.542	2.758	3.017	3.197
19	0.680	1.287	1.647	1.954	2.308	2.544	2.761	3.021	3.202
20	0.680	1.287	1.646	1.955	2.308	2.546	2.763	3.025	3.207
21	0.680	1.287	1.646	1.955	2.309	2.547	2.765	3.028	3.211
22	0.679	1.287	1.646	1.955	2.310	2.549	2.767	3.031	3.215
23	0.679	1.286	1.646	1.955	2.311	2.550	2.769	3.034	3.218
24	0.679	1.286	1.646	1.956	2.311	2.551	2.770	3.036	3.221
25	0.679	1.286	1.646	1.956	2.312	2.552	2.772	3.038	3.224
26	0.679	1.286	1.646	1.956	2.313	2.553	2.773	3.040	3.227
27	0.678	1.286	1.646	1.956	2.313	2.554	2.775	3.042	3.229
28	0.678	1.285	1.646	1.956	2.314	2.555	2.776	3.044	3.231
29	0.678	1.285	1.646	1.956	2.314	2.555	2.777	3.046	3.233
30	0.678	1.285	1.646	1.957	2.315	2.556	2.778	3.047	3.235
32	0.678	1.285	1.646	1.957	2.315	2.557	2.780	3.050	3.239
34	0.678	1.285	1.646	1.957	2.316	2.558	2.781	3.052	3.242
36	0.677	1.285	1.646	1.957	2.316	2.559	2.783	3.054	3.245
38	0.677	1.284	1.646	1.957	2.317	2.560	2.784	3.056	3.247
40	0.677	1.284	1.646	1.957	2.317	2.561	2.785	3.058	3.249
42	0.677	1.284	1.646	1.958	2.318	2.562	2.786	3.060	3.251
44	0.677	1.284	1.646	1.958	2.318	2.562	2.787	3.061	3.253
46	0.677	1.284	1.646	1.958	2.319	2.563	2.788	3.062	3.255
48	0.677	1.284	1.645	1.958	2.319	2.564	2.789	3.063	3.256
50	0.677	1.284	1.645	1.958	2.319	2.564	2.790	3.065	3.258
55	0.676	1.284	1.645	1.958	2.320	2.565	2.791	3.067	3.261
60	0.676	1.283	1.645	1.958	2.320	2.566	2.793	3.069	3.263
65	0.676	1.283	1.645	1.958	2.321	2.567	2.794	3.071	3.265
70	0.676	1.283	1.645	1.958	2.321	2.567	2.795	3.072	3.267
75	0.676	1.283	1.645	1.959	2.322	2.568	2.796	3.073	3.269
80	0.676	1.283	1.645	1.959	2.322	2.568	2.796	3.074	3.270
90	0.676	1.283	1.645	1.959	2.322	2.569	2.797	3.076	3.272
100	0.676	1.283	1.645	1.959	2.323	2.570	2.798	3.077	3.274
120	0.675	1.282	1.645	1.959	2.323	2.571	2.800	3.080	3.277
140	0.675	1.282	1.645	1.959	2.324	2.572	2.801	3.081	3.279
160	0.675	1.282	1.645	1.959	2.324	2.572	2.802	3.082	3.280
180	0.675	1.282	1.645	1.959	2.324	2.573	2.802	3.083	3.282
200	0.675	1.282	1.645	1.959	2.325	2.573	2.803	3.084	3.282
300	0.675	1.282	1.645	1.960	2.325	2.574	2.804	3.086	3.285
∞	0.6747	1.2818	1.6449	1.9598	2.3256	2.5747	2.8053	3.0877	3.2873

Source: J. H. Zar, *Biostatistical Analysis,* 4th ed. (Upper Saddle River, NJ: Prentice Hall, 1999).

Table J   Critical values of Rayleigh's z

n	α: 0.50	0.20	0.10	0.05	0.02	0.01	0.005	0.002	0.001
6	0.734	1.639	2.274	2.865	3.576	4.058	4.491	4.985	5.297
7	0.727	1.634	2.278	2.885	3.627	4.143	4.617	5.181	5.556
8	0.723	1.631	2.281	2.899	3.665	4.205	4.710	5.322	5.743
9	0.719	1.628	2.283	2.910	3.694	4.252	4.780	5.430	5.885
10	0.717	1.626	2.285	2.919	3.716	4.289	4.835	5.514	5.996
11	0.715	1.625	2.287	2.926	3.735	4.319	4.879	5.582	6.085
12	0.713	1.623	2.288	2.932	3.750	4.344	4.916	5.638	6.158
13	0.711	1.622	2.289	2.937	3.763	4.365	4.947	5.685	6.219
14	0.710	1.621	2.290	2.941	3.774	4.383	4.973	5.725	6.271
15	0.709	1.620	2.291	2.945	3.784	4.398	4.996	5.759	6.316
16	0.708	1.620	2.292	2.948	3.792	4.412	5.015	5.789	6.354
17	0.707	1.619	2.292	2.951	3.799	4.423	5.033	5.815	6.388
18	0.706	1.619	2.293	2.954	3.806	4.434	5.048	5.838	6.418
19	0.705	1.618	2.293	2.956	3.811	4.443	5.061	5.858	6.445
20	0.705	1.618	2.294	2.958	3.816	4.451	5.074	5.877	6.469
21	0.704	1.617	2.294	2.960	3.821	4.459	5.085	5.893	6.491
22	0.704	1.617	2.295	2.961	3.825	4.466	5.095	5.908	6.510
23	0.703	1.616	2.295	2.963	3.829	4.472	5.104	5.922	6.528
24	0.703	1.616	2.295	2.964	3.833	4.478	5.112	5.935	6.544
25	0.702	1.616	2.296	2.966	3.836	4.483	5.120	5.946	6.559
26	0.702	1.616	2.296	2.967	3.839	4.488	5.127	5.957	6.573
27	0.702	1.615	2.296	2.968	3.842	4.492	5.133	5.966	6.586
28	0.701	1.615	2.296	2.969	3.844	4.496	5.139	5.975	6.598
29	0.701	1.615	2.297	2.970	3.847	4.500	5.145	5.984	6.609
30	0.701	1.615	2.297	2.971	3.849	4.504	5.150	5.992	6.619
32	0.700	1.614	2.297	2.972	3.853	4.510	5.159	6.006	6.637
34	0.700	1.614	2.297	2.974	3.856	4.516	5.168	6.018	6.654
36	0.700	1.614	2.298	2.975	3.859	4.521	5.175	6.030	6.668
38	0.699	1.614	2.298	2.976	3.862	4.525	5.182	6.039	6.681
40	0.699	1.613	2.298	2.977	3.865	4.529	5.188	6.048	6.692
42	0.699	1.613	2.298	2.978	3.867	4.533	5.193	6.056	6.703
44	0.698	1.613	2.299	2.979	3.869	4.536	5.198	6.064	6.712
46	0.698	1.613	2.299	2.979	3.871	4.539	5.202	6.070	6.721
48	0.698	1.613	2.299	2.980	3.873	4.542	5.206	6.076	6.729
50	0.698	1.613	2.299	2.981	3.874	4.545	5.210	6.082	6.736
55	0.697	1.612	2.299	2.982	3.878	4.550	5.218	6.094	6.752
60	0.697	1.612	2.300	2.983	3.881	4.555	5.225	6.104	6.765
65	0.697	1.612	2.300	2.984	3.883	4.559	5.231	6.113	6.776
70	0.696	1.612	2.300	2.985	3.885	4.562	5.235	6.120	6.786
75	0.696	1.612	2.300	2.986	3.887	4.565	5.240	6.127	6.794
80	0.696	1.611	2.300	2.986	3.889	4.567	5.243	6.132	6.801
90	0.696	1.611	2.301	2.987	3.891	4.572	5.249	6.141	6.813
100	0.695	1.611	2.301	2.988	3.893	4.575	5.254	6.149	6.822
120	0.695	1.611	2.301	2.990	3.896	4.580	5.262	6.160	6.837
140	0.695	1.611	2.301	2.990	3.899	4.584	5.267	6.168	6.847
160	0.695	1.610	2.301	2.991	3.900	4.586	5.271	6.174	6.855
180	0.694	1.610	2.302	2.992	3.902	4.588	5.274	6.178	6.861
200	0.694	1.610	2.302	2.992	3.903	4.590	5.276	6.182	6.865
300	0.694	1.610	2.302	2.993	3.906	4.595	5.284	6.193	6.879
500	0.694	1.610	2.302	2.994	3.908	4.599	5.290	6.201	6.891
∞	0.6931	1.6094	2.3026	2.9957	3.9120	4.6052	5.2983	6.2146	6.9078

Source: J. H. Zar, *Biostatistical Analysis,* 4th ed. (Upper Saddle River, NJ: Prentice Hall, 1999).

Table K   Critical values of the Spearman rank correlation coefficient, r_s

$\alpha(2)$: n $\quad\alpha(1)$:	0.50 0.25	0.20 0.10	0.10 0.05	0.05 0.025	0.02 0.01	0.01 0.005	0.005 0.0025	0.002 0.001	0.001 0.0005
4	0.600	1.000	1.000						
5	0.500	0.800	0.900	1.000	1.000				
6	0.371	0.657	0.829	0.886	0.943	1.000	1.000		
7	0.321	0.571	0.714	0.786	0.893	0.929	0.964	1.000	1.000
8	0.310	0.524	0.643	0.738	0.833	0.881	0.905	0.952	0.976
9	0.267	0.483	0.600	0.700	0.783	0.833	0.867	0.917	0.933
10	0.248	0.455	0.564	0.648	0.745	0.794	0.830	0.879	0.903
11	0.236	0.427	0.536	0.618	0.709	0.755	0.800	0.845	0.873
12	0.217	0.406	0.503	0.587	0.678	0.727	0.769	0.818	0.846
13	0.209	0.385	0.484	0.560	0.648	0.703	0.747	0.791	0.824
14	0.200	0.367	0.464	0.538	0.626	0.679	0.723	0.771	0.802
15	0.189	0.354	0.446	0.521	0.604	0.654	0.700	0.750	0.779
16	0.182	0.341	0.429	0.503	0.582	0.635	0.679	0.729	0.762
17	0.176	0.328	0.414	0.485	0.566	0.615	0.662	0.713	0.748
18	0.170	0.317	0.401	0.472	0.550	0.600	0.643	0.695	0.728
19	0.165	0.309	0.391	0.460	0.535	0.584	0.628	0.677	0.712
20	0.161	0.299	0.380	0.447	0.520	0.570	0.612	0.662	0.696
21	0.156	0.292	0.370	0.435	0.508	0.556	0.599	0.648	0.681
22	0.152	0.284	0.361	0.425	0.496	0.544	0.586	0.634	0.667
23	0.148	0.278	0.353	0.415	0.486	0.532	0.573	0.622	0.654
24	0.144	0.271	0.344	0.406	0.476	0.521	0.562	0.610	0.642
25	0.142	0.265	0.337	0.398	0.466	0.511	0.551	0.598	0.630
26	0.138	0.259	0.331	0.390	0.457	0.501	0.541	0.587	0.619
27	0.136	0.255	0.324	0.382	0.448	0.491	0.531	0.577	0.608
28	0.133	0.250	0.317	0.375	0.440	0.483	0.522	0.567	0.598
29	0.130	0.245	0.312	0.368	0.433	0.475	0.513	0.558	0.589
30	0.128	0.240	0.306	0.362	0.425	0.467	0.504	0.549	0.580
31	0.126	0.236	0.301	0.356	0.418	0.459	0.496	0.541	0.571
32	0.124	0.232	0.296	0.350	0.412	0.452	0.489	0.533	0.563
33	0.121	0.229	0.291	0.345	0.405	0.446	0.482	0.525	0.554
34	0.120	0.225	0.287	0.340	0.399	0.439	0.475	0.517	0.547
35	0.118	0.222	0.283	0.335	0.394	0.433	0.468	0.510	0.539
36	0.116	0.219	0.279	0.330	0.388	0.427	0.462	0.504	0.533
37	0.114	0.216	0.275	0.325	0.383	0.421	0.456	0.497	0.526
38	0.113	0.212	0.271	0.321	0.378	0.415	0.450	0.491	0.519
39	0.111	0.210	0.267	0.317	0.373	0.410	0.444	0.485	0.513
40	0.110	0.207	0.264	0.313	0.368	0.405	0.439	0.479	0.507
41	0.108	0.204	0.261	0.309	0.364	0.400	0.433	0.473	0.501
42	0.107	0.202	0.257	0.305	0.359	0.395	0.428	0.468	0.495
43	0.105	0.199	0.254	0.301	0.355	0.391	0.423	0.463	0.490
44	0.104	0.197	0.251	0.298	0.351	0.386	0.419	0.458	0.484
45	0.103	0.194	0.248	0.294	0.347	0.382	0.414	0.453	0.479
46	0.102	0.192	0.246	0.291	0.343	0.378	0.410	0.448	0.474
47	0.101	0.190	0.243	0.288	0.340	0.374	0.405	0.443	0.469
48	0.100	0.188	0.240	0.285	0.336	0.370	0.401	0.439	0.465
49	0.098	0.186	0.238	0.282	0.333	0.366	0.397	0.434	0.460
50	0.097	0.184	0.235	0.279	0.329	0.363	0.393	0.430	0.456

n	$\alpha(2)$: $\alpha(1)$:	0.50 0.25	0.20 0.10	0.10 0.05	0.05 0.025	0.02 0.01	0.01 0.005	0.005 0.0025	0.002 0.001	0.001 0.0005
51		0.096	0.182	0.233	0.276	0.326	0.359	0.390	0.426	0.451
52		0.095	0.180	0.231	0.274	0.323	0.356	0.386	0.422	0.447
53		0.095	0.179	0.228	0.271	0.320	0.352	0.382	0.418	0.443
54		0.094	0.177	0.226	0.268	0.317	0.349	0.379	0.414	0.439
55		0.093	0.175	0.224	0.266	0.314	0.346	0.375	0.411	0.435
56		0.092	0.174	0.222	0.264	0.311	0.343	0.372	0.407	0.432
57		0.091	0.172	0.220	0.261	0.308	0.340	0.369	0.404	0.428
58		0.090	0.171	0.218	0.259	0.306	0.337	0.366	0.400	0.424
59		0.089	0.169	0.216	0.257	0.303	0.334	0.363	0.397	0.421
60		0.089	0.168	0.214	0.255	0.300	0.331	0.360	0.394	0.418
61		0.088	0.166	0.213	0.252	0.298	0.329	0.357	0.391	0.414
62		0.087	0.165	0.211	0.250	0.296	0.326	0.354	0.388	0.411
63		0.086	0.163	0.209	0.248	0.293	0.323	0.351	0.385	0.408
64		0.086	0.162	0.207	0.246	0.291	0.321	0.348	0.382	0.405
65		0.085	0.161	0.206	0.244	0.289	0.318	0.346	0.379	0.402
66		0.084	0.160	0.204	0.243	0.287	0.316	0.343	0.376	0.399
67		0.084	0.158	0.203	0.241	0.284	0.314	0.341	0.373	0.396
68		0.083	0.157	0.201	0.239	0.282	0.311	0.338	0.370	0.393
69		0.082	0.156	0.200	0.237	0.280	0.309	0.336	0.368	0.390
70		0.082	0.155	0.198	0.235	0.278	0.307	0.333	0.365	0.388
71		0.081	0.154	0.197	0.234	0.276	0.305	0.331	0.363	0.385
72		0.081	0.153	0.195	0.232	0.274	0.303	0.329	0.360	0.382
73		0.080	0.152	0.194	0.230	0.272	0.301	0.327	0.358	0.380
74		0.080	0.151	0.193	0.229	0.271	0.299	0.324	0.355	0.377
75		0.079	0.150	0.191	0.227	0.269	0.297	0.322	0.353	0.375
76		0.078	0.149	0.190	0.226	0.267	0.295	0.320	0.351	0.372
77		0.078	0.148	0.189	0.224	0.265	0.293	0.318	0.349	0.370
78		0.077	0.147	0.188	0.223	0.264	0.291	0.316	0.346	0.368
79		0.077	0.146	0.186	0.221	0.262	0.289	0.314	0.344	0.365
80		0.076	0.145	0.185	0.220	0.260	0.287	0.312	0.342	0.363
81		0.076	0.144	0.184	0.219	0.259	0.285	0.310	0.340	0.361
82		0.075	0.143	0.183	0.217	0.257	0.284	0.308	0.338	0.359
83		0.075	0.142	0.182	0.216	0.255	0.282	0.306	0.336	0.357
84		0.074	0.141	0.181	0.215	0.254	0.280	0.305	0.334	0.355
85		0.074	0.140	0.180	0.213	0.252	0.279	0.303	0.332	0.353
86		0.074	0.139	0.179	0.212	0.251	0.277	0.301	0.330	0.351
87		0.073	0.139	0.177	0.211	0.250	0.276	0.299	0.328	0.349
88		0.073	0.138	0.176	0.210	0.248	0.274	0.298	0.327	0.347
89		0.072	0.137	0.175	0.209	0.247	0.272	0.296	0.325	0.345
90		0.072	0.136	0.174	0.207	0.245	0.271	0.294	0.323	0.343
91		0.072	0.135	0.173	0.206	0.244	0.269	0.293	0.321	0.341
92		0.071	0.135	0.173	0.205	0.243	0.268	0.291	0.319	0.339
93		0.071	0.134	0.172	0.204	0.241	0.267	0.290	0.318	0.338
94		0.070	0.133	0.171	0.203	0.240	0.265	0.288	0.316	0.336
95		0.070	0.133	0.170	0.202	0.239	0.264	0.287	0.314	0.334
96		0.070	0.132	0.169	0.201	0.238	0.262	0.285	0.313	0.332
97		0.069	0.131	0.168	0.200	0.236	0.261	0.284	0.311	0.331
98		0.069	0.130	0.167	0.199	0.235	0.260	0.282	0.310	0.329
99		0.068	0.130	0.166	0.198	0.234	0.258	0.281	0.308	0.327
100		0.068	0.129	0.165	0.197	0.233	0.257	0.279	0.307	0.326

Source: J. H. Zar, *Biostatistical Analysis*, 4th ed. (Upper Saddle River, NJ: Prentice Hall, 1999).

Table L Critical values of the correlation coefficient, r

$\alpha(2)$: v / $\alpha(1)$:	0.50 0.25	0.20 0.10	0.10 0.05	0.05 0.025	0.02 0.01	0.01 0.005	0.005 0.0025	0.002 0.001	0.001 0.0005
1	0.707	0.951	0.988	0.997	1.000	1.000	1.000	1.000	1.000
2	0.500	0.800	0.900	0.950	0.980	0.990	0.995	0.998	0.999
3	0.404	0.687	0.805	0.878	0.934	0.959	0.974	0.986	0.991
4	0.347	0.608	0.729	0.811	0.882	0.917	0.942	0.963	0.974
5	0.309	0.551	0.669	0.755	0.833	0.875	0.906	0.935	0.951
6	0.281	0.507	0.621	0.707	0.789	0.834	0.870	0.905	0.925
7	0.260	0.472	0.582	0.666	0.750	0.798	0.836	0.875	0.898
8	0.242	0.443	0.549	0.632	0.715	0.765	0.805	0.847	0.872
9	0.228	0.419	0.521	0.602	0.685	0.735	0.776	0.820	0.847
10	0.216	0.398	0.497	0.576	0.658	0.708	0.750	0.795	0.823
11	0.206	0.380	0.476	0.553	0.634	0.684	0.726	0.772	0.801
12	0.197	0.365	0.457	0.532	0.612	0.661	0.703	0.750	0.780
13	0.189	0.351	0.441	0.514	0.592	0.641	0.683	0.730	0.760
14	0.182	0.338	0.426	0.497	0.574	0.623	0.664	0.711	0.742
15	0.176	0.327	0.412	0.482	0.558	0.606	0.647	0.694	0.725
16	0.170	0.317	0.400	0.468	0.542	0.590	0.631	0.678	0.708
17	0.165	0.308	0.389	0.456	0.529	0.575	0.616	0.662	0.693
18	0.160	0.299	0.378	0.444	0.515	0.561	0.602	0.648	0.679
19	0.156	0.291	0.369	0.433	0.503	0.549	0.589	0.635	0.665
20	0.152	0.284	0.360	0.423	0.492	0.537	0.576	0.622	0.652
21	0.148	0.277	0.352	0.413	0.482	0.526	0.565	0.610	0.640
22	0.145	0.271	0.344	0.404	0.472	0.515	0.554	0.599	0.629
23	0.141	0.265	0.337	0.396	0.462	0.505	0.543	0.588	0.618
24	0.138	0.260	0.330	0.388	0.453	0.496	0.534	0.578	0.607
25	0.136	0.255	0.323	0.381	0.445	0.487	0.524	0.568	0.597
26	0.133	0.250	0.317	0.374	0.437	0.479	0.515	0.559	0.588
27	0.131	0.245	0.311	0.367	0.430	0.471	0.507	0.550	0.579
28	0.128	0.241	0.306	0.361	0.423	0.463	0.499	0.541	0.570
29	0.126	0.237	0.301	0.355	0.416	0.456	0.491	0.533	0.562
30	0.124	0.233	0.296	0.349	0.409	0.449	0.484	0.526	0.554
31	0.122	0.229	0.291	0.344	0.403	0.442	0.477	0.518	0.546
32	0.120	0.225	0.287	0.339	0.397	0.436	0.470	0.511	0.539
33	0.118	0.222	0.283	0.334	0.392	0.430	0.464	0.504	0.532
34	0.116	0.219	0.279	0.329	0.386	0.424	0.458	0.498	0.525
35	0.115	0.216	0.275	0.325	0.381	0.418	0.452	0.492	0.519
36	0.113	0.213	0.271	0.320	0.376	0.413	0.446	0.486	0.513
37	0.111	0.210	0.267	0.316	0.371	0.408	0.441	0.480	0.507
38	0.110	0.207	0.264	0.312	0.367	0.403	0.435	0.474	0.501
39	0.108	0.204	0.261	0.308	0.362	0.398	0.430	0.469	0.495
40	0.107	0.202	0.257	0.304	0.358	0.393	0.425	0.463	0.490
41	0.106	0.199	0.254	0.301	0.354	0.389	0.420	0.458	0.484
42	0.104	0.197	0.251	0.297	0.350	0.384	0.416	0.453	0.479
43	0.103	0.195	0.248	0.294	0.346	0.380	0.411	0.449	0.474
44	0.102	0.192	0.246	0.291	0.342	0.376	0.407	0.444	0.469
45	0.101	0.190	0.243	0.288	0.338	0.372	0.403	0.439	0.465
46	0.100	0.188	0.240	0.285	0.335	0.368	0.399	0.435	0.460
47	0.099	0.186	0.238	0.282	0.331	0.365	0.395	0.431	0.456
48	0.098	0.184	0.235	0.279	0.328	0.361	0.391	0.427	0.451
49	0.097	0.182	0.233	0.276	0.325	0.358	0.387	0.423	0.447
50	0.096	0.181	0.231	0.273	0.322	0.354	0.384	0.419	0.443

ν	$\alpha(2):$ 0.50 $\alpha(1):$ 0.25	0.20 0.10	0.10 0.05	0.05 0.025	0.02 0.01	0.01 0.005	0.005 0.0025	0.002 0.001	0.001 0.0005
52	0.094	0.177	0.226	0.268	0.316	0.348	0.377	0.411	0.435
54	0.092	0.174	0.222	0.263	0.310	0.341	0.370	0.404	0.428
56	0.090	0.171	0.218	0.259	0.305	0.336	0.364	0.398	0.421
58	0.089	0.168	0.214	0.254	0.300	0.330	0.358	0.391	0.414
60	0.087	0.165	0.211	0.250	0.295	0.325	0.352	0.385	0.408
62	0.086	0.162	0.207	0.246	0.290	0.320	0.347	0.379	0.402
64	0.084	0.160	0.204	0.242	0.286	0.315	0.342	0.374	0.396
66	0.083	0.157	0.201	0.239	0.282	0.310	0.337	0.368	0.390
68	0.082	0.155	0.198	0.235	0.278	0.306	0.332	0.363	0.385
70	0.081	0.153	0.195	0.232	0.274	0.302	0.327	0.358	0.380
72	0.080	0.151	0.193	0.229	0.270	0.298	0.323	0.354	0.375
74	0.079	0.149	0.190	0.226	0.266	0.294	0.319	0.349	0.370
76	0.078	0.147	0.188	0.223	0.263	0.290	0.315	0.345	0.365
78	0.077	0.145	0.185	0.220	0.260	0.286	0.311	0.340	0.361
80	0.076	0.143	0.183	0.217	0.257	0.283	0.307	0.336	0.357
82	0.075	0.141	0.181	0.215	0.253	0.280	0.304	0.333	0.328
84	0.074	0.140	0.179	0.212	0.251	0.276	0.300	0.329	0.349
86	0.073	0.138	0.177	0.210	0.248	0.273	0.297	0.325	0.345
88	0.072	0.136	0.174	0.207	0.245	0.270	0.293	0.321	0.341
90	0.071	0.135	0.173	0.205	0.242	0.267	0.290	0.318	0.338
92	0.070	0.133	0.171	0.203	0.240	0.264	0.287	0.315	0.334
94	0.070	0.132	0.169	0.201	0.237	0.262	0.284	0.312	0.331
96	0.069	0.131	0.167	0.199	0.235	0.259	0.281	0.308	0.327
98	0.068	0.129	0.165	0.197	0.232	0.256	0.279	0.305	0.324
100	0.068	0.128	0.164	0.195	0.230	0.254	0.276	0.303	0.321
105	0.066	0.125	0.160	0.190	0.225	0.248	0.270	0.296	0.314
110	0.064	0.122	0.156	0.186	0.220	0.242	0.264	0.289	0.307
115	0.063	0.119	0.153	0.182	0.215	0.237	0.258	0.283	0.300
120	0.062	0.117	0.150	0.178	0.210	0.232	0.253	0.277	0.294
125	0.060	0.114	0.147	0.174	0.206	0.228	0.248	0.272	0.289
130	0.059	0.112	0.144	0.171	0.202	0.223	0.243	0.267	0.283
135	0.058	0.110	0.141	0.168	0.199	0.219	0.239	0.262	0.278
140	0.057	0.108	0.139	0.165	0.195	0.215	0.234	0.257	0.273
145	0.056	0.106	0.136	0.162	0.192	0.212	0.230	0.253	0.269
150	0.055	0.105	0.134	0.159	0.189	0.208	0.227	0.249	0.264
160	0.053	0.101	0.130	0.154	0.183	0.202	0.220	0.241	0.256
170	0.052	0.098	0.126	0.150	0.177	0.196	0.213	0.234	0.249
180	0.050	0.095	0.122	0.145	0.172	0.190	0.207	0.228	0.242
190	0.049	0.093	0.119	0.142	0.168	0.185	0.202	0.222	0.236
200	0.048	0.091	0.116	0.138	0.164	0.181	0.197	0.216	0.230
250	0.043	0.081	0.104	0.124	0.146	0.162	0.176	0.194	0.206
300	0.039	0.074	0.095	0.113	0.134	0.148	0.161	0.177	0.188
350	0.036	0.068	0.088	0.105	0.124	0.137	0.149	0.164	0.175
400	0.034	0.064	0.082	0.098	0.116	0.128	0.140	0.154	0.164
450	0.032	0.060	0.077	0.092	0.109	0.121	0.132	0.145	0.154
500	0.030	0.057	0.074	0.088	0.104	0.115	0.125	0.138	0.146
600	0.028	0.052	0.067	0.080	0.095	0.105	0.114	0.126	0.134
700	0.026	0.048	0.062	0.074	0.088	0.097	0.106	0.116	0.124
800	0.024	0.045	0.058	0.069	0.082	0.091	0.099	0.109	0.116
900	0.022	0.043	0.055	0.065	0.077	0.086	0.093	0.103	0.109
1000	0.021	0.041	0.052	0.062	0.073	0.081	0.089	0.098	0.104

Source: J. H. Zar, *Biostatistical Analysis,* 4th ed. (Upper Saddle River, NJ: Prentice Hall, 1999).

Glossary

Absolute value Mathematically speaking, a number's distance from zero. To the rest of us, it is simply the positive version of a number. The absolute value of -23 is 23; the absolute value of 6 is 6.

ANOVA (ANalysis Of VAriance) A powerful statistical test of the similarity between several sets of data or "treatments". *BioStats Basics Online* provides only the most basic ANOVA tests: a test of similarity means, and a two-way comparison of factors.

The ordinary ANOVA is similar to the t-test in that it determines the probability that parametric data sets with similar standard deviations are the same. Like the t-test, ANOVA requires that the data pass the F-test (similar standard deviations) before it may be used.

Bell curve The Gaussian or normal curve, which is shaped like a bell. It is given by the equation:

$$y = \frac{1}{\sqrt{2\pi}} \, e^{\frac{-x^2}{2}}$$

or something similar. This idealized unit curve has a mean of zero and a standard deviation of one and is called the perfect normal curve.

Actual data sets often fall into a bell-curve distribution. Though bell curves have the same shape as perfect normal curves, they may have any mean and standard deviation. These are called parametric data sets. If the data do not fall into a bell-curve distribution, they are called nonparametric.

The similarity of a curve to the bell curve is tested by a test for normality. In this and many other parametric tests the actual data curve is converted first into a perfect normal curve for mathematical convenience.

Best-fit line The line in regression analysis that minimizes the sum of the squares of the distance of each point from the line.

Best-fit plane The plane in multiple-regression analysis that minimizes the sum of the squared distances of each point from the plane.

Binomial data Data in which there are two mutually exclusive states, like male and female.

Categorized data A set of data in which each result or datum falls into a discrete category. Categorized data are necessary for the two chi-square tests.

A noncategorized ("continuous") data set may be broken up into categories when a categorized test must be used, but a great deal of precision (and thus certainty) may be lost.

Cause The phenomenon responsible for some observed distribution of data. Statistics exists to establish the probability that a given distribution differs from another distribution as a result of some cause, as opposed to by chance.

Chance As in "the chance of . . . ," the probability of a result occurring.

As in "due to chance," the random distribution of results. The alternative is causation—the presence of nonrandom effects in a data set.

Chi-square (goodness of fit) A test of "goodness of fit" between an expected and an observed number of results in several categories. The chi-square value is computed by the formula

$$\chi^2 = \sum \frac{(o - e)^2}{e}$$

where e is the expected number of results in a category, o is the observed number of results in the category, and $\sum$ means "the sum of."

Degrees of freedom in the chi-square (goodness of fit) test is equal to one fewer than the number of categories for the data.

Chi-square is pronounced "ki-square," and although it is represented as χ^2, it would be meaningless to take the square root and call it chi.

Chi-square (independence) A test of relatedness between sets of data that are divided into categories. The values are first put into a matrix, with each set in a row and each category a column. For instance, the number of results in the first data set falling into the third category would go in the first row, third column.

You then must compute the null distribution by the following method: first,

add the values in each row and in each column, as well as the total of all the values in the matrix. The expected value for any position in the matrix is equal to its row's sum multiplied by its column's sum, divided by the total sum of the matrix. Apply chi-square (goodness of fit) to the resulting numbers, taking each corresponding position in the matrix as a category.

Degrees of freedom in the chi-square (independence) test is equal to the number of columns minus one times the number of rows minus one.

Circular distribution A distribution of measurements in which the data are angles rather than ordinary numbers. Circular distributions differ from other distributions because they cycle back on themselves; thus, a result of 2° differs from 358° by 4° rather than 356°. The angles may represent true bearings, or any other periodic data (like days in a year) divided into 360.

Cluster A group of data close together, with the most points near the center and the fewest near the edges. A group of results obtained from a parametric distribution will form a cluster.

Continuous data Data that can take any value along a range, as, for example, height.

Conversion (scaling) The process by which a normal curve that has been widened, compressed, or moved left or right is transformed into a unit curve with a mean of zero. It is accomplished by subtracting the mean from each datum, then dividing each datum by the curve's standard deviation. This process is also called *scaling*.

Correlation Correlation is a measurement of the degree to which two sets of data are related. A perfect correlation (1.0 or −1.0) indicates that for the data sets given, the highest value always corresponds to the highest (or lowest, in the case of a negative correlation), the second highest to the second highest (or lowest), and so on. A zero correlation indicates that the two variables show no relationship, or are distributed randomly in relation to each other.

Correlation does not assume a cause–effect link; regression analyzes putative cause–effect data.

The square of a correlation is the proportion of the variance in one variable that is due to variation in the other.

Correlation is the critical factor dealt with by path analysis, which determines what part of a correlation is due to cause–effect links between variables.

Cosine A mathematical function often denoted as "$\cos(x)$". Its graph has a distinctive wavy pattern.

The cosine of an angle is the length of the horizontal side of a right triangle with that angle and a hypotenuse of length one.

The x coordinate of a point is equal to the cosine of the angle which a line to it makes with the x axis times its distance from the origin (0,0).

Datum (pl. data) A unit of information. Typically a number, though "it was green" constitutes a qualitative (rather than quantitative) datum and could be the basis of a category ("fifteen results [insects] were in the category of 'it was green' ").

Calling a single number "data" (it's a datum) or saying "this data" or "the data is" is almost as annoying to a statistics-literate person as pronouncing chi-square "chee-square." Nevertheless, even professionals

sometimes slip up—in which case they merely forgot to add the word "set" after "data," thus singularizing the term. Mortals, on the other hand, make this error out of boundless ignorance.

Degrees of freedom (df) A degree of freedom is a set of alternatives not actually excluded. For instance a piece of paper on a bulletin board has three degrees of freedom (height, left–right position, and angle), assuming that it doesn't fall off of the bulletin board like most unanchored pieces of paper. If you put a thumbtack into it, only its angle is free to change, and if you put several thumbtacks into it, it has no degrees of freedom left.

For a statistical example, take the chi-square test. The number of degrees of freedom is one fewer than the number of categories. This is because if there is only one category, it would be entirely certain that all of the data will be in that category, so one category leaves no freedom.

The number of degrees of freedom is typically written as v (though, for clarity, we have often used df in this text).

Dependent variable The "effect" in a cause-and-effect relationship. The "cause" is the independent variable. The value of the dependent variable "depends" on the value of the independent variable. This distinction is important in regression and multiple-regression analysis.

Discrete data Data that can take only certain specific values along a range, as for instance the number of students in a class, which can only be a whole number.

Distribution A description of the frequencies of various data. Distributions usually can be represented with a histogram.

ne various types of distributions discussed in this text (sample distributions, null distributions, parent distributions, normal distributions) are all descriptions of the frequency with which different possible sorts of data appear.

In general, distributions may be divided into samples and populations; a population is an ideal set of all possible results (like null distributions), while a sample is just a few data observed (sampled) from a parent population.

Exponential Described by a function similar to $y = e^2$, where e is the base of the natural logarithm ($e = 2.718282$). Exponential functions occur in nature in situations where things exhibit growth or decay, and in systems where the probability of something depends upon how many times it has occurred in the past (e.g., radioactive decay is less likely after most of the substance has already decayed, children are more likely to be born if their parents were born, etc.).

A typical exponential function of the form $y = e^2$ stays close to zero for a long time, then rapidly accelerates upward at an ever-increasing rate. The bell curve is an unusual and special kind of exponential function.

Note: exponential is *not* pronounced "expodential."

F-test This test is used to determine the probability that the standard deviations of two sets of data agree with the hypothesis that the sets were derived from the same parent curve.

The main time you actually use the F-test is as a preparation for using the t-test or ANOVA. In order to use most versions of the t-test, you must have data consistent with the hypothesis described above, so if the F-test fails you may use only the t-test for unequal SDs.

The F-test operates by finding the standard deviations of the two sets, squaring them (to obtain the variance), and then dividing the larger by the smaller. The resulting number is called F and must be compared to a table of F-values.

The other use of the F-test is to determine if two distributions have the same variance as a test of whether the data in one of them are independent.

Friedman test The Friedman test is used for nonparametric data broken up into sets (groups; often a single individual under several treatments constitutes a "set") and treatments (conditions under which data were obtained from each set). It is one of the less powerful tests and is quite tedious to perform.

Gauss test If a distribution is known to be normal (parametric), then it may be compared to normal null distribution by means of the Gauss test. This test is very accurate but mathematically difficult.

BioStats Basics Online offers the Gauss test in two flavors: an all-purpose Gauss test and a simplified version for two-choice data. The two-choice version is more precise but is limited to data derived from a two-choice situation (e.g., a rat is put in a two-arm maze 10 times and its preferred arm recorded 10 times; the test is then repeated with 100 different rats).

The mathematical foundation of the Gauss test is that the probability of a result (in a one-tailed situation) is equal to the area under the bell curve to the left of that result. Because the distribution is normal, all the data may be considered as located at the mean for the purposes of this test.

Grouped data Data that are broken up into groups, each group containing a datum for each of the different treatments used. If only two treatments were used (or one and a control), the data are instead said to be paired. Grouped data may only be used with multiple-sample tests. Such groups are called *blocks* in many statistics texts.

Histogram A chart showing the relative frequencies of different results, with the data categories on the horizontal axis and the data frequency on the vertical axis.

Drawing a histogram is one of the easiest ways to test for normality: many distributions are obviously normal or nonnormal. A carefully analyzed histogram may also reveal a way to rescue nonnormal data.

Imperfect normal distributions A distribution which, though normal, is not centered at zero, or has a standard deviation other than one, or (usually) both. The process of converting an imperfect normal distribution to a perfect one is called conversion.

Independent Uncorrelated.

Independence may be tested for in parametric data by the chi-square test for independence, and more generally by any test for correlation or for relatedness of data sets. For example, the number of car accidents per year and the wobble of the magnetic north pole are utterly independent.

Independent variable The "cause" variable in a cause-and-effect relationship. Its value

determines the value of the dependent variable. This distinction is important for regression and multiple-regression analysis.

Interpolation Approximating an intermediate value between two known values (e.g., the test gives critical values for $v = 20$ and $v = 30$, and you want to know it for $v = 23$). A zillion methods exist, but the most common is linear interpolation (all-purpose and fast): if A is the lower value, B is the higher value, F(A) is the result from A (the critical value for $v = A$), F(B) is the result from B, and C is the value for which you want a result, then F(C) = [(B − A) (F(B) − F(A)) + F(A)]/(C − A).

Kruskal-Wallis test The Kruskal-Wallis test is used to analyze the difference between the effects of more than two treatments on ungrouped samples, even if they are of different sizes.

Linear Forming a line if graphed.

Paired data are linear if the difference in x-values between two points relates to the difference in y-values by the same proportion no matter which points are chosen.

The attributes of the line (and of the paired parent distribution) may be determined by linear regression analysis if there is a clear cause–effect relationship between the variables, the values of the cause variable are well known, and the values of the effect variables are parametrically distributed for any cause-variable value.

Matrix A group of numbers arranged into rows and columns so as to form a rectangle.

Grouped data from multiple treatments are arranged into matrices for statistical tests; chi-square (for independence) uses a matrix.

Mean The average value. The mean is obtained by summing all the values and dividing by the number of values.

The mean of a perfect normal curve is at zero.

The mean is meaningful only for parametric data; the nonparametric equivalent is the median.

Mean vector The mean of circular data. Its angle is the mean angle (or mean bearing), and its length is inversely related to the standard deviation.

The mean vector is obtained by finding the means of the cosines of the angles and the means of the sines of the angles. The mean vector's x and y coordinates will be given by the mean summed sines and cosines, respectively. Its angle is the arctangent of y divided by x. Its length will be the square root of x squared plus y squared.

Mean bearing The angle of the mean vector. It indicates the average angle of a group of angles.

Unlike ordinary distributions, which must have a mean, circular distributions may have no mean; there may be no significant mean bearing at all. To test for the presence of a mean, use the z-test.

Median The middle value in a list of data ordered by value. The typical measure of the average value in nonparametric data.

Monte Carlo simulation The steps are simple for the experienced computer programmer:

First, program a function which will perform the test and give the test statistic.

Second, program the computer to generate random data of the appropriate sort.

Third, have the computer run the test and record the results a few million times. This may take a while.

Finally, graph the resulting numbers and determine what fraction of them are less than or equal to (in the one-tailed case) the value you got from your actual data. This is your one-tailed P-value.

Multiple regression A technique for assessing the relative contribution of two independent variables on a dependent variable when the independent variables are themselves correlated. The technique generates a best-fit plane. It is a special (but more useful) case of path analysis.

Multiple-sample test A test that analyzes the similarity of several different sets of data (or a set of grouped data), as opposed to a two-sample test, which analyzes only two sets of data (or a set of paired data).

Negative correlation A correlation of the smallest of one thing to the largest of another and vice versa. For instance, larger animals tend to have slower heart rates, so size and heart rate show a negative correlation.

Noise Randomness, usually a result of random factors that obscure a real trend. Data from an experiment with a great deal of noise will have a large standard deviation, and analysis will require more data to establish significant results than from a less "noisy" experiment.

Nonlinear Not conforming to a line. Whether data are linear or not is determined by linear regression analysis.

If data are not linear but still show a correlation, they may follow a mathematical function other than a line. The best way to guess at this is to plot the data on a graph and see what it looks like. If you can determine the type of function, apply the inverse of that function to the data and then test them again. This is a nonlinear regression analysis.

For instance, if the data showed an exponential curve you would do linear regression analysis on the data after taking the logarithm of each y-value, thus analyzing the data's exponential regression (but only if the exponentiated data are still normal).

Nonparametric Not conforming to a bell-curve distribution. Some such data may be normalized ("rescued"); some are unsalvageable. Nonparametric data yield much less certainty, and should be rescued whenever possible. Data are shown to be parametric or nonparametric by a test for normality.

Normal distribution A normal distribution is a distribution that fits a bell curve. Parametric data have a normal distribution; nonparametric data do not. Some nonnormal distributions may be normalized to become normal distributions.

Normalize Making nonparametric data parametric for the purpose of testing. Normalizing is very tricky, but can vastly improve certainty in an experiment. Only continuous data can be normalized; never try to normalize categorized data. Normalization is accomplished by applying some mathematical function to the data–for instance, squaring every result.

Null distribution The null distribution is the distribution predicted by the null hypothesis. Typically, the null distribution will be derived from some well-known law (such as Mendel's ratios for genetic phenotypes).

Null hypothesis The hypothesis of no difference. The null hypothesis is that your samples will show no discernable difference or trend except that due to chance. The goal of most experiments is to test the null hypothesis.

If the null hypothesis is disproven by showing a statistically significant difference between the distribution which it predicts (the null distribution) and the observed distribution, then some variable in your experiment (probably) had an effect on the data you collected. If the null hypothesis is not disproven, then nothing is proven.

One-tailed The probability measurement of how likely it would be for a result to be more extreme in a known direction. If you simply want to know whether one group gives higher (or lower) results consistently, you would use a one-tailed probability. If you want to know whether one group gives different results consistently, you would use two-tailed probability measurement of how likely it would be for a result to be more extreme.

P-value The P-value of a result is the sum of the probabilities of that result and all of those results more extreme than it in some direction (the one-tailed case), or of any result with a lower probability (the two-tailed case).

The P-value is the general test for statistical significance. Typically, a set of data with a $P < .05$ is considered significantly different from the null distribution. More careful experimenters use $P < .01$ or even $P < .005$ as the cutoff to avoid error.

Paired data Data that consist of many pairs of results, with each pair containing one result from one treatment and one result from another (or a control). Paired data may be graphed, correlation-analyzed, regression-analyzed, or put through any of the two-sample tests. You may also extract pairs from grouped data and analyze them, but to do so after analyzing the grouped data first is to run the risk of performing post hoc analysis.

Parametric Data whose distribution conforms to the bell curve. Parametric data may be analyzed with far more certainty than nonparametric data because of the many well-known mathematical qualities of that curve.

Parent distribution All the data there are to be gathered about some parameter; sample distributions are some subset of a parent distribution.

Path analysis A method for determining the correlations between variables in attempting to work out the cause-and-effect interactions between various independent variables on one or more dependent variables.

Perfect normal Data whose distribution fits a bell curve, has a mean of zero, and has a standard deviation of one. Perfect normal data are rare, but imperfect normal data may be made perfect through conversion, which is necessary for many tests.

Poisson analysis A method for dealing with binomial data when the alternative outcome in which you are interested is quite rare compared to the other possibility. The method predicts the distribution of occurrences at the rare-is-common end of the binomial without the need for direct calculation of all of the rare : common combinations, which is otherwise tedious and computationally difficult.

Positive correlation A correlation in which the highest of one data set is paired with the highest of the other, and the lowest with the lowest.

Post hoc analysis The analysis of data with respect to a null distribution derived from the same data (either a null distribution chosen to agree with or disagree with those data). Results derived by such analysis are worthless because they merely measure the tendency of the data to follow itself.

A classic example of post hoc analysis may be found in the following (purely hypothetical) example: in studying some natural phenomenon, an experimenter obtains some paired data. Looking at these data, the experimenter saw a line. So he performed linear regression analysis and found that, yes indeed, the data set was linear.

Though the experimenter may well be right, he also could be wrong. The researcher has analyzed the data against a null hypothesis (they're linear) derived from the data themselves. The only possible conclusion: the data resemble themselves. Post hoc analysis is probably the ultimate statistical sin. Its name is derived from the logical fallacy that it is related to, *post hoc ergo procter hoc* ("after this, therefore because of this").

Probability The fraction of the total tests that will yield a particular result if a very large number of tests is conducted. A probability of 1.0 is a certainty (will always happen) and a probability of 0 is an impossibility (will never happen).

Probabilities are always between zero and one.

Rank-sum (Mann-Whitney) method A two-sample test of unpaired nonparametric data. This test is not very powerful and may require a great deal of data to establish significance.

All the data are ranked, with ties being broken by assigning each an intermediate rank. Then each of the two data sets is summed. If the two data sets are of unequal size, a scaling step is involved. The smaller of the two sums is then compared to tables.

Regression analysis Linear regression analysis is used to determine whether or not a set of paired data conforms to a line (that is, whether there is a tendency toward $y = mx + b$ for some constant m and b, for $m \neq 0$).

BioStats Basics Online provides a linear regression analysis only; if you wish to test data for nonlinear regression, there are hints on the "nonlinear" card.

Linear regression analysis is done by means of a lot of complicated equations. If you are devoted enough to want to know these, check any comprehensive statistics book.

Linear regression assumes that your variables are related as a cause and an effect—an independent and a dependent variable. If this is not the case, you must restrict yourself to correlation analysis.

Sample distribution The set of observed data. Statistics exists to compare observed data to another sample distribution, a theoretical null distribution, or an all-inclusive parent distribution.

Sample mean The mean of a small sample of data. For instance, if several rats are run 10 times each through some experiment, each rat would give a 10-data-point sample, and the mean of this sample would be a sample mean. By comparing the sample means, you could find the particular bias of each rat.

One occasion to use a sample mean is when a data set is hopelessly nonparametric. If the data can be broken up into samples, and the mean of each sample taken, then the sample means will often distribute themselves parametrically.

Sample size The number of data points in a sample. The larger the sample size, the more likely it is that the sample will reflect the actual distribution, and the chances of getting an accurate result are increased.

In the case of taking small samples from a larger distribution (for instance, in the technique of analyzing sample means), larger sample sizes will give nicer, more parametric results but will require more data to establish significance.

Scatter How much the data points are spread out. Scatter is measured by the standard deviation.

More-scattered data give more uncertainty and require more data to establish significance.

Scatter is often a result of random noise in the experiment and can often be reduced by careful experimental design and technique.

Standard deviation (SD) A measure of how scattered the data points are. The standard deviation of a perfect normal curve is 1.0, and approximately 34.1% of the data falls between the mean and the standard deviation on one side of a bell curve.

The standard deviation is equal to the square root of the variance:

$$\sqrt{\frac{\sum(d^2)}{n-1}},$$

where d is the difference between a result and the mean, and n is the total number of results.

Standard error The range within which the mean is found 68.26% of the time. It is computed by taking the standard deviation and dividing it by the square root of the sample size.

Signed rank (Wilcoxon) test A nonparametric test of paired data. The null hypothesis is that there will be no difference except that due to chance between the two data sets.

The signed rank is computed by subtracting the second number in each pair from the first, and ranking the numbers by their absolute values (so, for instance, 4, -2, -7 would get ranks of 2, 3, 1). Those ranks corresponding to negative differences (second number greater than first) are summed, and then those corresponding to positive differences are summed. The smaller sum is then compared to a table.

Significance level The P-value that is considered sufficiently unlikely to disprove the null hypothesis. Different people use different significance levels in their analyses, but the most common ones are $P < .05$, $P < .01$, and $P < .005$. Higher P-values are easier to attain, but also increase the chance of "false positives," or significance from random data. Remember that 1 time in 20,

$P < .05$ will occur by chance, while only once in 100 times will $P < .01$ occur (and only once in 200 times will $P < .005$ occur) from chance alone.

Significant Beyond the significance level; almost certainly not due to chance. If data yield a significant result from the proper statistical test, then the null hypothesis may be discarded. If results are not significant, then it is possible for the null hypothesis to be true. Determining significance (or lack thereof) is the ultimate object of almost every statistical test.

Sine A mathematical function often denoted as "sin(x)." Its graph has a distinctive wavy pattern.

The sine of an angle is the length of the vertical side of a right triangle with that angle and a hypotenuse of length one.

The y coordinate of a point is equal to the sine of the angle which a line to it makes with the x axis times its distance from the origin (0,0).

Tail Either of the ends of the bell curve, which extends out to infinity.

All P-values must be either one-tailed (high or low) or two-tailed (high *and* low). This merely indicates whether the P-value is the area under the curve to the left of the result, to the right of the result, or simply more extreme in either direction. By considering what exactly you are testing for, you can decide between these.

Test of normality A test to determine whether a given set of data is normal. It uses a known normal curve and compares it to the distribution of your data into 10 evenly spaced categories. Although typically any data set with a $P > 0.5$ of normality may be considered normal, the test for normality cannot tell you that a distribution is normal, only that it is "more normal" than similar randomly generated distributions. A better test for normality is often to plot your data and see if the graph looks normal.

The test for normality in *Biotats Basics Online* is based on the Kolmogorov-Smirnov test for goodness-of-fit for continuous distributions, comparing your data to the normal curve after scaling your distribution to fit. The critical values used are produced by a Monte Carlo simulation.

Transformation The process by which a nonparametric data set is made parametric by application of some mathematical procedure to each datum in the set.

t-**Test** A simple (paired or unpaired) two-sample parametric test. It is the second easiest test to perform (after chi-square) and probably the most common. The only drawback is that the *F*-test must be used first. But if the *F*-test shows a significant difference then you must use a version of the *t*-test for distributions with differing SDs.

Depending on the type of data, three different *t*-tests may be used:

1. The one-group *t*-test tests the similarity between a set of data and a null distribution with known mean.
2. The two-group paired *t*-test tests the null hypothesis of no difference for paired parametric data.
3. The unpaired *t*-test tests the null hypothesis of no difference on two unpaired sets of data.

Tukey-Kramer method One of several imperfect methods for establishing which data set in an ANOVA test is anomalous, and estimating a *P*-value. We chose this

method because it makes the most conservative estimate and thus is least likely to yield misleading results.

Two-choice tests Tests in which there are only two alternatives (e.g., coin flipping). In such cases, the *P*-value of any given result is easily calculable.

One common variation on two-choice tests, in which many small samples are taken (e.g., run 12 rats 10 times each, yielding twelve ten-result sets), can be analyzed easily with *BioStats Basics Online*.

Two-choice data distribute themselves exactly by the binomial distribution, and therefore are easy to deal with.

Two-sample test A test that analyzes the similarity of two different sets of data (or a set of paired data), or one set of data against a null distribution. The opposite is a multiple-sample test that analyzes many sets of data (or a set of grouped data).

Two-tailed The probability measurement of how likely it would be for a result to be more extreme in either direction from the mean. If you wanted to know whether one group gives different results consistently, you would use two-tailed probability. If you simply want to know whether one group gives higher (or lower) results consistently, you would use a one-tailed probability.

Unit curve The perfect normal curve: a bell curve with a mean of zero and SD of one.

Unpaired data Data that, though they fall into sets, do not have any correspondence between sets. For instance, if two different groups were used in measuring something, then the results would be unpaired because there is no relationship between individuals in Group 1 and Group 2.

Variance The square of the standard deviation. A measure of the scatter of a group of results.

The variance is important because when your data result from the addition of several factors, the total variance is the sum of the variances of all the contributing causes; this is called *additivity* and is the basis of ANOVA.

x **axis** The horizontal axis on a graph, corresponding to all points $(x,0)$, where x is any number. The distance from a point to the x axis is its y coordinate. The X axis is the axis along which the independent variable (or cause) is shown if a cause–effect relationship is suspected.

y **axis** The vertical axis on a graph, corresponding to all points $(0,y)$, where y is any number. The distance from a point to the y axis is its x coordinate. The Y axis is the axis along which the dependent variable (or effect) is shown if a cause–effect relationship is suspected.

Bibliography
More Information on Data Used in the Text

Alessandro, D., J. Dollinger, J. Gordon, S. Mariscal, and J. L. Gould. The ontogeny of the pecking response in herring gull chicks. *Animal Behaviour* 37, 372–82 (1989).

Arduino, P., and J. L. Gould. Is tonic immobility adaptive? *Animal Behaviour* 32, 921–22 (1984).

Bischoff, R. J., J. L. Gould, and D. I. Rubenstein. Female-choice sexual selection in guppies. *Behavioral Ecology and Sociobiology* 17, 253–55 (1985).

Canty, N., and J. L. Gould. The hawk-goose experiment: Sources of variability. *Animal Behaviour* 50, 1091–95 (1995).

Gould, J. L. The locale map of honey bees: Do insects have cognitive maps? *Science* 232, 861–63 (1986).

Gould, J. L. How do bees remember flower shape? *Science* 227, 1492–94 (1985).

Gould, J. L. Pattern learning in honey bees. *Animal Behaviour* 34, 990–97 (1986).

Gould, J. L., and K. P. Able. Human homing: An elusive phenomenon. *Science* 212, 1061–63 (1982).

Gould, J. L., and C. G. Gould. *The Honey Bee*. New York: W. H. Freeman, 1995.

Gould, J. L., and C. G. Gould. *Sexual Selection*. New York: W. H. Freeman, 1996.

Gould, J. L., M. Henerey, and M. C. MacLeod. Communication of direction by the honey bee. *Science* 169, 544–54 (1970).

Haines, S. E., and J. L. Gould. Female platys prefer long tails. *Nature* 370, 512 (1994).

Keeton, W. T. The mystery of pigeon homing. *Scientific American* 231 (6), 96–107 (1974).

Margolis, R. A., S. Mariscal, J. Gordon, J. Dollinger, and J. L. Gould. The ontogeny of the pecking response in laughing gull chicks. *Animal Behaviour* 35, 191–202 (1987).

Neft, D. S., and R. M. Cohen. *The Sports Encyclopedia: Baseball*. New York: St. Martin's, 1994.

Nightingale, F. *Notes on Matters Affecting the Health, Efficiency, and Hospital Administration of the British Army*. London, 1858.

Pavlov, I. P. *Conditioned Reflexes*. Oxford: Oxford University Press, 1927.

Tinbergen, N., G. J. Broekhuysen, F. Feekes, J. C. W. Houghton, H. Kruuk, and E. Szulc. Egg shell removal by the black-headed gull. *Behaviour* 19, 74–117 (1963).

Wainer, H. Visual revelations. *Chance* 8 (3), 52–56 (1995).

More Inclusive or More Detailed Texts of Note

Anderson, T. W., and J. D. Finn. *The New Statistical Analysis of Data.* New York: Springer-Verlag, 1996.

Box, G. E. P., W. G. Hunter, and J. S. Hunter. *Statistics for Experimenters.* New York: Wiley, 1978.

Cabrera, J., K. Schmidt-Koenig, and G. S. Watson. The statistical analysis of circular data, vol. 9. In *Perspectives in Ethology,* ed. P. P. G. Bateson and P. H. Klopfer. New York: Plenum, 1991.

Freedman, D., R. Pisani, R. Purves, and A. Adhikari. *Statistics,* 2nd ed. New York: W. W. Norton, 1991.

Pearson, E. S., and H. O. Hartley, eds. *Biometry Tables for Statisticians.* Cambridge: Cambridge University Press, 1956.

Rohlf, J. F., and R. R. Sokal. *Statistical Tables.* San Francisco: W. H. Freeman, 1973.

Siegal, S. *Nonparametric Statistics for the Behavioral Sciences.* New York: McGraw-Hill, 1956.

Snedecor, G. W., and W. G. Cochran. *Statistical Methods.* Ames: University of Iowa Press, 1989.

Sokal, R. R., and J. F. Rohlf. *Biometry.* New York: W. H. Freeman, 1981.

Young, H. D. *Statistical Treatment of Experimental Data.* New York: McGraw-Hill, 1962.

Zar, J. H. *Biostatistical Analysis.* Upper Saddle River, N.J.: Prentice Hall, 1996.

Index

(Pages in *italics* refer to end-of-chapter boxes; topics in Exercises are not indexed; "ff" indicates that the topic is treated through the end of the chapter.)

Absolute value, 213
Accuracy, 80, 86
Additivity, *136*, *148*, *186*, *286*
Admissions (college), 295–302
Amphibians, 275–6
ANOVA, 154ff, 313, 332
  one-way ANOVA, 161–5, *183*, 332
  two-way ANOVA, 165–71, *183*, *185*, 332
  model I, II, and III ANOVA, *186–7*
  three-way ANOVA, 172
  nested ANOVA, 174
  MANOVA, 171
  ANCOVA, 303
Arbuthnot, J., 215
Assortative mating, 278–9

Ballistic missile accuracy, 81
Bar graph, 20
Baseball, 14, 117, 119, 123–5
Basketball, 315
Bayes, T., 61
Bayesian analysis, 60–3
Bees. *See* Honey bees.
Beherens, W. V., 125
Bell curve, 25, 47, 79, *106*
Best-fit line, 267–72
Best-fit plane, 292–3

Bias, 27, 117
Bimodal distribution. *See* Multimodal distribution.
Binomial, 41–60, 63–5, 86, 91–3, 194–7, 328, 329
Binomial theorem, 63–5
Blackbirds, 20–21
Blue jays, 1–2, 7
Bonini correction, 177

Cancer, 177, 262–4
Categorical data, 16, 189ff, 326
Cause and effect, 2, 5–6, 14–5, 169, 255, 257–9, 267, 277–9, 329
Central-limit theorem, 130–1
Challenger disaster. *See* Space shuttle.
Chance, 2, 13, 56
Chickens, 25, 215–20
Chi-square test
  for goodness of fit, 190–99, *211*, 310, 330
  for independence, 199–204, *212*, 310, 330, 331
Cigarettes, 262–4
Circular distribution, 25, 235ff, *252*, 313, 327, 328

Classical conditioning, 3
Clock shifting, 246–7
Clumping, 81, 120–1
Clustering, 120, 237–42
Coefficient of dispersion, 67, 121
Coincidence, 3. *See also* Post hoc analysis.
College admissions, 295–302
Column dropping, 193, 222
Common causes, 277
Confidence interval, 126–30
Confounding variables, 28
Continuous data, 16, 79ff, 92, 111ff, 140ff, 154ff, 215ff, 235ff, 326
Control tests, 7
Correlation, 257, 260–6, *286*, 292, 298, 329, 331. *See also* Regression analysis.
Correlation coefficient, 257–9, 261–3, 265, 267, 272–3
Cosine, 238–9
Covariance, 278–9
Crickets, 41–2, 57–8, 81, 119

Daisy fleabane, 151–3, 158–61
Darwin, C., 256

Data, 16–8
Data types, 16–8, 326
Death feigning, 25, 213–9
Degrees of freedom, 97
Dependent variable, 265,
  291ff
Discrete data, 16–8, 92,
  326
Distribution, 18ff, 120, 318,
  325, 326–8
Dogs, 3

Eggs, 200–3
End effect, 270
Error bars, 128
Estimate of value, 13–5

False positive (Type I error),
  52, 70, 76–7, 152–4,
  174–8, 275, 332
False negative (Type II
  error), 52, 71, 76–7,
  260, 275
Fish, 88, 177–8. *See also*
  Guppies; Salmon.
Fisher, R. A., 115, 125, 154,
  204
Fisher's Exact Test, 202–3,
  212–3
Flattened distribution, 328
Food-avoidance learning,
  1–2, 7, 9
F-test, 115, 154, 161, 166,
  182–3, 329
Friedman test, 220–2, 332

Galton, F., 255, 265
Gauss, K. F., 79, 270
Gauss test, 98, 329
Gosset, W. S., 98, 112

GPA, 15, 32, 257–60,
  295–302, 316–20
Grouped data (sets), 18, 41,
  86, 144, *149*, 155,
  220–2
Growth (plants), 141ff, 173
G-test, 197
Gulls, 4, 200–203
Guppies, 28, 173–4,
  223–4, 278

Height, 13, 17–20, 25–6,
  31, 34, 82, 84, 93–94,
  125–7, 199, 256,
  265–8, 272, 278,
  289–92
Heritability, 289–94
Histogram, 20
Homing pigeons, 237–8,
  241–2, 246–7
Honey bees, 3, 48, 242–4,
  310–12
Human orientation, 242–3
Hypothesis, 5–7
Hypothetical distribution.
  See Null distribution.

Imprinting, 193
Independence, 44–47, 66,
  82, 117
Independent variable, 267,
  291ff
Intermediate causes, 278
Interpolation, 125

Job status, 293

Kolmogorov-Smirnov test, 320
Kruskal-Wallis test, 225–6,
  233, 332

Language
  of honey bees, 310–12
  human, 5
Learning, 1ff, 48–54
Least-squares method, 270
Legendre, A. M., 270
Likelihood ratio, 61–3
Linear regression, 267–73
Lung cancer, 262–4

Mann, H. B., 223
Mann-Whitney test. *See*
  Rank-sum test.
MANOVA, 171
Matrix, 201
Mealworms, 1–2, 7
Mean, 31, 33, 79, 84, *107*,
  111, 126, 326, 328
Mean vector, 237–9
Mean bearing, 236–8
Measurement error, 80
Median, 31, 326
Mendel, G., 190
Missile accuracy, 81
Mode, 31
Monte Carlo simulation,
  314–21, 333
Mosquitofish, 177–8
Multimodal distribution, 25,
  32, 328
Multiple regression, 294ff,
  307, 332
Multiple sample
  comparisons, 151,
  174–8, 332
Multiple-sample test, 151ff,
  217ff, 332

Negative correlation, 258
Nests, 200–2

Nightingale, F., 236–7
Noise as an element in experiments, 2, 8
Nonlinear regression, 275–7
Nonnormal distribution. *See* Nonparametric distribution.
Nonparametric distribution, 25, 139ff, 215ff, 260–64, 329
Normal distribution. *See* Parametric distribution.
Normalization. *See* Transformation.
Normality, testing for, 89–90, 141, 329
Null distribution, 23, 48, 83, 188, 314–20, 325
Null hypothesis, 7, 13, 23, 188

Odds. *See* Probability or odds.
Odds in Bayesian analysis, 61–3
One-tailed test, 58–60, 95
Operant conditioning, 3
Orientation, 25–6
O-rings. *See* Space shuttle.

Paired data, 35, 121–3, 172, 215–20
Parameters, 29, 83–4, *107*
Parametric distribution, 25, 29, 79ff, 111ff, 139ff, 154ff, 235ff, 265ff, 294ff, 326
Parent distribution, 23, 48, 83, 93, 190, 325
Pascal, B., 46–7
Pascal's triangle, 46–47

Path analysis, 289–94, 309, 331
Pavlov, I., 3
Pearson, K., 190, 256 265, 278, 289, 291, 298
Peaked distribution, 327. *See also* Multimodal distribution.
Peas, 190–1
Perfect normal, 58
Pigeons, 237–8, 240–1, 246–7
Plants, 151–3, 159–60, 173, 188–90
Poisson, S., 65
Poisson analysis, 65–71, 77, 313, 329
Population, 23. *See also* Parent distribution.
Positive correlation, 259
Post hoc analysis, 8, 59–60, 172, 277, 312–3
Power, 122
Precision, 80
Probability or odds, 13, 48–54
Probability analysis (in learning), 2
Product law, 44–7, 69, 152, 312
P-value, 52, 77, 95, 152, 279, 318

Quadrat, 120

Rank sum (Mann-Whitney) test, 223–4, *232–3*, 331
Rats, 9
Rayleigh, Lord, 237

Rayleigh *u*-test, 242–5, *252*, 331
Rayleigh *z*-test, 240–1, *252*, 331
Redwing blackbirds, 20
Regression analysis, 257–60, 267ff, *287*, 331–332
Regression line, 267–72
Regression to the mean, 131, 256, 267
Reptiles, 275–6
Root-mean-square (RMS) method, 34

Salmon, 190–1
Sampling, 27–29
Sample distribution, 8, 22, 48, 83, 93, 112ff, 325
Sample mean, 143–6, 155
Sample size, 8, 33, 55–7, 93, 111, 319, 326
Sample variance. *See* Variance.
SAT scores, 20, 22, 121–2, 255–8, 295–302
Scatter, 261. *See also* Noise.
Scatter plot, 20–22
Second-order statistics, 87, 127, 241–2
Secular trend, 256, 267
Sets. *See* Grouped data.
Shoe sizes, 15, 24, 34, 265–6
Sign test, 215–7
Significance, 53, *76*, 178, 279, 320
Signed-rank (Wilcoxon) test, 217–9, *232*, 331
Sine, 238–9

Skewed distribution, 25, 32, 63, 142–3, 327
Smoking, 262–3
Space shuttle, 67–71, 273–5
Spearman, C. E., 261
Standard deviation (SD), 30, 33, 84, *108*, 126–7, 136, 326
Standard error (SE), 126–31, *136–7*, *149*, 263–4, 265, 270–1, 327
Statistics, 9, 29, 325
Suicide, 66–7
Superstition, 3
SUVs, 130
Synergy, 165, 291
Syllable number, 15

Tail, 32, 57–9, 95, 140
Test of normality, 89–90, 141, 329
Test statistic, 316
Theory of island biogeography, 276–7
Tinbergen, N., 200–3
Tonic immobility, 215–20

Transformation, 139ff, *148–9*, 275–7, 330
Trend, 56
t-test, 99–101, 112–25, 140, 330
  one sample, 98–101, *109*, 330
  paired samples, 121–3, 124, *135*, 330
  unpaired samples, 123, 124, *135*, 330
  unpaired samples with different SDs, 123–4, *136*, 330
Tukey-Kramer method, 164–5, *183*, 332
Two-choice tests. *See* Binomial.
Two-sample test. *See* Chi-square test for independence; Rank-sum test; Signed-rank test; *t*-test; Watson-Williams test.
Two-tailed test, 58–60, 95
Type I error. *See* False positive.

Type II error. *See* False negative.

U-test, 223, *233*, 331
Unpaired data, 123, 222–6

Variance, 33, 88, *107*, 115–7, 124, *136*, 289, 326, 328, *306*. *See also* ANOVA.

Wages, 140
Watson, G. S., 223
Watson U^2-test, 246–7
Watson-Williams test, 247, *253*, 331
Weeds, 151–3, 158–60
Weight, 151–3, 158–60, 272
Weighted average, 162
Whitney, D. R., 223
Wilcoxon, F., 217, 222
Wilcoxon test. *See* Signed-rank test.

x axis, 20, 91, 238–9, 267

y axis, 20, 92, 238–9, 267
Yule, G. U., 298

Choosing the Right Test

I. What is the **Question?**

    ✔ ***Means and Medians*** are measures of the center of a parametric or nonparametric distribution, respectively.

    ✔ ***Variance*** is a measure of the scatter in a distribution.

II. What **Type** of **Data** do you have?

    ✔ ***Circular*** data are measured on a repeating scale (e.g., direction).

    ✔ ***Parametric*** data are distributed as a bell curve.

    ✔ ***Continuous*** data can take any intermediate value in a range.

    ✔ ***Categorical*** data exist in mutually exclusive categories (e.g., male and female).

III. What kind of **Comparison** are you making?

    ✔ ***Sample*** distributions are small subsets of parent distributions.

    ✔ ***Null*** distributions are theoretical.

    ✔ ***Binomial*** data are categorical data with only two categories.

    ✔ ***Parent*** distributions contain all the data for a particular population.

IV. **How Many Comparisons** are you making?

V. Are the **Samples Related?**

    ✔ ***Paired*** data measure the same individual under two sets of circumstances.

    ✔ ***Grouped*** data measure the same individual under three or more sets of circumstances.

VI. Is there a **Cause-and-Effect** relationship?

    ✔ An ***independent*** datum does not depend on the value of any other datum.

VII. **Test to Use:**

Key

G = grouped
NP = not paired
P = paired

PD = paired dependent
PI = paired independent
S:B = sample to binomial

S:N = sample to null
S:P = sample to parent
S:S = sample to sample